Terrain Evaluation

Second Edition

Terrain Evaluation

Second Edition

An introductory handbook to the history, principles, and methods
of practical terrain assessment

Colin W. Mitchell

Longman
Scientific &
Technical

Copublished in the United States with
John Wiley & Sons, Inc., New York

Longman Scientific & Technical,
Longman Group UK Ltd,
Longman House, Burnt Mill, Harlow,
Essex, CM20 2JE, England
and Associated Companies thoughout the world.

Copublished in the United States with
John Wiley & Sons, Inc., 605 Third Avenue, New York, NY 10158

First published 1973

Second edition 1991

British Library Cataloguing in Publication Data
Mitchell, C. W. (Colin Ware)
  Terrain evaluation.–2nd. ed.
  1. Terrain evaluation
  I. Title II. Series
  551

ISBN 0–582–30122–X

Library of Congress Cataloging-in-Publication Data
Mitchell, Colin W.
  Terrain evaluation : an introductory handbook to the history,
  principles, and methods of practical terrain assessment / by Colin
  W. Mitchell. -- 2nd ed.
    p.   cm. -- (The World's landscapes)
  Includes bibliographical references and index.
  ISBN 0–470–21697–2
  1. Landforms.   I. Title.   II. Series.
  GB401.M57   1991
  551.4'1--dc20   90–13468
                                                              CIP

Transferred to digital print on demand 2002

Printed and bound by Antony Rowe Ltd, Eastbourne

To Dr Robert M. S. Perrin

Contents

Contents

Contents

Preface to second edition

The aim of this book is to meet the need, expressed by T. G. Miller (1967) and others, for a brief, popular, and inexpensive introduction to terrain evaluation. From a start in the years following the Second World War, many individuals and organizations caused a wide expansion of the field. Some workers published summaries of their own and related contributions (e.g. Christian and Stewart, 1968; Beckett and Webster, 1969; Webster and Beckett, 1970), and the proceedings of an important symposium, edited by G. A. Stewart, assembled summaries of the advances to 1968. The first edition of the present work, published in 1973, sought to provide a simple collation and summary of these and related principles, methods and examples for the benefit of those professionally involved in the use and management of land, students, and general readers.

Since that time public awareness of the need for environmental conservation and planning has continually grown. This has contributed to extensive developments in most aspects of the subject, especially in classifications of the earth's surface, remote sensing and data processing technologies, and geographical information systems.

The preparation of a second edition gives the opportunity to incorporate these changes, to extend the range of topics covered, notably in including terrain processes and the place of vegetation, and to include a wider range of land user interests. The author especially wishes to record here his debt to Dr John Howard through whom he obtained a greater understanding of the place of vegetation.

The emphasis in this book, however, differs from that in their joint book *Phytogeomorphology*. Although the land units defined are the same, greater emphasis is placed on the geomorphological than the ecological aspects. This is for two reasons. First, it is written from the point of view of the earth scientist. Second, many of the concepts and examples quoted are from areas where the vegetation pattern may be an unreliable guide to landscape classification, either because it is induced, as in the settled parts of North America and western Europe, or because it is absent, as in deserts.

To achieve a synthesis within the compass of a relatively short book, it has been necessary to handle many topics with less depth than would be desirable for the specialist. However, it is contended that the value of a general synthesis outweighs this disadvantage, and that consistency and continuity are aided if done by a single hand.

Mathematical formulae have been kept to a minimum, and most practical techniques have been outlined rather than being described in detail. To compensate for these restrictions, there are many references to the literature.

The emphasis throughout is on the physical, rather than the economic, social or legal aspects of the subject.

C. W. Mitchell
University of Reading,
1990

Acknowledgements

I am grateful to Dr Jim Houston, the editor of this series, for inspiring this book.

I thank Drs John Hardy, Russell Thompson, John Whittow, and Graham Yapp for advice on particular chapters, Professor Jim Douglas for his comprehensive review of the manuscript, and my wife for her patience and help during the time of preparation.

We are grateful to the following for permission to reproduce copyright material:

American Institute of Biological Sciences for fig. 16.9 (Phillips, 1965); American Society of Agricultural Engineers for fig. 7.2 (Wischmeier et al., 1958); American Society of Agronomy, Inc., Crop Science Society of America, Inc., & Soil Science Society of America, Inc. for fig 7.6 (Hoffman & van Genuchten, 1983); American Society of Landscape Architects for fig 26.3 (Lewis, 1964); Association of American Geographers for figs. 6.1 (Peltier, 1950), 16.3 & 16.4 (Hammond, 1962), 16.5 (Hammond, 1964), 16.6 (Murphy, 1968), 16.7 (Savigear, 1965) & 25.2 (Johnson, 1921); Association Nationale Des Geographes Marocains for tables A1-3 (Joly, 1957); Auckland University Press for table 23.3 (Fox, 1956); the author, Dr. P. Beckett for fig. 15.2 (Beckett); Collins (William) Sons & Co. PLC for table 18.6 (Manley, 1962); CSIRO (Division of Geomechanics) for fig. 23.2 (Australia, CSIRO, 1967); CSIRO (Division of Wildlife & Ecology) for fig. 23.3 (CSIRO); Gebrüder Borntraeger for fig. 3.2 (Dalrymple et al., 1968); Geological Society for fig. 16.1 (Geological Society Working Party, 1982); Controller of Her Britannic Majesty's Stationery Office for figs. 12.2 & A1-10 (Perring & Mitchell, 1970); Hunting Technical Services Ltd. for fig. 8.1 (Hunting Technical Services Ltd., 1956); The Institute of British Geographers for figs. 10.4 & 10.6 (Cooke & Harris, 1970); International Institute for Land Reclamation & Improvement (ILRI) for fig. B2 (van Beers, 1958); International Institute for Aerial Survey & Earth Sciences (ITC) for fig. 16.2 (Verstappen & van Zuidam, 1968); the author, C. Jesty for fig 16.12 (Jesty & Wainwright, 1978); the author, Prof. H. Jenny for fig. 21.4 (Jenny, 1941);

Landscape Research Group for fig. 26.1 (Tandy, 1967); Longman Group UK Ltd. for fig. 10.1 (Curran, 1985); the author, W.P. Lowry for table 18.3 (Lowry, 1989; Thompson *et al.*, 1986); The Macauley Land Use Research Institute for fig. 21.2 (Ragg, 1960); Macmillan Company of Australia Ltd. for fig. 25.6 (Parry *et al.*, 1968); Masson S.A. for fig. 23.6 & table 23.4 (Long, 1974); McGraw-Hill Publishing Co. for fig. 16.8 (Raisz, 1962); Methuen & Co. Ltd. for figs. 7.5 (Bagnold, 1941, 1965), 18.3, 18.5 & tables 7.2, 18.2 (Oke, 1978) & 19.1 (Waltz, 1969); the author, Dr. T. Partridge for fig. 24.2 (Brink & Partridge, 1967); Pergamon Press PLC for fig. 18.7 (Verstappen & van Zuidam, 1970) Copyright 1970 Pergamon Press PLC; Royal Geographical Society for table 5.2 (Mitchell *et al.*, 1979); Royal Meteorological Society for table 18.4 (Thompson, 1973); Sudan Government (Ministry of Agriculture) for fig. 8.2 (Smith, 1949); Soil Survey & Land Research Centre for fig. 2.1 (Jarvis *et al.*, 1979); Soil & Water Conservation Society for fig. 7.1 (Wischmeier *et al.*, 1969); the author, A. Strahler for fig. 6.2 (Shrahler, 1969); United States Army Engineer Waterways Experimental Station for figs. 10.4, 10.6 (Cooke & Harris, 1970) & 25.4 (USAEWES, 1959); United States Dept. of Agriculture (USDA) for fig. 10.5 (Olson, 1970) & tables 7.1 (Wischmeier & Smith, 1978) & 23.2 (Klingebiel & Montgomery, 1961); Friedr. Vieweg & Sohn for fig. 18.1 & table 18.1 (Geiger, 1965); the author, D. Way for figs. 6.24 & 6.25 (Way, 1968); John Wiley & Sons Ltd. for fig. 19.5 (Dunne, 1978) Copyright © 1978 John Wiley & Sons Ltd., Williams & Wilkins for fig. 21.3 (Thorp, 1931) © 1931 Williams & Wilkins.

Whilst every effort has been made to trace the owners of copyright material, in a few cases this has proved impossible and we take this opportunity to offer our apologies to any copyright holders whose rights we may have unwittingly infringed.

We are grateful to the following for permission to reproduce copyright material:

Editions Technip for fig. 15.1; Ministry of Defence (Air Force Department) for figs. 10.2 & 10.3; Weather and the Royal Meteorological Society for fig. 18.5.

I am happy to acknowledge the following non-copyright illustrations from the US Government (11.3, 11.5, 11.7 & 11.8).

Part I

Principles of terrain evaluation

1

What is terrain evaluation?

1.1 Outline and objectives of this book

Terrestrial life depends on a surface mantle of rock and soil not more than a few metres deep and the associated plant cover. The character and behaviour of this mantle under the influence of climate determine its suitability for all types of land use. Terrain represents one of the triad of factors of production: land, labour, and capital. It differs from the others in being relatively fixed in location and extent and in being more amenable to geographical forms of analysis. Accelerating population growth and earth-transforming technologies are changing the environment at an unprecedented rate, often for the worse. At the same time, modern methods of data processing make it possible to gather and manage information much more efficiently and rapidly than hitherto. There is an urgent need to harness this capability in order to improve land use and management.

Terrain evaluation is an important technique in achieving this. It integrates other land resource factors, notably surface materials, soils, water, and vegetation on a common readily comprehensible basis, such that a map of terrain can be used as a framework for the others. For this reason, it forms the basis for the interdisciplinary approach known as 'integrated survey'.

Part I of this book considers the general principles of how and why an intelligence system based on terrain forms a valuable foundation for land use planning. The latter part of the present chapter defines terms, gives an account of the scope and requirements of an environmental intelligence system, and outlines the reasons for using a natural classification of terrain as its basis. Chapter 2 discusses the scale spectrum, the conceptual development of regionalizations, and boundary delimitation, and Chapter 3 outlines terrain and landscape classifications from international to local scale. Chapter 4 describes the way in which these have been developed into parametric and physiographic schemes, and Chapter 5 outlines a scalar hierarchy of physiographic units.

Part II outlines the type of information about the earth's surface required in a practical intelligence system. Chapter 6 considers the variations imposed by climate, structure, and lithology on surface form, and Chapter 7 the natural processes causing erosion, degradation, and pollution of the environment. Chapter 8 discusses the place of vegetation both as an aid in the recognition of terrain units, and in schemes where it plays a part in defining them, and considers the relevance of different types of vegetation classification to terrain evaluation.

Part III (Chs 9–15) deals with the practical aspects of data collection and analysis. This includes the techniques of project planning (Ch. 9), remote sensing (Chs 10 and 11), sampling (Ch. 12), field observation (Ch. 13), geographical information systems and data processing (Chs 14 and 15). Since these techniques have wide applications in all environmental sciences, the emphasis is on aspects specifically relating to terrain surveys, giving literature references to enable the reader to follow up specialist aspects in detail.

Part IV covers the techniques of display, reporting, and mapping of terrain data, Chapter 16 giving the principles of terrain mapping, and Chapter 17 showing how this and other forms of data presentation can be developed as an output from a geographical information system.

Part V (Chs 18–27) describes the applications of terrain evaluation to each of the main practical disciplines concerned with land uses. These are: environmental climatology, hydrology, geology, soils, archaeology, agriculture, civil and military engineering, and environmental planning. The general format of each chapter is first to discuss the relevance of terrain to the land use in question and then to describe the systems which have been developed for analysing and classifying the terrain for it. The final chapter synthesizes these methods and systems in relation to likely developments in the 1990s.

1.2 Definition of terms

The term 'terrain evaluation', used as the title of this book, follows the precedent of research carried out under the auspices of the Military Engineering Experimental Establishment (now the Military Vehicles and Engineering Establishment) (Beckett and Webster, 1969). The subject has developed in response to the need for an understanding of terrain by the increasing number of disciplines concerned with its use. These are both the pure sciences, such as geology, hydrology, geography, botany, zoology, ecology, pedology, and meteorology; and the applied sciences, such as agriculture, forestry, civil and military engineering, and landscape planning. This range of interest underlines the need for explaining terminology which may be understood in different senses in different disciplines.

'Terrain' is defined by the *New English Dictionary* as a 'tract of country considered with regard to its natural features and configuration'. This is preferable to similar terms because its meaning is more strictly confined to the surface of the earth and has fewer academic and practical connotations. It has now been used in the titles of a number of publications in this sense, e.g. Beckett and Webster (1969), Way (1973), and Townshend (1981). 'Environment' and 'milieu' have meanings which are somewhat too general and extend well beyond the confines of geography. 'Physiography' is an older term which includes not only surface form and geology but also climatology, meteorology, oceanography, and natural phenomena in general. 'Geomorphology' has the advantage of being more narrowly confined to landforms but is too strongly involved with considerations of process. 'Microrelief' is too exclusively geometric and does not comprehend earth materials or structure. 'Regolith' and 'soil', conversely, relate only to materials and not to geometry, regolith being restricted to the loose overburden mantling undecomposed bedrock and soil still more narrowly limited to its upper and biophysically weathered part. 'Landscape' or 'land' are perhaps the closest equivalents, but both are somewhat wider concepts than terrain. The former rather too strongly connotes the visual and artistic aspects, and the latter has been given the specific technical meaning of the total physical environment (including climate, relief, soils, hydrology, and vegetation) to the extent that it influences land use (FAO, 1976a). Terrain must also be distinguished from the term 'terrane' used by geologists for three-dimensional blocks of crust characteristically separated from each other by tectonic action. These may be large such as the Guyana and Saharan Shields separated by the Atlantic Rift, or smaller fragments of crust in younger mountain ranges such as the North American Cordillera.

'Evaluation' is defined as the 'act or result of expressing the numerical value of; judging concerning the worth of' an object. This double meaning makes it somewhat more inclusive and thus preferable to such terms as 'analysis', 'quantification', 'assessment', or 'appraisal'.

1.3 The scope of terrain evaluation

The scope of the subject is wide. It begins with the user's need and the whole problem of acquiring information about the terrain both from old records and new field surveys, laboratory studies, and statistical analyses. It therefore includes the study of these techniques. Secondly, it is concerned with the abstraction, classification, storage, and reproduction of such information to make it available quickly and cheaply to users. Thirdly, it considers the means by which these ends can be achieved.

Land is not static but experiences continual change both of its physical characteristics and also of the socio-economic conditions governing its use. The absolute and relative values placed on different tracts vary between societies and over time in the same society. Many examples of this could be quoted. The sugar-growing islands in the West Indies were highly valued in the eighteenth century but became much less important after sugar-beet became the main source of sugar in Europe as a result of the British blockade during the Napoleonic Wars. Similarly, the changed patterns of leisure in the last generation have dramatically increased land values in popular recreational areas such as the Alps, Mediterranean, Caribbean, and east Australian coast. There is now great pressure on land and ecosystems in specially favoured and accessible locations such as the shores of Lake Geneva, the Costa do Sul, southern Florida, and the Great Barrier Reef islands.

Even in the same society, land evaluations must take account of different perceptions of the environment, especially concerning rural areas. These perceptions have been altered and sharpened in recent years by the wider appreciation of the fragility of our planetary ecosystem and of the need for conservation. We can discern two types of approach today. The first, which could be called the 'environmental management approach', sees the existing pattern of land use as being in approximate equilibrium with the environment. It favours leaving the existing system approximately as it is and altering it only to maximize production and minimize wastage. The second, or 'socio-economic planning approach', seeks all-round improvement by replacing the old ecosystem with a planned new one. The difference between these two approaches is more than academic because we are already faced in some areas with pressures to manipulate land resources in ways which cannot be harmonized with maintaining their short-term equilibrium.

There are three types of terrestrial phenomenon which are generally excluded from consideration in this book as they do not fall within the normal definition of terrain:

1. *The atmosphere.* This is too variable and ephemeral to be assigned to sufficiently small and closely definable tracts of the earth's surface.
2. *Large expanses of water beyond the range of riparian uses.* These are subject to different types of analysis than are applicable to terrain.
3. *That part of the earth's crust which lies at a greater depth than about 6 m.* Terrain evaluation is not concerned with operations such as mining and well drilling except in so far as they involve the exploitation of the immediate surface.

1.4 Information requirements of a terrain evaluation system

Although a wide range of professions is concerned with land uses and the aggregate of their information requirements is large, it is not infinite. There is considerable overlap between the interests of different users. Most would agree on a central core of important information about land which would include such properties as gradient, soil texture, and moisture regime. Each would, however, require additional data of less general interest. Only agriculturists and foresters, for instance, would need detailed information on soil nutrient status, and only engineers would require it on some aspects of soil strength.

Nor is the range of land use intensities infinite. Terrain must always be evaluated within an economic context, and the importance of biophysical factors relative to others diminishes at very large and very small scales. Planning at international or national levels is concerned with macro-economic considerations of which terrain forms only a small part. At the other extreme, the construction of buildings or monuments or the layout of gardens involve so much input per unit area that terrain costs become a relatively insignificant part of the total. Thus terrain evaluation aims primarily at intermediate levels of land use intensity where investment per unit area is neither so small that land characteristics become unimportant, nor so large that their control becomes a relatively small proportion of total cost. Experience has shown that this intermediate level is approximately that of the agriculturist choosing a farming system, the engineer constructing a 'B' road, the battalion commander concerned with the movement of about a dozen vehicles, or a similar degree of land use interest in other fields. Such a level has been defined as 'moderately extensive'.

The information requirements for the main specialisms concerned with terrain evaluation are:

1. *Earth and related biological sciences.* These include geology, geomorphology, zoology, botany, ecology, and pedology. Such disciplines differ from others concerned with terrain in that their requirements are wider ranging and more variable. They require information both on past processes and on future trends.

2. *Meteorology and climatology.* These are concerned with the effect of terrain on weather and climate. Slope, aspect, exposure, and the nature of the soil surface influence climate both directly through their effect on winds, insolation, fog, cloud, and rain, and indirectly through the vegetation.

3. *Hydrology.* This requires knowledge of terrain in a number of ways, especially those relating to surface and subsoil water in defined territorial areas such as river catchments. Specifically, it is concerned with runoff regimes and quantities, stream flows, infiltration, and groundwater depths and movements with practical application to water supplies, erosion, and flood hazards.

4. *Agriculture, pasture, and forestry.* These include all specialisms concerned with raising economic plants and animals, and thus comprehend farming, pastoralism, ranching, horticulture, forestry, and related disciplines. Three groups of land properties are normally important: (a) soil fertility resulting from nutrient status, texture, moisture regime, and an absence of soil limitations or hazards; (b) soil manageability consisting of tilth, hardness, permeability, and slope; and (c) the nature of existing vegetation.

7

5. *Civil engineering.* Most engineering operations require knowledge of terrain factors. They can be categorized as concerning (a) long but relatively narrow corridors of ground for roads, railways, canals, etc.; (b) more compact areas for large construction works such as airfields, factory complexes or large dams; and (c) smaller sites for minor works such as individual buildings or quarries. The terrain information required is generally of two types: the sources of materials for borrow and construction, and the physical conditions of the site for load bearing, drainage, and liability to flooding or erosion.

6. *Military activities.* These are essentially civil engineering operations, but with emphasis on less permanent works and always requiring very rapid response times. Military interest in terrain also focuses on artillery lines of sight, the suitability of the ground for excavating trenches and fortifications, holding posts, tent pegs, and mines, accepting parachute drops, and sustaining the passage of troops and vehicles. This last is known as 'going' or 'trafficability', and depends on soil strength, stickiness, and the frequency of gradients exceeding certain critical figures. Vehicle designers require, in addition, information on the distribution and areal extent of features that pose special design problems, such as marshes or sand dunes.

7. *Urban and rural residential and recreational planning.* Terrain is an important determinant of landscape aesthetics and must be considered in all planning schemes. Often, in developed areas, the less valuable land is for agriculture the more it is for residential and recreational purposes, requiring separate evaluations of each. Specifically, while agriculture prefers flat, fertile land, which usually has undramatic scenery, residential and recreational developments favour the proximity of hilly areas covered with forest or moorland, and containing rivers, lakes, sea coasts, and sandy beaches in so far as their exploitation can be harmonized with social and economic costs.

8. *Nature conservancy and wildlife planning.* These interests require detailed knowledge of the environment, particularly in relation to the habitats of wild plants and animals. Such habitats have scientific interest and educational value to the extent of their geological or ecological typicality or rarity. Demonstration sites for these purposes have enhanced value if they also possess aesthetic attractions.

9. *Archaeology.* This is concerned with past land uses such as quarries, buildings, tombs, monuments, and earthworks. It depends on recognizing them from the way they have modified the natural land surface.

The preparation and execution of works which change land uses require terrain data. Most come from published sources and remotely sensed imagery. Much more exist in unpublished files and notebooks which are neither readily available nor indexed. Still more are unwritten and either lost altogether or preserved only in the memories of individuals whose whereabouts is unknown. Therefore, in any intelligence system, there is a need for a 'bank' of environmental data which can: (a) anticipate information needs, (b) acquire and store data about terrain and its past and potential uses, and (c) translate these data into usable form.

Further reading

Goodall and Kirby (1979), Howard and Mitchell (1985), Stewart (1968).

2

Principles of physical regionalization

2.1 The scale spectrum

The first stage of terrain classification is scale definition. Landscape features vary in size by several orders of magnitude, from continents at one extreme to particles of sand or clay at the other. Tricart (1965b) has suggested an eight-level scalar hierarchy and Haggett et al. (1965) a subdivision of the earth's total surface area successively by powers of 10 to give 'G_a scale values' or 'G'. The area of the globe is 5.098×10^8 km^2. This is assigned a value of 0. One-tenth of this (5.098×10^7) is 1, one-hundredth is 2 and so on. G can be determined from formula [2.1] for any given area (R_a) being considered.

$$G = 8.7074 - \log R_a \text{ (km}^2) \qquad\qquad [2.1]$$
$$G = 10.7074 - \log R_a \text{ (ha}^2).$$

Tricart orders and G_a scale values for a scalar hierarchy of land regions are shown on Table 2.1.

Table 2.1 Landscape Scales (after Tricart 1965 and Haggett et al. 1965)

Order	Area (km^2)	G^a scale value	Example
I	10^7	1.71	Australian continent, humid temperate zone
II	10^6	2.71	American Piedmont, North European Plain
III	10^4	4.71	Lowland Britain, Florida Peninsula
IV	10^2	6.71	Weald, Cape Cod
V	10	7.71	North Downs, Nantucket
VI	10^{-2}	9.71	Terrace, scarp, fan
VII	10^{-6}	13.71	Soil polygon, tussock, runnel
VIII	10^{-8}	15.71	Pebble, weathering detail on rock

Most terrain classes of practical significance lie in the middle of the G_a range. They reflect the 'natural scale' of a landscape, deriving from such factors as distance to the horizon, relief contrasts, and internal articulation. It is, for instance, 'wide' in the Great Plains, 'grand' in the Alps, and 'domesticated' in densely populated parts of Europe and Asia. Even within one area, the natural scale can vary with land use intensity, so that a variety of mapping scales may be appropriate. The planning of individual buildings, for instance, requires larger-scale mapping than township development, and this in turn larger-scale mapping than, say, regional afforestation.

Terrain classifications have traditionally tended towards the recognition of a nested hierarchy of units. There are a number of reasons for this. Descending scalar sequences such as that on Table 2.1 suggest themselves almost naturally to the observer. Hierarchies are the most widely used system of classification in everyday life and are universally understood. They can be constructed on the basis of a few characteristics and are easily memorized because they minimize the number of individuals in each class. Different classes can be simply related to each other via the first higher category in which they both fall. Where defined without categorical overlapping, the nested hierarchy can suggest and illuminate generalizations for each level and its relations to neighbouring levels.

Some natural features often possess a remarkable simplifying invariance under changes of magnification. This is true of landforms such as coastlines and mountains, which often have statistically similar contour and shoreline patterns at a wide variety of scales, despite having totally different detail at each. The property of objects whereby magnified subsets look like the whole and each other is known as 'self-similarity', and a scalar series of self-similar phenomena is known as a 'fractal series' (Mandelbrot, 1982; Peitgen and Saupe, 1988). It differs from a series of the more traditional Euclidean shapes, such as cones, wedges, and cubes, by not becoming ever smoother on magnification. The statistical similarity between the members of a fractal series can exist over quite a wide, although finite, range of sizes. The largest topographic variations are limited by the dimensions of the planet or the force of gravity (the materials may not be strong enough to support excessively high mountains). The smallest may be limited by the smoothing of erosion, the basic grain size of the rock and sand, or at the very smallest, by the atomic nature of the particles. In some senses, therefore, a scalar hierarchy of normal landforms can be thought of as a fractal series.

2.2 Historical development of regional concepts

Classical geography distinguished between 'formal' regions based on biophysical factors, and 'functional' regions based on human factors. This distinction is still useful, although many regionalizations, like the French *pays*, are a combination of the two.

Functional regions, often in the form of cadastral or administrative units, have been used in social sciences such as cultural geography (Unstead, 1933), agriculture (Whittlesey, 1936), and economics (e.g. by Isard, 1975). Today some governmental authorities collect environmental information on the basis of individually owned parcels, and several Australian states base their geographical information systems on cadastral units. Such units serve administrative and economic convenience, but have two main disadvantages. First, because they are subject to change, data from one period or locality may be difficult to compare with that from another. Second, because each cadastral unit is a composite of terrain types, some of which may be contrasting, it is difficult to make valid generalizations or extrapolate data from one area to another on the basis of physiographic analogy. For these reasons, administrative and cadastral subdivisions of the landscape have not been considered as terrain classes in this book.

Formal regions based entirely on 'natural' criteria have been used widely in the environmental sciences, notably in ecology, soil science, and climatology. Some are based on single land attributes and others synthesize two or more. Early examples of the former were the schemes of de Candolle (1856) and Schimper (1903) for plants, Wallace (1876) for animals, Dokuchaiev (1899, quoted by Mishustin, 1983) for soils, and Koppen (1931) for climate, and of the latter that by Herbertson (1905) who subdivided the whole world into 'natural regions' based on climate, configuration, and vegetation.

Regions can be 'specific' or 'generic' (Unstead, 1937). Specific regions are unique localities identified by a name, such as the Paris Basin or Appalachia. Generic regions are generalized categories which recur. This recurrence has always been recognized in lay parlance by the abundance of generic terms. These include common words for large landforms like 'mountain', 'plateau', and 'plain', for medium-sized features like 'cliff', 'tor', and 'marsh', and for small components like 'ledge', 'cave', and 'hummock'. The fact of recurrence is vital in classification and is used, for instance, by planetary scientists looking at the surface of the moon or Mars. Faced with an unknown terrain, they first of all delimit areas of similar features, recognizing recurrent patterns. The recognition of a region as belonging to a generic type allows analogies to be drawn between its different occurrences. For example, the 'Chilterns' are a specific region, but 'chalk cuesta' is a generic term which could relate to other regions besides the Chilterns.

Some regions, such as volcanoes or sand dunes, have clear sharp boundaries, while others, such as the subdivisions of alluvial or glaciated tracts, have more indefinite ones. Most regions consist of core areas surrounded by transitional zones, which may be larger than the cores. Plant geography provides a useful terminology, calling whole units 'formations', their central cores 'nodes', and their transitional surroundings 'ecotones' (Poore, 1956).

The regions considered thus far have been 'uniform' in the sense that their boundaries define areas to which a single description can be applied. Maps of such regions are generally termed 'choropleth maps', although some cartographers restrict this term to maps displaying census tracts or administrative areas, using 'chorochromatic' for those displaying natural units. Others use the term 'dasymetric maps' for maps combining natural units with administrative districts (Burrough, 1986). Where the definitive character of regions is the internal interrelationships of their parts, rather than their internal homogeneity, they are designated 'nodal'.

Although originally derived from the relations between a city and its hinterland (Dickinson, 1930; Christaller, 1933), the nodal concept applies to many physical regions. One example is where zones of sediment or plant diffusion radiating from a source merge into those from another. A dissected anticline exposing concentric layers of different rocks may be radially eroded to leave a nested series of alluvial annuli, or plants may diffuse in patterns dictated by wind speed and direction. The anticline and the plant source constitute the nodes around which the ecotones develop.

Regions are also continually changing. This is especially true of the internal organization of nodal regions where the node itself has a changing role. To extend the previous example, the erosion of the anticline will yield a continually changing pattern of sediments as its inner layers are consecutively exposed by erosion, and the plant diffusions will change with the colonization sequence.

2.3 Boundary delimitation

Since terrain units are defined in terms of a number of properties which are not geographically coterminous, compromise is necessary in delimiting their boundaries.

Four types of delimiting criteria can be recognized: 'general', 'differentiating', 'diagnostic', and 'intrinsic' (Vink, 1983). Most regional classifications begin with the use of obvious general physical features such as beaches, cuestas, river valleys, etc. The exact delimitation of their boundaries must normally be based on more precisely defined criteria, such as slope angle or water-table depth, selected because of their importance to land use. Diagnostic criteria are those which are particularly efficient in mapping. Their selection will automatically exclude others, such as differences in soil texture or structure which, although important, are less efficient for this purpose. Intrinsic soil properties such as base status or cation exchange capacity, may also be important but are not normally used in boundary delimitation. A scientific classification for mapping purposes will apportion land properties into these four categories, all of which should be referred to on the legend.

Once these criteria are selected, it is necessary to harmonize them for boundary delimitation. When there is only one quantified attribute, the degree of similarity between neighbouring areas can be indicated by the technique of 'cross-boundary similarity'. This statistically compares the attribute for each pair of contiguous mapped areas according to a standard similarity index such as the 'product moment correlation coefficient', and then draws the boundary between them with a thickness inversely proportional to their degree of similarity. The resulting pattern of boundaries emphasizes the main interregional contrasts (Smith, 1975).

When the classification is based on two or at most three criteria, simple two-axis or three-axis graphs can show how they can be grouped according to similarity (Fig. 2.1). A graphical solution is Maull's 'girdle' method (Grigg, 1967).

13

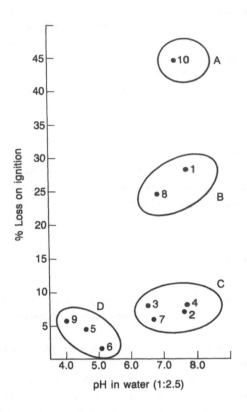

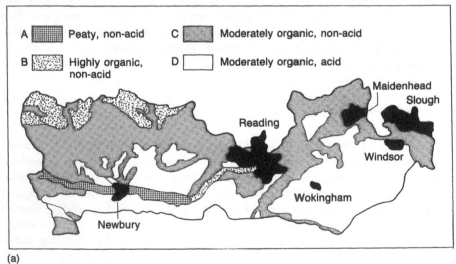

(a)

Fig. 2.1 Two classifications of soils in Berkshire:
(a) using two attributes (pH and percentage loss on ignition);

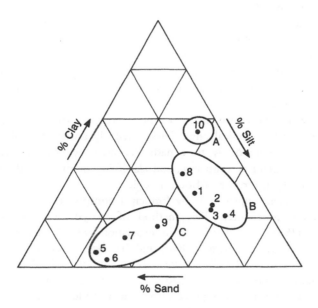

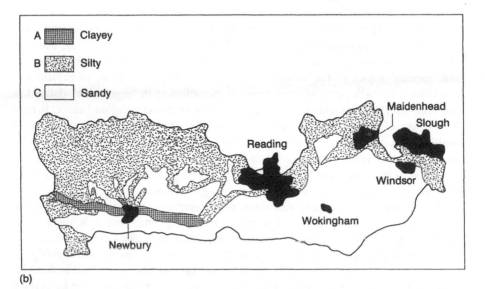

(b)

Fig. 2.1 cont.
(b) using three attributes (mechanical fractions – sand > 60 μm, silt 2–60 μm, clay < 2 μm).
Soil classes: (1) rendzine; (2) brown calcareous earth; (3) argillic brown earth; (4)
palaeoargillic brown earth; (5) podzol; (6) gley podzol; (7) argillic gley; (8) alluvian gley; (9)
stagnogley; (10) humic alluvial gley. (Data from Jarvis *et al.*, 1979)

This proceeds by drawing maps of the regional limits of each land property in question, superimposing them, and selecting the line where more than a given number of boundaries coincide. The method can be refined by drawing in a calculated mean boundary line, or by incorporating calculations of the sizes of overlapping areas multiplied by the number of properties involved in the overlap. But this approach becomes increasingly complicated and unsatisfactory as the number of criteria is enlarged. Although it can give an acceptable compromise amenable to computer automation, it is generally better to resolve such differences by adjusting the boundaries, and if necessary the legend, by further reference to the physical factors involved.

When there are more than three definitive criteria, statistical methods usually become preferable to graphical ones. These normally assume that all sites or areas can be assigned unequivocally to a particular class by choosing two or more attributes, treating them as 'crisp sets', and classifying them by methods which can be referred to under the name of Boolean logic.

The most efficient approach is by 'principal components analysis' (PCA) and related grouping procedures. This method of analysis was first suggested by Hagood *et al.* (1941), was developed in association with the technique of 'direct factor analysis' (Cole and King, 1968), and has been used in terrain study by Mather and Doornkamp (1970), Cadigan *et al.* (1972), and Mather (1976). The method reduces the number of mapped attributes by grouping them into sets that are more or less similar. A computer carries out a multiple correlation analysis in such a way that the correlation coefficients between every pair of attributes at each location are assigned locations in multidimensional space, where they will tend to form clusters. These clusters can be interpreted in terms of a small number of axes, called 'principal components', at different 'angles' to one another. These axes are not themselves the attributes, but represent contributions of various sizes from all and 'explain' a part of the variations between them. The amount of explanation contributed by each principal component is quantified as its 'eigenvalue', defined as the value of each characteristic or as a latent root of the correlation matrix. It may be thought of as the length of vector which passes through a scatter of points in multidimensional space so that it reduces to a minimum the distance between the points and the vector. Two or three eigenvalues normally explain most of the variations within the matrix, and these, rather than the original land attributes themselves, can be used as the basis for mapping boundaries.

The complexity of natural variations, the limitations of field sampling, and the uncertainties of cartographic delimitation, however, often make it difficult or unsatisfactory to regard sites or areas as crisp sets. In these circumstances, it is sometimes useful to be able to recognize them as having a 'partial membership' in a particular class, and to quantify the level of this membership. This is done by considering them as 'fuzzy sets' rather than as crisp sets. A fuzzy set can be defined as a generalization of a crisp set in a situation where the class boundaries are not, and cannot be, sharply defined. The method applies a mathematical formula to all relevant properties of the site or area such that its membership in a particular class can be numerically graded. It is designed to solve questions of the following form:

16

'If certain standard values for soil depth, texture grade, and gradient are critical for a certain land class, can one include a site which may be adequate in two but falls somewhat short in the other?' It answers such questions by deriving an overall 'composite membership function' for the site between 0 and 1 showing how well it accords with the central concept of the class (i.e. how closely the function approaches to 1). The formula compares the value of each property at each site with the value required for that property to belong to the class. It then weights the result from each property according to its importance and sums the results of all properties for the site. If this sum is over 0.5 (the 'cross–over point'), the site is assigned to the class, if less, it is excluded. The method is useful but can only be employed after field experience in the area in question makes it possible to determine which properties should be definitive and to quantify: (a) the relative importance of each; (b) the critical values of each required for class membership; and (c) the numerical penalty to be assigned to deviation from this critical value. The method has the advantage of reproducibility and exactness, but appears time-consuming for routine use. Details of the method of calculation are given by Kandel (1986) and are simply summarized with a worked example by Chang and Burrough (1987).

The delimitation of boundaries should not obscure the interdependence of regions. The character of land cannot be understood fully in terms of local controls acting in isolation but is in part determined by relationships with adjoining areas from which they receive and to which they give runoff, groundwater, microclimatic influences, and sedimentation. They are 'open' rather than 'closed' systems. This is emphasized when an area is subjected to such hazards as flooding, erosion or burial from forces originating elsewhere. When extreme these can drastically alter its character.

2.4 The reality of regions

In the light of the foregoing discussion, it is pertinent to ask whether regions have a 'real ' existence or whether they are just arbitrary spatial subdivisions of multithetic continua. This debate has parallels in other disciplines. In history or sociology the question arises whether communities or nations really exist as 'organisms' or are merely collections of individuals. In ecology, the same query applies to climax associations, and in soil science to soil classes. The debate is that between 'realism' and 'nominalism', first discussed by Aristotle, and put in the context of soil science by Robinson (1949). The realist maintains that abstractions like 'table', 'region', 'community', 'podzol', etc. have a real existence of which the individual examples are expressions. The expression may deviate more or less from the ideal, but the ideal would still exist in the absence of any example of it. The nominalist maintains, to the contrary, that abstractions are mere names, and have no essence apart from the examples which compose them. There is thus no such thing as 'mountain' in the abstract, but 'mountain' is merely the name we arbitrarily assign to a group of objects having certain recognizable characteristics in common. This controversy has an echo in regional science. Some workers, including notably the Russians, claim that there are absolute physical and human regions and that it is the aim of science to discover and elucidate them. It is hard, however, to reconcile this with the difficulty of defining particular regions and more specifically with the concept of nodal regions which grade imperceptibly into one another. On the whole, the nominalist view is the more practical and terrain units are most conveniently viewed as assemblages of properties rather than as truths to be discovered.

Further reading

Douglas (1977), Grigg (1967), Smith (1975), Vink (1983).

3
Types of physical regionalization

3.1 Worldwide classifications

Large-area classifications have been attempted since the mid nineteenth century in a number of environmental sciences. More recent contributions have included those by Thornthwaite (1948), Meigs (1957), and Unesco (1979) for climate, Holdridge (1947, 1966) and Holdridge and Toshi (1972) for 'life zones', and FAO for 'agro-ecological zones (Higgins and Kassam, 1981).

The earliest general classification of landforms seems to have been that by Passarge (1919). His system was hierarchical and remains one of the most comprehensive produced. It included five categoric levels of descending importance: 'type', 'class', 'order', 'family', and 'kind'. The two types represented land and coastal forms. The classes within each of these were those mainly suffering erosion and those mainly experiencing aggradation. Orders identified the main types of processes: tectonic, volcanic, and eruptive. Families subdivided these on the basis of the main forms in which processes are expressed, e.g. the faulted, flexured, and fractured families of the tectonic order. Finally, the kinds represented the degree to which the characteristics of the family were expressed, e.g. the distinction between symmetrical, asymmetrical, and overthrust kinds of the faulted family.

Other classifications include the comprehensive schemes of Howard and Spock (1940), von Engeln (1942), and the military classifications of coastal environments by Putnam et al. (1960) and of deserts by Perrin and Mitchell (1970). These tend to follow a pattern of first separating landforms into destructional types dominantly experiencing erosion and constructional types dominantly experiencing aggradation, subdividing the former on form and lithology, and the latter on formative process.

3.2 National and regional classifications: geomorphological

National and regional classifications based primarily on geomorphology have

19

been undertaken since the mid nineteenth century mainly by earth scientists and engineers. They have been reviewed by Wright (1967), Christian and Stewart (1968), Beckett and Webster (1969), Perrin and Mitchell (1970), Schneider (1970), Howard and Mitchell (1980, 1985), Erol (1983), Vink (1983), and Bailey *et al.* (1985). Some of the categories proposed are summarized in Table 3.1. It is significant that this table differs from Table 2.1 in the levels recognized. It excludes Order VIII which represents details, such as small particles and rock pitting, too small to be significant for land use, but includes practical intergrades between V and VI and between VI and VII.

The earliest work was done in the USA in the late nineteenth century, stimulated by the challenge of rapid expansion into new lands. Bowman (1911) subdivided the country into 'physiographic types' related to land uses, and in 1914 Joerg reviewed its regionalizations to that time. This influenced many surveys between the wars (Heath, 1956). In 1916, Bowman showed the need for detailed landscape classification for the study of land uses in the Andean altiplano and, in the same year, the Association of American Geographers established a committee under Fenneman to define the physiographic regions of the country, which were further elaborated in 1928. They used 'section', 'order', and 'division' as their higher units. Perhaps the earliest systematic attempt at regional land evaluation was the Michigan Economic Land Survey of 1922 (Mabbutt, 1968). This, followed by Veatch's study (1933), classified the agricultural areas of the state into 'natural land types', which have had as permanent a value as geological or soil maps of the area. Landscapes were subsequently delineated as physiographic provinces in the southeast by Hodgkins (1965), in the Pacific northwest by Franklin and Dyrness (1973), and in coastal areas by Terrell (1979). Lotspeich and Platts (1982) proposed that climate should be combined with physiography to form a single theme.

The pioneering study in Great Britain was that by Bourne (1931) who defined a 'site' as a unit which would for all practical purposes provide throughout its extent similar conditions of climate, physiography, geology, soils, and edaphic factors. Sites recurred in associations he called 'regions'. He also pointed out the help which could be obtained from aerial photographs. Wooldridge (1932) described 'flats' and 'slopes' as the ultimate units of relief and defined the generic term 'facet'. Unstead (1916, 1933) recognized a hierarchy of units by defining the terms 'feature', 'stow', and 'tract' in order of increasing size for the smaller subdivisions of landscape. Linton (1951) combined Unstead's with Fenneman's terms (replacing feature with 'site') into a hierarchy of morphological regions. Jenny (1958), in the USA, subsequently suggested the term 'tesserae' for units which seem to be somewhat smaller than facets. More recently, soil surveys in mountainous areas such as the Scottish Highlands have used a system which is nearer to landscape ecological mapping than to traditional soil mapping (Vink, 1983).

The Australian work is considered in more detail in connection with agriculture and engineering in Chapters 23 and 24. It is sufficient to state here that the contribution has come since 1946 from two departments of the

Table 3.1 Major heirarchical classifications of terrain units (approx. nationally & chronologically arranged)

Order Tricart, 1965b	I	II	III	IV	V		VI		VII
Dimension Schneider, 1970	←— Planetarisch —→			Regionisch	←— Chorisch —→		←— Topisch —→		
UK–CSIRO WR Howard and Mitchell, 1985	Land zone	Land division	Land province	Land region	Land system	Land catena	Land facet and Land clump	Land subfacet	Land element
CSIRO AG Grant, 1973/74					Pattern		Unit		Component
Netherlands vanZuidam, 1985					Terrain system (pattern)		←— Terrain unit —→		←— Terrain component —→
China Ni Shao Xiang, 1985		Realm (0)	Temperature belts (1)	Region (2); subregion	Zone (3); subzone	Natural provinces (4) Natural prefectures (5)	Natural county (6); Land type		
USA Vink, 1983; Bailey et al., 1985			Domain; division	Subzone; province	Section; district	Land type association	Land type	Land type phase	Site; tessera (Jenny, 1958)
Canada Vink, 1983; Bailey et al., 1985			Ecozone; ecoprovince	Ecoregion	Ecodistrict	Ecosystem	Ecotype	Ecophase	
Mexico Quinones-Garza (1983)		Physiographic province	Physiographic subprovince	Physiographic discontinuity	System of topoforms		Topoform		Element
Turkey Erol, 1983	Yerkure	Kusak; Ülke	Bolge	Bolum	Yöre	Gevre	Kesim; Alt Yöre		Alan
Indonesia Desaunettes, 1977				Physiographic region	Land system; complex of land catenas	Land catena	Land facet		Land element
Germany Schmithusen, 1948, 1963, 1976	Geogr. zone	Idiochore	Region; Synergotyp	Grosseinheit; Synergen	Haupteinheit; Synergie	Grundeinheit; Geotop	Grossfliese		Fliese; Choreose
Socava, 1974	Gürtel	Kontinente	Landschafts zone	Subzone; Provinz	Naturbezirke Landschaft				

Table 3.1 cont.

Order Tricart, 1965b	I	II	III	IV	V	VI	VII
Herz, 1973	Gürtel	Megachore	Makrochoren Gefüge	Makrochore	Mesochore; Mesochoren Gefüge	Mikrochore; Mikrochore Gefüge	Physiotop; Physiotop Gefüge
Schultze, 1955, 1966			Gruppe von Grosslandschaften	Landschaftsgruppe; Grosslandschaft	Landschaft	Gruppe v. Zellen Kompleks; Landschaftsteil; Landschafts-Zellen Kompleks	Landschafts zelle
Richter, 1967, 1968	Gürtel; Kontinent	Megachore; Makroregion	Mesoregion	Mikroregion	Mesochore Oberstufe	Mesochore Unterstufe; Mikrochoren Gruppe; Mikrochore	Physiotop; Ökotop
Klink, 1966	Zone	Grossregion	Region	Gruppe v. Haupteinheit	Naturräumliche Haupteinheit	Untereinheit; Teileinheit	Fliese; Ökotop; Physiotop
Haase, 1964				Makrochore	Mesochore Oberstufe	Mesochore Unterstufe; Mikrochore; Mikrochoren Gruppe; Ökotop Gefüge	Ökotop
Neef, 1963	Geosphäre	Georegion	Megachore	Makrochore	Mesochore	Mikrochore	Physiotop; Ökotop
Naturräumliche Gliederung Deutschlands, 1954		Makroregion; Mesoregion	Mikroregion	Megachore Übereinheit; Makrochore Grosseinheit	Mesochore Haupteinheit	Teile v. Haupteinheit	
Gellert, 1959/60	Megaregion		Grossregion Ordnung-suafel	Grossregion Ord: 2, 3	Naturml. Ord: 4	Mikrochore Teileinheit	
Müller-Miny, 1958					Naturml. Ord: 5, 6	Naturml. Ord: 7	
Paffen, 1948, 1953	Landschafts-zone	Landschafts-region	Gross Landschaftsgruppe	Gross-landschaft	Einzel-landschaft	Klein-landschaft; Landschafts-Zellen Kompleks	Teil Kompleks Landschaftszellen
Otremba, 1948			Region	Gruppe der Phys. Geog. Einheit	Phys. Geog. Raumeinheit	Unter-gliederung der PGR	
Carol, 1957					Flurregion Kompleks	Flurregion; Gruppe der Flurkompleksen; Flurkompleks	Flur; Flur-stuck

Table 3.1 cont.

Order Tricart, 1965b	I	II	III	IV	V	VI	VII
Troll, 1939 (Schneider, 1970)							Klein-landschaft; Ökotopen-kompleks; Physiotop; Biotop; Ökotop.
USSR Isacenko, 1965	Zone	Subzone	Provinz;	Unterprovinz; Okrug	Rayon Landschaft	Mestnost	Urochishche; Sub urochishche
Gvozdetskiy, 1962			Type class	Subtype	Group	Species	Fazies
Poland Kondracki, 1964	Territorium	Subzone	Provinz	Unterprovinz; Makroregion; Makrorayon	Mesoregion; Mesorayon	Mikroregion; Mikrorayon	
Hungary Pecsi and Somogyi, 1967	Geozone	Kontinent; Megaregion	Grosslandschaft; Makroregion	Mittel-landschaft; Mesoregion	Klein Ls.gruppe; Mikroreg. syst.	Klein-landschaft; Mikroregion	Fazies system; Ökotop-system; Fazies; Ökotop
Japan Nakano, 1962		Division	Province	Association	Section		Series; Landform type
UK Unstead, 1916, 1933; Bourne, 1931; Wooldridge, 1932.	Major region	Minor region	Subregion	Tract group	Tract	Subgroup; (region, Bourne)	Stow (site, Bourne) (flat, slope, facet, Wooldridge); Feature; Site

CSIRO: the Division of Water Resources (formerly Land Research and Regional Survey) and the Division of Applied Geomechanics (formerly Soil Mechanics). The former introduced the terms 'land unit' as well as land system (Christian and Stewart 1968), and the latter used the words 'pattern', 'unit', and 'component' for describing physiographic divisions of decreasing magnitude (Aitchison and Grant, 1967).

The Soviet approach derives from the pioneer work of Dokuchaiev, Beng, Ramensky, and others, and was described by Solntsev (1962) and Vinogradov *et al.* (1962). The hierarchy of land units appears to be remarkably similar to those from other parts of the world, notably the Australian. Definitions are in terms of relief, soil, bedrock, microclimate, and habitat conditions such as moisture, salinity, and vegetation. Economic significance is also noted. The smallest unit is the *fazies*, with constant site and habitat conditions, and a single 'biocoenosis' repeated in regular patterns. Prokaiev (1962) opposed the too rigid definition of *fazies* because of the variety of land uses that were related to them and their inevitable internal variability. The next larger units in order of increasing size are *zveno, sub-urochishche, urochishche,* and *mestnost.* The *urochishcha* (plural of *urochishche*) are usually clearly interpretable on aerial photographs and are defined as associations of *fazies* with a homogeneous substrate and a common pattern of drainage and transport of suspended and dissolved materials. They can be simple or complex. They are regarded as complex if they have an association of *fazies* related to a single relief form with sharply changing substratum, or an alternation of small relief forms, such as sand hillocks.

In East Germany, landscape regionalizations were inspired by Troll (Leser, 1978), Neef (1963), and Haase (1968), and were strongly influenced by the needs of agrarian planning in state-run agriculture. The approach was an 'agricultural habitat mapping based on landscape ecological investigations' (Vink, 1983). In West Germany, earlier concepts of *Fliese, Physiotop,* and *Landschaftszelle* were systematized into a hierarchical arrangement by the Institut für Landeskunde at Bad Godesberg (Schneider, 1966).

As indicated in Table 3.1, the comparable series of units in Japan are, in order of increasing size: 'landform type', 'series', 'section', 'association', 'province', and 'division' (Nakano, 1962), and similar systems have been evolved more recently in other countries, including Indonesia (Harrop, 1974; Desaunettes, 1977), Turkey (Erol, 1983), and China (Zhao Songqaio, 1984, 1986; Lin Chao, 1984; Ni Shao Xiang, 1985).

3.3 National and regional classifications: ecological

Ecologically based regionalizations have been largely the work of foresters, plant ecologists, and agricultural scientists. They have emphasized the characteristics and interrelations of plant populations, and are likewise summarized in Table 3.1.

24

Their development can be traced through the evolution of a terminology in the science relating plants to their environment. The introduction of the term 'physiography' to describe vegetal as well as geomorphic aspects of the environment is ascribed to Linnaeus. Although less used today and superseded partly by 'biogeocoenose', it retains this wide sense. In the early nineteenth century Alexander von Humboldt ascribed the term 'geobotany', previously referring to traditional plant taxonomy, to plant and animal geography. His *Kosmos* appeared in English translation in 1856. In Europe geobotany has come to be applied to field botany (Mueller-Dumbois and Ellenberg, 1974) and the study of communities especially in relation to the higher levels of organization (Dansereau, 1951). Since about 1960 it has also been used for the science concerned with the recognition of geological phenomena from their plant cover (Brooks, 1983). The term 'ecology' was first suggested by Haeckel in 1866 for the study of plant–habitat relationships. This was subdivided by Schroter (quoted by Troll, 1971) into 'autecology' and 'synecology', relating to the habitats of single plant species or of whole communities respectively. This usage is still followed in Continental literature, while American and British publications use 'ecology' as equivalent to synecology but extend it to include animal as well as plant ecology.

Passarge (1919) popularized the term *Landeskunde*. Although in English-speaking countries it has tended to be used mainly when referring to the German and Russian approaches to the science, its translation 'landscape science' has remained in general use. Troll saw the potential importance of the idea behind these words. He suggested the term *Landschaftsökologie* and as early as 1939 advised that the science it represented be used in conjunction with the interpretation of aerial photographs as a means for exploring little known landscapes, and foresaw that this would be one of the most important directions for future geographical research (1971). Leser has used the term as the title for a comprehensive book on the subject (1978). The term 'physiographic plant geography' has also been used in this sense (e.g. Zimmerman and Thom, 1982).

A number of terms refer to areal units. The 'ecoregion' concept was advanced by Dokuchaiev (1899) and refined by Passarge (1919). Clements and Shelford (1939) used 'biome' to refer to a large 'climatic–faunistic unit with a particular type of vegetation'. Unesco's 'Man and Biosphere' programme (1973b) recognized two hierarchical levels: realms and provinces in their international classification of vegetation, leaving the establishment of lower entities to local experts. Walter and Box (1976) introduced a scheme for classifying ecosystems based on biomass.

In German and Russian literature the terms 'biocoenose' and 'biogeocoenose' are widely used. Biocoenose is a synonym for the plant community at any rank and so can include the large-scale 'plant panformations' of Du Rietz (1936) as well as smaller units such as the 'plant formation'. It does, however, exclude the habitat, known as the 'biotope'. This closely parallels Tansley's 'ecosystem' (1935), which consists of the organic biome in combination with the inorganic factors of the habitat. The equilibrium system resulting from the combination of these two factors is the biogeocoenose, which can be defined as a:

> combination on a specific area of the earth's surface of homogeneous natural phenomena (atmosphere, mineral strata, vegetable, animal and microbic life, soil, and water conditions) possessing its own specific type of interaction of these components and a definite type of interchange of their matter and energy among themselves and with other natural phenomena, and representing an internally contradictory dialectical unity being in constant movement and development (Sukachev and Dylis, 1966).

Morozov (1931) in his foundation work on 'forest biogeocoenology' recognized the individual tree stand as the biogeocoenose. A considerable number of terms have developed from this, notably those extending it to cover different scales, e.g. 'micro–', 'meso–', and 'macro–landscapes' for units of increasing size in forest and rangeland, and small 'epimorphs' within larger 'epigenema' in bogs (Sukachev and Dylis, 1966). When human uses are included the units are called *Kulturbiogeoceonoses*.

A more theoretical approach is through the science of 'chorology' which treats of the laws of distribution of living organisms over the earth's surface. It appears to owe its original inspiration to Kant's lectures on physical geography (quoted 1922), and was developed by Hettner (1934) and others. De Jong (1962) gave it a more philosophical form, emphasizing that its meaning was broad enough to embrace the whole study of differentiations within the distributions of phenomena. Isacenko (1965) distinguished between the 'typological' (i.e. thematic) and 'regional' subdivisions of landscape components, equivalent to the 'general' and 'special' types of chorological differentiation described by de Jong.

Many countries now have national schemes of ecological classification and mapping. Those in North America have been summarized by Bailey *et al.* (1985). In the USA they began with Allen (1892) and Merriam (1898) and were developed in Dice's map of biotic provinces (1943), Austin's map of land resource regions (1963), and Aldrich's concept of 'life areas'(1966). At the regional and state level, Braun mapped the deciduous forest regions of eastern North America (1964), Zimmerman a part of southeast Arizona (1969), and Arno the forest regions of Montana (1979). The aquatic resources of some states are being inventoried by the US Environmental Protection Agency within a framework of 'aquatic ecosystem regions'. Ecological land classifications at more detailed levels include a hierarchy in the sense of Fenneman by Godfrey (1977) and 'climate and vegetation zones' in the 'life-zone system' of Holdridge and Toshi (1972). A map of 'Ecoregions of the United States' at a scale of 1:7 000 000 was produced (Bailey, 1980, 1983), making an inland counterpart of the regionalization of marine and estuarine systems proposed by Cowardin *et al.* (1979). This is combined with the scheme of the Rocky Mountain Forest and Range Experiment Station at Fort Collins, Colorado, as simplified by Vink (1983) in Table 3.1.

Since Canada is largely covered with natural vegetation, land classifications of the country have been mainly ecological. Forest regionalizations began as early as 1937 and were presented as *Forest Regions of Canada* by Rowe in 1972, while Zoltai *et al.* (1975) published a map of wetland regions. At the regional or provincial level, Krajina (1965) in British Columbia, Hills (1942, 1949, 1950, 1960) and Hills and Portelance (1960) in parts of Ontario, and Loucks (1962) in the Maritimes mapped regionalizations in general use for planning developments in their respective areas. Their underlying concepts were defined by Lacate (1961, 1969). At a wider scale, Hare (1959) described a reconnaissance survey of Labrador–Ungava in which the sample areas were studied on aerial photographs to obtain a key of surface types, and then large areas mapped using the key. More recently, landscape ecological surveys have been carried out within the scope of Environment Canada, the national organization for environmental planning. The Canada Land Inventory (1965) produced many maps of lands in agricultural use. When it was nearly completed a new system, known as the 'biophysical land classification', was developed for the inventory and evaluation of northern forests and wildlands (Rubec, 1979). This is significant as the first major classification which includes water areas as well as climatic, land, and vegetal factors. The special interest in wetlands is shown by Welch (1978) and guidelines were produced to aid surveys for environmental impact analysis (Ecological Land Survey Task Force, 1980). Examples of its use are surveys of the Yukon (Oswald and Senyk, 1977), Quebec (Jurdant *et al.*, 1977), the Northwest Territories (Wiken *et al.*, 1981), and Labrador (Lapoukhine *et al.*, 1978). The work has been reviewed by Wiken (1979) and Vink (1983) and has a rapidly growing literature.

Two concepts are evident in the Canadian work which find echoes elsewhere. The traditional and more established is the 'environment controlling system'. This assumes the primary influence of one dominant component of the ecosystem, such as macroclimatic regime, and maps this. The alternative is the 'environmental synthesis system' used in the Canada Land Inventory. This is holistic integration of all environmental controls in any given ecoregion, the mapped units reflecting the relative weights given to each. Both schemes have used the hierarchy given in Table 3.1, which approximately parallels that of the USA, but with somewhat fewer levels.

In France, Long (1974) recognized a hierarchy of five ecological units, ranging in size from the *élément de station écologique* to the *zone écologique* which closely parallels the terrain hierarchy considered in Chapter 4. A system of environmental mapping at a scale of 1:50 000 has been based on this (Journaux, 1978).

In the Netherlands, landscape ecological surveys derive from soil mapping established by Edelman in 1950. They are based on photo-interpretation with scale and legend tied to the practical evaluation of the mapped units, adding landscape phenomena such as mass wasting, vegetation, artefacts, and aquatic features. Soil maps remain as a special type of landscape ecological map appropriate where soils provide the best differentiating characteristic. Illustrations of both types are given by Vink (1983).

27

3.4 Theoretical landform associations

Landforms are often geographically grouped into genetically linked associations. These associations can recur in the same way as can the individual landforms. Therefore, both can be generic as well as specific. Various authors have suggested models of such generic associations, mainly of landform sequences on hillslopes.

Among the earliest was Milne's use of the term 'catena' to describe recurring toposequences of soils on common parent materials on hillsides in East Africa (1935). Penck (1927, as quoted by Thornbury, 1954) considered landscape within the context of the earth's crustal instability and viewed hillslopes as being convex, concave, or linear, depending on the dominant process they were undergoing. If uplift was dominant and erosion unable to keep pace with it, slopes were convex; if erosion was dominant they were concave; while if the two processes were in balance they were linear. In a landscape under the second of these conditions (the *absteigende Entwicklung*) Penck recognized two parts of the hillslope, an upper relatively steep slope (*Boesche* or *Steilwand*) and a lower relatively gentle slope (*Haldenhang*), which generally met one another at a steep angle. Meyerhoff (1940) called the former the 'gravity slope' and the latter the 'wash slope'.

King (1962), from South African experience, suggested a subdivision of hillslopes into four units: 'crest', 'scarp', 'debris slope', and 'pediment', each associated with distinctive processes of mass movement and water flow, to which Ruhe (1975) added a lower unit known as a 'toeslope' (Fig. 3.1). A more closely articulated scheme, based on data from New Zealand, not only subdivided the hillslope into a nine-unit surface, but showed the genetic relations between the units (Fig. 3.2) (Dalrymple *et al.*, 1968; Conacher and Dalrymple, 1977).

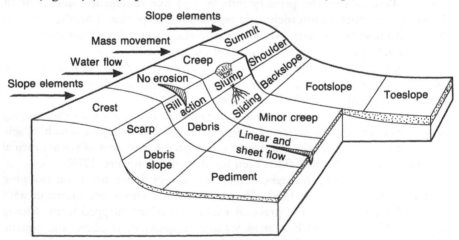

Fig 3.1 The morphological elements of a hillslope formed under the action of running water and of mass movement under gravity. Foreground (King, 1962); background: elements of a fully developed hillslope (Ruhe, 1975)

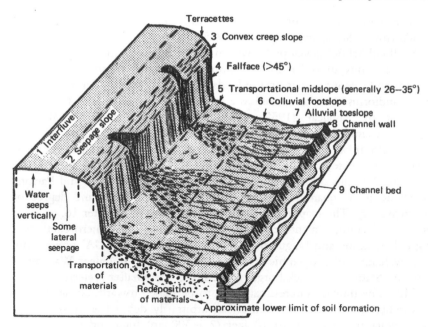

Fig 3.2 Hypothetical nine-unit land surface model (after Dalrymple *et al.*, 1968)

In order to avoid the subjectivity inherent in most hillslope schemes, Young (1971) proposed partitioning the slope profile into units in such a manner that the coefficient of variation of the gradient within each does not exceed specified maximum values and each measured length is allocated to the longest acceptable slope unit of which it is a part.

Theoretical associations of a few landforms restricted to narrower distances and smaller ranges of lithological variety are more frequent. Indeed, they are used in most geomorphological and soil surveys. A few examples may be quoted. Sand dunes in arid areas have been considered to be repetitions of only three surfaces: sand sheets, pack faces, and slip faces (Mitchell and Perrin, 1966). Coral reefs have been subdivided into types and constituent units on the basis of Australian examples by Fairbridge (1946–47) and volcanoes and volcanic landforms generally by Cotton (1944).

3.5 Geometrical systems for small landforms

Geometrical classifications differ from those discussed in previous sections in that they consider only the smallest land units, take no account of soil or lithology or any other feature except surface geometry, and are essentially parametric rather than physiographic. They explore means of expressing microrelief on maps and mathematically. Notable examples are the morphometric mapping of the 'Sheffield School' and the general analysis of small landforms by Stone and Dugundji (1965).

The Sheffield School developed a system for recognizing and mapping small morphological units in order to determine the origins of small geometric subdivisions and discontinuities on the landscape and to train undergraduates to recognize landforms. This originated as a body of techniques evolved for slope measurement by Savigear (1952, 1956, 1962) and Young (1963), and developed into morphological mapping by Waters (1958), Savigear (1960), Bridges and Doornkamp (1963), and Curtis *et al.* (1965). The techniques were summarized and extended by Savigear (1965). Young (1971) placed the definitions of the units on a more objective and quantitative basis, and Parsons (1978) extended the quantification to include also their shape and degree of surface irregularity. The mapping technique is discussed in Chapter 16.

The classification, quantification, and mapping of microrelief was investigated by Stone and Dugundji (1965) as part of the 'MEGA' programme (military evaluation of geographic areas) of the US Army Engineer Waterways Experiment Station at Vicksburg, Mississippi, hereinafter referred to as USAEWES. The military interest caused it to be biased towards considerations of trafficability. Microrelief was somewhat arbitrarily defined as consisting of surface irregularities 3 inches to 10 feet (7.5–300 cm) high and 4 to 64 feet (1.2–20 m) in lateral extent. In California 22 micro-terrains were examined, ranging from a wave-cut terrace to a boulder-free dry wash. These were mapped at the very large scale of 1:120 and a series of radial profiles were constructed through each area at bearing intervals of 15°. The profiles were then evaluated to give parameters which could be regarded as indicative of surface roughness. A combination of four, each measured along every profile within each micro-terrain and then quantified by Fourier analysis, gave a 'roughness vector'. These were: average number of changes in level (M), average height (An) and steepness (P) of major relief features, and the extent to which there was periodicity (K). These permitted two further quantities to be determined: the 'cell length' (CL) which indicated the minimum distance along the traverse required to encounter all its normal variations, and the 'avoidance ratio' (ρ), equal to $P \times M$. The latter proved to be the most generally useful value as it gave a quantitative index of surface roughness which was inversely related to vehicle trafficability.

Further reading

Bailey *et al.* (1985), de Jong (1962), Leser (1978), Stone and Dugundji (1965), Sukachev and Dylis (1966).

4

Parametric and physiographic systems of terrain evaluation

4.1 Introduction

In approaching the practical classification of terrain at the scale required for most land use planning, two alternative approaches are possible.

First, one can consider the terrain from the point of view of the uses envisaged, devise a list of the relevant land attributes and the class limits required within each, and then map each one. A superposition of the resulting maps will give a complete classification. As an illustration, one could say that soils in Britain become more suitable for the cultivation of wheat with decreasing altitude and gradient and increasing clay content. If one then classifies each of these attributes into acceptable and unacceptable classes, one can map each of them separately and then overlay them to obtain a composite map which shows the favourability or unfavourability of each parcel in terms of the three attributes. As the number of land attributes or land uses increases, for example if one also wants to know about soil depth and barley yields, the operation becomes more complex. This is known as the 'parametric' method.

Alternatively, one can develop a scheme based on the natural classifications of terrain outlined in Chapter 3. This is achieved by recognizing natural units, mapping them, and then measuring their properties quantitatively to relate them to land uses. It is known as the 'genetic' method at small map scales, and as the 'physiographic' or 'landscape' method at large map scales. The latter two terms are sometimes used to include genetic methods as well.

4.2 Parametric systems

4.2.1 *Definition*

Parametric land classification is the subdivision of land on the basis of selected attribute values. This yields, for each attribute considered, an array of numerical values or quantitative statements related to sample points. These data can be handled manually or by computer to give tables, histograms, graphs, maps, or a combination of these.

When the objective is cartographic, the simplest result is obtained by dividing a single factor into classes at certain critical values, then drawing contours known as 'isopleths' around them, as in a hypsometric map of altitude classes. Additional quantitative parameters, such as isohyets or isotherms, can be contoured in the same way and then superimposed to give increasingly complex maps. Alternatively, numerical criteria can be added to qualitative or descriptive maps to give them a more quantitative framework.

Parametric classifications can be either of the relatively simple 'coordinate' type or of the somewhat more complex 'multivariate' type. Coordinate classifications use a limited number of environmental attributes to produce a closed legend quantifying and grading each attribute and then mapping their combinations which occur. The method is demonstrated graphically in Figs 4.1–4.4 by showing how single terrain attributes can be combined to give a composite parametric map. Figure 4.1 shows hypsometry, Fig. 4.2 rainfall, and Fig. 4.3 geology of southeastern England. Figure 4.4 is a composite parametric map of the three. Multivariate systems assume that sites are 'polythetic individuals' that possess a number of properties. Each property is viewed as a dimension, and the individual site is placed in multidimensional space according to the grading of each of its properties in each dimension. It is then necessary to classify the sites by a mathematical analysis of their positions in the multidimensional space, grouping them into selected units according to their grade in all properties.

4.2.2 *Application to terrain evaluation*

Although parametric classifications are old, they have only been developed for terrain evaluation since the 1950s parallel to, but largely independently of, landscape systems. The method has been most fully developed for military purposes. The USAEWES has been an important agency (1959), and different aspects of the same general approach were used in other postwar US military research such as that by the Quartermaster Research and Engineering Centre at Natick, Mass. (Wood and Snell, 1959), the Air Force Cambridge Research Laboratories (e.g. Ta Liang, 1964), the Office of Naval Research (Melton, 1958), and Cornell Aeronautical Laboratory Inc. (1963). It has also been used in military land evaluation by the Canadian army (Parry *et al.*, 1968); for agricultural land evaluation in Canada and Britain (Bibby and Mackney, 1969), in the USA (US Department of the Interior, 1951; Olson, 1974), the USSR (Ignatyev, 1968), and eastern Europe (Teaci and Burt, 1974); and for urban site and recreational planning, especially in the USA (e.g. Kiefer, 1967).

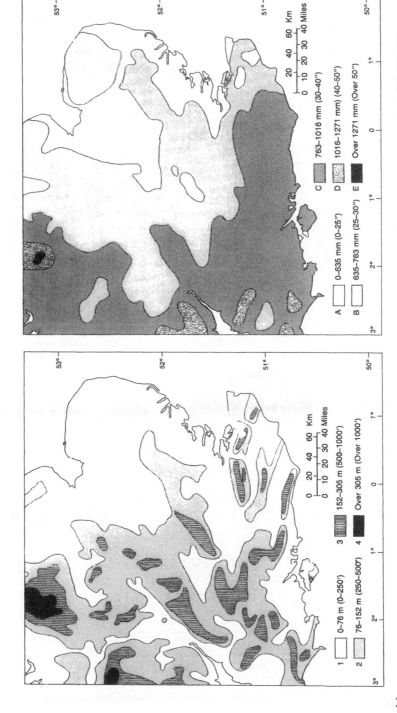

Fig. 4.2 Southeast England: average annual rainfall 1901–31

Fig. 4.1 Southeast England: hypsometry

33

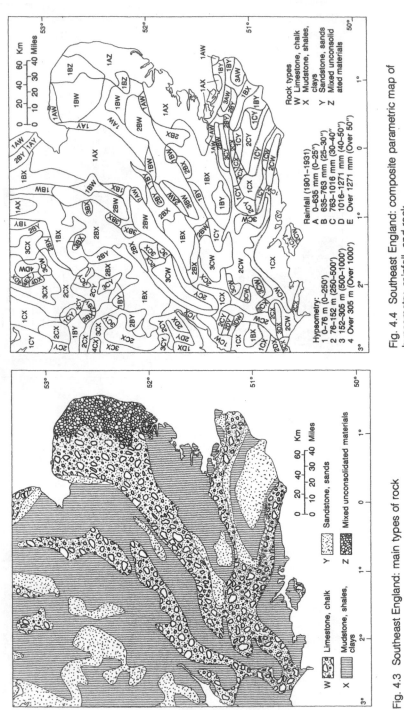

Fig. 4.4 Southeast England: composite parametric map of hypsometry, rainfall, and rock

Rock types
W Limestone, chalk
X Mudstone, shales, clays
Y Sandstone, sands
Z Mixed unconsolidated materials

Rainfall (1901–1931):
A 0–635 mm (0–25″)
B 635–763 mm (25–30″)
C 763–1016 mm (30–40″)
D 1016–1271 mm (40–50″)
E Over 1271 mm (Over 50″)

Hypsometry:
1 0–76 m (0–250′)
2 76–152 m (250–500′)
3 152–305 m (500–1000′)
4 Over 305 m (Over 1000′)

Fig. 4.3 Southeast England: main types of rock

W Limestone, chalk
X Mudstone, shales, clays
Y Sandstone, sands
Z Mixed unconsolidated materials

When parametric classifications are used for terrain, certain inherent problems must be considered. The chief of these are the choice of attributes to be mapped, their subdivision into classes, and the recognition of these classes on the ground. This last requirement is often difficult to meet because of the precision of class definitions and their usual exclusion of recognition criteria.

4.2.3 *Choice of attributes and their subdivision*

Terrain attributes in a parametric map must be recognizable and measurable in the field, and must define the land units at scales relevant to the land uses being considered. National schemes generally agree not only in recognizing a scalar hierarchy of land units, but also in according priority to those at two levels of magnitude: larger grouping units, giving areal coverage and a correlative framework, and smaller basic units which are homogeneous for most practical purposes. The former are hereinafter called 'land systems', and are appropriate to mapping scales of 1:250 000–1:1 000 000. The latter are called 'land facets' and are suited to mapping scales of 1:100 000 and larger, which accord to those of aerial photography.

The objectives of these two scales are somewhat different. At the smaller scale of the land system, representative mapped attributes are selected to record the general character of an area in terms of a single parameter such as the 'most commonly occurring slope class' or of a quantified combination of two or more. Examples are the 'sodium value', combining soil texture and alkalinity to characterize Sudan soils (Jewitt, 1955), and the quantification of desert relief in terms of plan–profile, relief, and slope (USAEWES, 1959). At the larger scale of the land facet, the dominant need is to find a minimum number of parameters which can define individual landforms closely enough to ensure that they have internal homogeneity or at least internal predictability.

The recognition of parametric units requires fairly detailed field sampling unless it is possible to use attributes which are quantifiable by visual observation alone. These are rare and so reliance must normally be placed on the quantification of natural breaks in the landscape which are significant to land use. In the last resort, it is necessary to use arbitrary mathematical ranking of the values of each attribute.

4.3 Landscape systems

4.3.1 *Types of landscape system*

As Beckett (1962) pointed out, we may start from the commonplace that every country may be divided into a finite number of physical regions, each with a characteristic landscape. This is recognized in lay parlance by the use of such terms as Cotswolds or Fens in Britain, Champagne or Jura in France, Great Plains or Appalachia in the USA. Such regions are clearly distinct from their neighbours and usually recognizable both on the ground and from the air. Their unity is due to a common tectonic and climatic history acting on broadly uniform rock, and comparable conditions elsewhere will tend to give similar regions. It is the basic assumption of both genetic and landscape systems that this gives rise to analogies between separated areas close enough for predictions between them to be useful.

4.3.2 *Selection of criteria in defining land classes*

The most useful type of classification, as pointed out by J. S. Mill (1891), is that whose classes permit the widest range of generalizations to be made. For this reason, classifications of natural phenomena should be genetic, selecting criteria which are the causes of other things, or at least sure marks of them. In general, therefore, the best scheme for terrain is one based on origin, process, and form. Its units will reflect the interaction of these factors and have the practical advantage of integrating the many attributes of the landscape into a single whole.

Secondly, the criteria used in defining units should be chosen from the most enduring characteristics of the landscape, giving landforms priority over plants except where the latter are dominant, as in mangrove swamps, some wetlands, or where other distinctions are not clear. Climatic factors can be used to define broad zones, while fauna and the works of man are better considered as attached information rather than as definitive of classes.

Thirdly, landscape units have the characteristics of both uniform and nodal regions. While their internal homogeneity generally consists in a uniformity of natural properties, they also often have a 'sense' of internal variation, deriving from an organization of properties around a node. Neither 'uniformity' nor 'nodality' are absolutes, but vary both with the complexity of different landscapes and with the scale of mapping.

These natural criteria must be balanced by the overriding practical need for simplicity. This is achieved by keeping the number of landscape classes to a minimum while ensuring that each is a valuable vehicle for data storage, i.e. that it is both internally homogeneous and distinct from others.

4.3.3 *Large units: the genetic approach*

Large units, derived from subdivisions of landscape on the basis of causal environmental factors, are appropriate at mapping scales of 1:1 000 000 or smaller which can show areas of continental size on a single sheet. Such large units have, as Mabbutt (1968) pointed out, three practical advantages. First, they provide a useful coordinating framework. Second, they economize the effort of field sampling in the reconnaissance of large areas by providing an overall view of genetic relations. Third, they have educational value in explaining world patterns and regional history, and in broad-scale planning.

On the other hand, they have a number of disadvantages. Because the units are large, usually extending to thousands of square kilometres, they have both great internal complexity and vague boundaries. The lower size limits to practicable genetic subdivision cannot be reduced enough to give units homogeneity for most types of land use. This caused an early search for smaller and more precise units.

4.3.4 *Small units: the landscape or physiographic approach*

The importance of two basic sizes of unit has been emphasized. Both sizes recur in the landscape so that it is possible to recognize analogies between similar, but separated, areas. In this they differ somewhat from the genetic units whose greater size and complexity make them less easily matched. Both land systems and land facets are part natural in that they are based on obvious subdivisions of the landscape, but are also part artificial in that their definitive criteria are based on practical utility.

4.3.5 *Reproducibility versus recognizability of terrain units*

It is useful to make predictions about areas without access to them. Such predictions must be reliable not only where sampling is possible but also where it is not. They may be long range, correlating between land types in widely removed settings; medium range, between examples within a single region; or short range, between sites within a single occurrence of a land type. Accuracy can be expected to deteriorate with the distance over which they are made. Such predictions require the recognition of the terrain type in the unknown area, and the description, if possible in quantitative terms, of its analogue in a known area. Only when these are combined is it possible to use information about the first to make predictions about the second. Two types of capability are therefore needed: recognition of types of landscape, and knowledge of the properties of these types, i.e. recognizability and reproducibility (Beckett and Webster, 1965a).

The recognizability of a type of terrain can be defined as the proportion of it which can be recognized, without ground check, out of the total area it covers, with the tools and resources generally available. These consist of published literature, topographic, geological, and soil maps, and aerial and satellite photographs. In developed regions, comprehensive topographic map coverage may be at scales larger than 1:10 000 and geological maps at larger than 1:100 000. In developing countries 1:50 000 and 1:2 000 000 respectively are more typical, although larger scales are often available of limited areas. Soil maps are generally at similar scales to geological maps, but, apart from worldwide syntheses such as the FAO/Unesco *Soil Map of the World* (1974–), detailed mapping covers a smaller percentage of the earth's surface. It can also normally be assumed that workers exist with some experience of the terrain and climate in question.

The use of aerial and satellite photography for recognizing terrain types underlines the difference between features used in their definition and those used in their recognition. The former are few, fundamental, invariable, and always present, the latter are visible, but not necessarily practically important or always the same. Induced vegetation or the colour of the ground surface, for instance, are among the most useful clues to the recognition of land units on aerial photographs, but are too ephemeral to be normally definitive.

The reproducibility of a type of terrain, on the other hand, is the degree to which it is possible exactly to delimit it in any area. The closer the match required between two occurrences, the more precise must be the definition of attributes, because this will allow the most exact description and boundary delineation. To take an example: a flood plain in a Mediterrranean climate with a slope of less than 2°, silty soil, and a water-table normally below 1 m, is more 'reproducible' than one where only the climate and slope are specified, and this in turn more reproducible than just any flood plain.

On the other hand, the recognizability of a terrain unit, which can be defined as the proportion of its total distribution that can be recognized with the means usually available, will not increase, but will decrease with the precision of its definition. This is because it is easier to recognize broad categories of landform than precisely defined units. In the example quoted, it is easier to recognize any flood plain than one where climate and slope limits are specified, and this in turn easier to recognize than one where more exactly definitive criteria are specified.

Therefore, reproducibility and recognizability are more or less in inverse relation and it is necessary to reach a compromise which optimally combines the two. The principle established by Beckett and Webster (1965a) has been to give roughly equal weight to each, so that land units are not defined to such a degree of precision that a significant proportion of their occurrences become unrecognizable.

4.4 Parametric and physiographic systems compared

4.4.1 *Advantages of parametric systems*

Parametric systems make an objective linkage between sites which eliminates the danger of assuming hierarchical relationships. They permit affinities to be seen between sites that might not otherwise appear. They can provide a useful tool in the analysis of landscape processes. For instance, a combination of maps of topographic benches and of subsoil organic layers could indicate the evolution of a flight of river terraces containing palaeosols.

They emphasize computation, which makes possible a more statistically reliable means of measuring variance, formulating rational sampling policy, and expressing the probability limits of findings. The new techniques of scanning and computing enable them to give an increasingly rapid and complete picture. They are favoured by two new developments: (1) sensors which are able to scan directly attributes which have hitherto had to be inferred from associated features, and (2) electronic data handling and image processing which favours information in quantitative form.

4.4.2 *Advantages of physiographic systems*

First, they help explain the causes of landscape differentiation, both spatial and temporal. Viewing landscape as a whole assists understanding of origins, processes, and genetic relationships. This in turn calls attention to exploitable natural resources, such as springs, mineral veins, and placer deposits. On a smaller scale, surface features such as slope and moisture can give useful indications of soil profile. Landscape features such as the drainage network, the character of the vegetation, and the distribution of snow can lead to an understanding of temporal variations, important in predicting natural hazards such as floods or erosion.

Parametric classifications, on the other hand, can give little information about unsurveyed areas, subsoils, or temporal changes. It is necessary to correct this by sampling in greater detail and more frequently, using more refined scanning techniques, and economizing effort by making a narrower selection of properties which experimentation shows to be most closely related to the land uses envisaged. To predict temporal changes, a parametric system must sample attributes specially chosen to reveal them. It is especially important to include these when surveying areas with extreme seasonal changes of temperature or moisture, particularly when the two occur in combination, as in circumpolar or monsoonal regions.

Second, landscape units govern the distribution of other environmental features such as water, soils, and land uses, and can form a useful framework for their survey and representation. This can reduce duplication of field effort by providing a common basis for many land users. Even in areas where a substantial amount of information already exists but needs reassessment for planning purposes, there are advantages in groups of specialists undertaking it jointly and on a common basis. The greater the number of land uses involved the more the physiographic method will tend to be favoured.

Third, landscape units are visible and readily comprehensible both on the ground and from the air. This is clear from the extent to which they are everywhere recognized in lay language. Local people in rural areas can sometimes recognize more types of land than professional soil surveyors and terminologies for small differences can be surprisingly detailed. In the Sudan Gezira, for instance, the colloquial name for the black cracking clays is *badob*, but they are known as *dahrat* near the Blue Nile, *fuda* where the surface is granular mulch, *mayaa* in basin sites, and *gardud* around the base of hills where they are non-cracking (Tothill, 1952). This ready recognizability is increased by today's wide reliance on remotely sensed imagery for thematic mapping. It contrasts with the less ready comprehensibility of parametric units, especially when based on more than two or three land attributes.

Fourth, a physiographic classification is more speedy and economical in the present state of technology. If information is required over the whole range of needs in agriculture, engineering, and planning, the number of parameters required becomes too large for it to be practicable to measure them all and very complicated to combine them.

Fifth, because landforms are recurrent, there are great advantages in using physiographic analogy to predict land attributes between different areas. Such predictions are not possible with parametric systems because, with very few exceptions, exact measurements of the earth's surface cannot be extrapolated beyond their place of measurement. Because of this limitation, parametric systems require more detail than landscape systems. This necessitates more ground measurement and greater detail in mapping. As information on most attributes tends to be sparse, maps must therefore either be based on slow and costly ground surveys which tend to be restricted to small areas, or else be at scales that are too small for most users. Finally, because information is based on point samples only, accuracy tends to fall off rapidly between the points unless remote scanning of attributes has been possible to fill the intervening areas.

Sixth, physiographic systems make it easier to grade the relative importance of different land attributes. With parametric systems the mathematical process may assign equal weight to all definitive land attributes even if they are unequal in importance. Those that are both indicative of genesis and important to land use, such as gradient or soil texture, should count for more than those of lesser importance such as soil colour or some trace elements.

Seventh, physiographic units lend themselves readily to a variety of scales because they are normally both composite and clearly divisible. For example, Mount Monadnock in New Hampshire, USA, is a recognizable unit made up of an open summit, wooded slopes, and stony gullies, and for detailed studies of flora or fauna, each of these subdivisions could be further subdivided. The open summit, for instance, consists of bare rock, boulders, and fissures, while the gullies contain small waterfalls, steep undercut banks, and pools. By contrast, parametric surveys at one scale are based on a sampling density which may be inappropriate at another.

To summarize, with the standard equipment and expertise of today, the landscape approach offers the possibility of more rapid survey at lower cost than does the parametric. Its reliability is adequate for reconnaissance and, with moderately close sampling, for semi-detailed surveys, and has the advantage of combining all survey work into a single operation. For detailed surveys or in areas without visible internal differentiation, or for the quantitative analysis of important features, the parametric approach adds precision and reliability. The two systems should, therefore, be thought of as ultimately complementary.

Further reading

Christian (1958), Howard and Mitchell (1985), Mitchell (1988), USAEWES (1959).

5

The terrain hierarchy

5.1 The units

The terrain hierarchy discussed in this chapter was developed by a number of bodies, notably including the Australian CSIRO (Christian and Stewart, 1968), the UK Oxford–MEXE–Cambridge Group (Beckett and Webster, 1969), the South African National Institute of Road Research (Brink *et al.*, 1968), and the University of Melbourne (Howard, 1970c). The units represent every scale from global to local, and thus parallel other scientific classifications such as those of soils by the US Department of Agriculture (USDA) (1976) and of vegetation by Long (1974). The internal homogeneity and mutual distinguishability of the units have been tested statistically at all the main categoric levels. The categories are given in Table 5.1, together with the most relevant mapping scales and remote sensing platforms for the recognition and representation of each level.

To some extent, the hierarchy described here can be thought of as a fractal series. All sizes of unit have a genetically controlled assemblage of relief and drainage features, sometimes bounded by a coastline, and islands, for instance, exist which accord in size to each of the lower levels. However, the definitions of the units, especially the larger ones, depend on climatic, structural, and lithological criteria, as well as those of form. This makes the recognition of fractal series within the terrain hierarchy highly complex.

5.2 The land zone

Land zones can either be viewed as representing the major continents and ocean basins or else as representing the earth's major climatic zones. Their boundaries differ except in the Arctic and Antarctic. The subdivision of the land surface most significant to its uses, however, is climatic. This has been emphasized by true colour photographs from satellites which show the way continents have latitudinal zones: green in humid regions, brown in deserts

Table 5.1:   Hierarchical classification of terrain, soil, and ecological units

Terrain unit	Definition	Soil unit	Vegetation unit	Mapping scale (approx.)	Optimal remote sensing platform
Land zone	Major climatic region	Order	—	<1:50 000 000	
Land division	Gross continental structure	Suborder	Plant pan-formation; *zone écologique*	1:20 000 000 –1:50 000 000	Meteorological satellites
Land province	Second-order structure or large lithological association	Great group	—	1:5 000 000 –1:20 000 000	
Land region	Lithological unit or association having undergone comparable geomorphic evolution	Subgroup	Sub-province	1:1 000 000 –1:5 000 000	Landsat SPOT ERS
Land system (simple) *	Recurrent pattern of genetically linked land facets	Family	*Région écologique*	1:200 000 1:1000 000	Landsat SPOT, ERS, and small-scale aerial photographs
Land catena	Major repetitive component of a land system	Association	*Secteur écologique*	1:80 000– 1:200 000	
Land facet	Reasonably homogeneous tract of landscape distinct from surrounding areas and containing a practical grouping of land elements	Series	Sub-formation; *station écologique*	1:10 000– 1:80 000	Medium-scale aerial photographs, occasionally SPOT and Landsat
Land clump	A patterned repetition of two or more land elements too contrasting to be a land facet	Complex	Sub-formation: *station écologique*	1:10 000– 1:80 000	
Land subfacet	Constituent part of a land facet where the main formative processes give material or form subdivisions	Type	—	Not mapped	Large-scale aerial photographs
Land element	Simplest homogeneous part of the landscape, indivisible in form	Pedon	*Elément de station écologique*	Not mapped	

* A land system can be complex if it represents a combination of two or more geomorphogenetically related simple land systems, or compound if the combination is not geomorphogenetic. Complex and compound land systems are appropriate to the mapping scale of the land region.

Sources: soil units, USDA (1976); vegetation units, Howard (1970c), Long (1974).

and white around the poles. Thus, climate is the most useful basis for defining land zones. The 11 units shown in Fig. 5.1 are broadly internally homogeneous in the environmental processes they experience. Land zones are appropriately mapped at scales smaller than about 1:50 000 000 suitable for single-page representations of the whole world. The units are too large to be covered by single frames of earth resources satellites and can only be appreciated when these are assembled into a mosaic or viewed on small-scale satellite imagery, such as that taken from the National Oceanographic and Atmospheric Administration advanced very high resolution radiometer (NOAA/AVHRR) meteorological satellite system (Justice *et al.*, 1985). Land zones accord moderately well, both in degree of conceptual generalization and in geographical distribution, with the 'soil orders' of the USDA system and the highest category mapping units of the FAO/Unesco *Soil Map of the World* (1974–) and with the Unesco scheme for the classification and mapping of vegetation (1973a).

5.3 The land division

The land division is the first subdivision of the land zone, and represents the major continent-sized structures within it, appropriately delimited at mapping scales of 1:20 000 000–1:50 000 000, so that the whole world can be shown on a single page. The cool temperate oceanic land zone, for example, includes a number of land divisions, such as the European, northeast American, northwest American, east Asian, and New Zealand. The arid land zone can similarly be subdivided into the detached constituent areas also visible on Fig. 5.1: Sahara, Kalahari, Arabia, Chinese Inner Asia, etc.

Land divisions within the same land zone differ considerably in landforms and soils. For instance, deserts in the Old World are largely on relatively level geological shields while much of those in the New World lie around mountain chains at the junctions of tectonic plates. Even within the shields, there are profound differences. The soils of North Africa and the Middle East, for example, differ from those in Australia in being generally more calcareous. Likewise, although the physiognomy of plants and the ecology of animal populations are similar in all parts of the same land zone, the actual species compositions can be almost totally different, as evidenced by the contrasts between American, Eurasian, African, and Australasian flora and fauna. Within each land division, the uniformity of vegetation is at the level of plant pan-formations and of Long's *zone écologique* (1974).

5.4 The land province

Land provinces are subdivisions of the land division based on a broad homogeneity of structure and rock type, such as a dominance of igneous mountains, folded sedimentary hills, or wide detrital plains, and may have

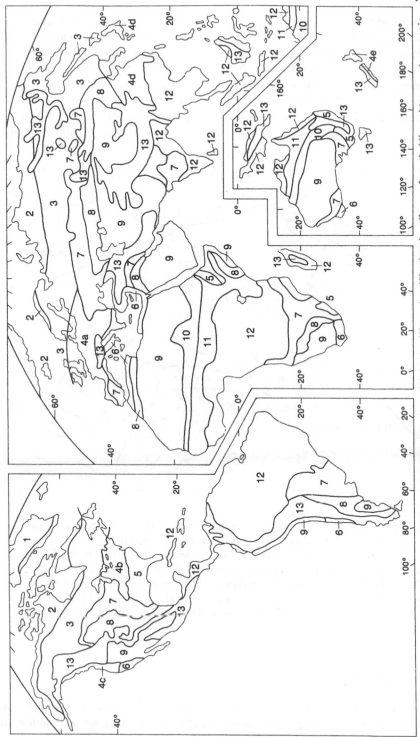

Fig. 5.1 World land zones and some land divisions. World land zones: (1) Arctic; (2) tundra; (3) taiga (coniferous forest); (4) cool temperate oceanic; (5) warm temperate oceanic; (6) Mediterranean; (7) prairie and steppe; (8) temperate semi-desert; (9) desert; (10) sahel; (11) savana; (12) humid tropical; (13) montane. Land divisions of (4): (a) N European; (b) NE American; (c) NW American; (d) E Asian; (e) New Zealand

boundaries which are coterminous with those of plant formations. Their extent is generally measured in hundreds rather than thousands of kilometres, and they are appropriate to the mapping scales of about 1:5 000 000 to 1:20 000 000 used in atlases for single continents. The North American or the European cool temperate oceanic land divisions, for instance, are readily divisible into land provinces. In the former, examples would be the Appalachians, the Piedmont, the Great Lakes Basin, and in the latter, lowland Britain, the Paris Basin, and the Pyrenees. Figure 5.2 shows the subdivision of the European land division of the cool temperate oceanic land zone.

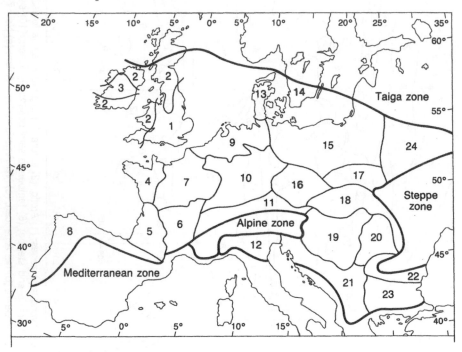

Fig. 5.2   Land provinces in European division of cool temperate oceanic land zone: (1) lowland Britain; (2) highland Britain; (3) lowland Ireland; (4) Armorica; (5) Aquitaine; (6) Massif Central; (7) Paris Basin; (8) pluviose Iberia; (9) Dutch–West German Lowland; (10) Franco-German Uplands; (11) Alpine Foreland; (12) Po Plain; (13) Danish Lowland; (14) Swedish Lowland; (15) Baltic Lowland; (16) Bohemian Plateau; (17) South Polish Uplands; (18) Carpathians; (19) Pannonian Plain; (20) Transylvania; (21) Dinaric Alps; (22) Romanian Pediment; (23) Bulgarian Platform; (24) Russian Plain

5.5   The land region

Land regions are subdivisions of land provinces on the basis of structure and lithology. They have a smaller range of surface form, a closer association of parent rocks, and have had a more homogeneous geomorphic evolution. They are appropriately shown on single-page maps of medium-sized countries such as Britain, Thailand, Zambia, or Venezuela, or of major regions of large countries such as the North American Piedmont or the Gangetic plain in India, according to mapping scales in the approximate range 1:1 000 000–1:5 000 000. Examples of land regions in the former would be the New England and Eastern New York Upland or the Northern Coastal Plain, as defined by Austin (1963). A provisional subdivision of England and Wales is given in Fig. 5.3. Because the land region is essentially geological and

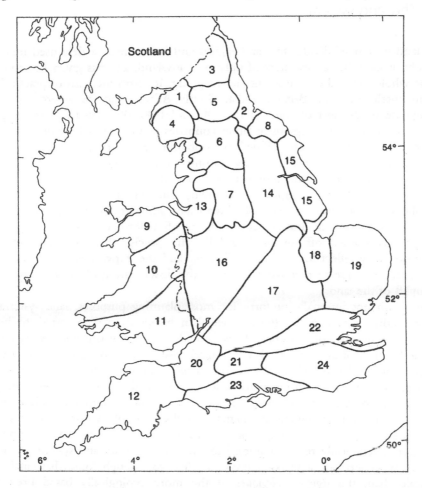

Fig 5.3 Land regions of England and Wales: (1) Cumbrian Lowlands; (2) Durham–Northumberland Lowland; (3) Cheviots; (4) Lake District; (5) north Pennines; (6) central Pennines; (7) south Pennines; (8) North York Moors; (9) North Wales; (10) Central Wales; (11) South Wales; (12) the Southwest; (13) Lancashire–Cheshire Plain; (14) Vale of York–Lincoln; (15) Wolds; (16) Midlands triangle; (17) scarplands; (18) Fenlands; (19) East Anglia; (20) Somerset Lowland; (21) Salisbury Plain; (22) Thames Valley; (23) Hampshire Basin; (24) Weald.

geomorphological in definition, it is sometimes hard to recognize on satellite imagery without considerable ground data, especially in vegetated areas, and a more ecologically based unit, the land subprovince, has been used. This corresponds to Christian and Stewart's (1968) synthesized compound or complex land system, and approximately also to Holdridge's 'ecological life zone' (1947) which is essentially climatic and often equates with plant formations in Latin America.

5.6 The land system

The land system subdivides the land region and is the most widely used unit. Broadly, it identifies a local tract of upland or lowland, and has given its name to the whole method of assessment and mapping. It forms the basis of practical terrain intelligence in Britain, India, South Africa, and elsewhere. It is appropriate to the sort of scales (1:150 000 to 1:1 000 000) used for planning the main administrative regions of larger countries and for the whole of smaller countries. Land systems are defined as containing a recurrent assemblage of land facets, which have a 'genetic' relationship to one another and everywhere occur in roughly the same proportions. They thus reflect an aspect of the 'systems' approach in geography in that many land systems are linked to their neighbours by a single pattern of climatically generated processes. Specific examples illustrate such linkage. Lowlands are often formed by the deposition of materials washed off surrounding hills, the precise shape of sand dune complexes may reflect local wind oscillations, and the stepwise arrangement of river terraces is due to regional changes in the downcutting of the river or to the uplift of the land.

Although not internally uniform for most planning purposes, land systems are generally of such a size (tens rather than hundreds of kilometres) to be useful for a number of purposes. They provide an appropriate framework for practical information, are readily comprehensible and traversable on the ground, identifiable on all aerial and most satellite photography, and amenable to mapping at a useful and inexpensive scale while at the same time permitting a relatively complex legend giving most of the detail required for planning purposes. The intricate geomorphogenetic combination of two or more simple land systems is a 'complex land system' and an equivalent non-geomorphogenetically related combination is a 'compound land system'. Land systems in general are approximately equivalent to, though somewhat smaller in scale than, the *régions écologiques* of the more ecologically based French system (Long, 1974).

Land systems may either be abstracts or local forms, the former being used in mapping and extrapolating between different areas, the latter for local planning. Local forms may sometimes accord with areas identified by a local term. Such terms are abundant everywhere, but can only be useful as land systems if their particular area has a practical significance. Also, because they are smaller than land regions, it is rarer to find examples whose names are household words outside their immediate localities. Nevertheless, some examples could be quoted. In the Weald land region of southern England, instances are the Vale of Kent, the High Weald, and the South Downs (Fig. 5.4), and in the Sonoran Desert land region of the southwestern USA, the Mohawk Mountains and Death Valley. These represent local forms respectively of land system abstracts which could be called clay vale, sandstone ridge, chalk fold hills, river valley on clay, and playa basin.

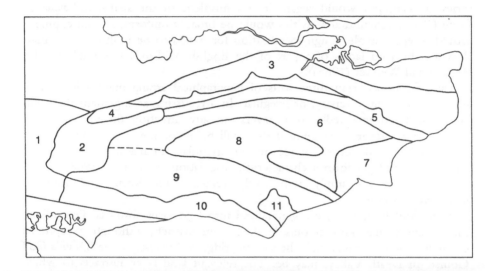

Fig 5.4   Land systems of the Weald land region: (1) Hampshire Downs; (2) Western Heights; (3) North Downs; (4) Vale of Holmesdale; (5) Greensand Ridge; (6) Vale of Kent; (7) Romney Marsh; (8) High Weald; (9) Vale of Sussex; (10) South Downs; (11) Pevensey Levels. (After Stamp and Beaver, 1933)

5.7 The land catena

It is sometimes necessary to recognize a unit intermediate in size between the land system and the land facet, and the term 'land catena' was suggested for this by Howard (1970a). He defined it as a major repetitive component of the land system consisting of a chain of geographically related land facets (e.g. hilltop to valley bottom). A transect of the land facets in a land catena would be recorded at right angles to the topographic grain of the country. The land catena generally contains a characteristic soil association, and an assemblage of several plant subformations, and is equivalent in scale to Long's *secteur écologique*.

5.8 The land facet and land clump

The land facet is the main subdivision of the land system, and is the most widely used planning unit because it can be regarded as homogeneous for moderately extensive land uses (Beckett and Webster, 1965a). It is defined in terms of its geology, water regime, topography, and in unimproved or uncultivated areas, sometimes also on its vegetal structure. Its terrain features are simple. It lies on generally homogeneous rocks, with a single water regime (including both surface and groundwater) and a constant 'sense' of internal variation. A pedologist would map its soils as one or an association of soil series, an ecologist would regard it as equivalent to the *station écologique* of Long (1974), agricultural activities would be broadly uniform, and an engineer would accept a single design specification for works to be built on it. A map showing the land facets of a part of the English Midlands was published by Beckett and Webster (1965a).

Land facets occur in three sorts of geographical arrangement within land systems: recurring, catenary, or irregular. In a recurring arrangement they will be encountered repeatedly in a traverse in any direction through the land system. In a catenary arrangement they will be encountered only on traverses in a particular direction, while in an irregular arrangement they occur haphazardly and yet form a characteristic association (Haantjens, 1968). These three arrangements imply respectively regular, directional, and multiple formative processes.

Land facet local forms are generalized into land facet abstracts. Few of the former are widely known outside their own districts, although exceptional examples, such as Dover cliff, the summit ridge of Everest, or the 'Devil's Golf Course' in Death Valley, may be. The relevant land facet abstracts to which these belong would be cliffs, arêtes, and wet salt flats. Other examples of land facet abstracts would be small river terraces, alluvial fans, or dunes.

Some writers have regarded the difference between land facets and land systems as only a matter of scale, e.g. Christian (1958) and Vinogradov *et al.* (1962). Others such as Beckett and Webster (1962) have recognized a fundamental difference in kind, maintaining that land systems are basically morphogenetic, and land facets basically physiographic, units. Mabbutt (1968) favoured the latter view, but for the reason that land facets are characterized by an unbroken continuity of internal properties that land systems cannot have.

Although both land systems and land facets are called 'natural' units, their definitions emphasize attributes of practical utility and those from which they can be recognized in the field and from the air, so they are also in part 'artificial'. This is especially true of land facets whose value derives from an optimal compromise between their reproducibility and recognizability.

An intermediate level of generalization between the land facet abstract and local form can also be recognized, the 'variant'. This represents a type of land facet distinguished by a particular lithology or a distinct type of intensity of genetic processes. Examples of variants of the land facets just quoted would be chalk cliffs, high altitude glaciated arêtes, and wet salt flats with seasonally self-churning surfaces.

Both local forms and variants can be regarded as 'cognate' subtypes of land facet abstracts if they are relatively predictable parts of the land systems in which they occur, i.e. if they derive from mere local differences in lithology or the intensity of the genetic processes. If they are relatively unpredictable, e.g. due to the presence of isolated areas of vulcanism or exotic deposits, they are considered as 'non-cognate' local forms or variants of the land facet.

Sometimes, for instance in much dissected terrain, an area of land at the same scale as the land facet consists of a repetition of two or more contrasting land elements whose individual occurrences are too small to constitute separate land facets. In this case, a distinct unit known as the 'land clump' is recognized, having the same scale and status in mapping as the land facet. The relation between them is therefore analogous to, though often different in scale from, that between the soil series and the soil complex in soil mapping. Examples of land clumps would be badlands, areas of hummock dunes, and the exposed beds of braided streams.

5.9 The land subfacet and land element

Land subfacets are parts of the land facet caused by differences in the intensity, but not the nature, of the dominant genetic processes (Beckett and Webster, 1965a). They are therefore distinguishable on the basis of material or surface form. An example would be the convex top and the concave bottom of a hillslope too short to be a land catena. These components could have gradients or development potential which differed enough to be distinguished but not to constitute separate land facets.

The land element is the smallest unit of the landscape. It represents Wooldridge's 'morphological electron' (1932) and is indivisible on the basis of form. It is uniform in lithology, soil, and vegetation structure at the level of the plant subformation and of floristics at the level of the plant consociation or association, and is equivalent to Long's *élément de station écologique*. Examples of land elements would be small rock outcrops, gullies on hillsides, or swampy patches in fields. In regions of the world where there are negligible differences in topography, as on some alluvial plains and former lake beds, land elements may extend over considerable areas.

5.10 The testing of the terrain hierarchy

A terrain unit within this hierarchy can only have practical value if (a) its internal variations are less than those of the landscape as a whole, i.e. having a coefficient of internal variation below a statistically significant level, and (b) it differs significantly from others, i.e. has an adequate proportion of inter-unit comparisons showing statistically significant difference. These questions were considered in a series of papers between 1965 and 1979 (Beckett and Webster, 1965a, b, c, and d; Mitchell and Perrin, 1966; Perrin and Mitchell, 1970; Mitchell 1971; Mitchell *et al.*, 1979).

Beckett and Webster showed that in a 5000 km^2 area of the English Midlands, centred on Oxford, land facets were sufficiently homogeneous and different from each other to be used for practical predictions within the study area (1965a, b, c, and d). On the strength of these results, a comprehensive terrain classification was tested over a whole climatic zone, the hot deserts.

Data on soil particle size distributions from the southwest USA, Libya, and the Arabian area, distributed so as to represent every level of the hierarchy, were analysed. The mean coefficient of variation (the standard deviation expressed as a percentage of the mean, hereinafter abbreviated to CV) for 36 land facets was 62.5 per cent when they were considered globally, 57.6 per cent when pooled within land system abstracts, and 45.6 per cent when pooled within land system local forms.

Because a mere statement of within land facet variability has little value if all land facets are alike, it was necessary to supplement these measures with tests to show how often inter-land facet comparisons showed significant difference. These showed that when each land facet is compared with every other within the arid land zone, about one-third of the comparisons showed statistically significant differences; when such comparisons occurred within single land system abstracts, about one-half showed such difference, and when they occurred within single land system local forms, about two-thirds did (Mitchell, 1971).

The rate at which this predictability increased with each categoric confinement of the land area was then determined. The procedure was to calculate a mean and CV over the whole land zone for each of 62 quantified land attributes, and then to calculate them within the land division (pooling the results from all land divisions) and then within the land province and so on down to the land system local form. The median value of the CVs at each level of the hierarchy was then used to calculate the percentages by which the total variance was reduced at each level of the hierarchy, i.e. the degree to which predictive power increased with each diminution of the geographical area over which the predictions were made. The result is indicated in Table 5.2.

The figures are cumulative so that, for instance, the second level reduces the variance by 6.2 + 12.3 = 18.5 per cent, the third by 6.2 + 12.3 + 8.2 = 26.7 per cent and so on. They do not total 100 per cent because each is a median of the percentage values obtained from all 62 properties, and so they should be used for internal comparison only. The last figure shows that about one-third of the total variance of terrain properties remains within the facet local form and thus is unexplained by the classification. Although each level of the hierarchy 'reduces' the variance, the largest single contribution is from the local land facet level, and the next largest from the land province.

Table 5.2 Median values of the variance components for 62 soil properties expressed as percentages of the total variance (from Mitchell *et al.*, 1979)

Between land divisions within land zone	6.2
Between land provinces within land divisions	12.3
Between land regions within land province	8.2
Between land system local forms within land regions	10.1
Between land facet local forms within land system local form	18.2
Between sites within land system local form	32.7
Total	87.7

These findings are generally supported by available evidence from other climatic zones. There is a prima-facie case for assuming that the world can be subdivided into a hierarchy of land units containing a finite number of land system local forms. This number would not exceed a few thousand even if Millard's estimate of 2000 (1967) is probably too low. Each of these land systems contains an assemblage of land facets which are internally homogeneous and mutually distinct for moderately extensive land uses so that valid predictions can be made of site conditions between the different parts of each, and with somewhat less, but often acceptable, accuracy between comparable land facets in closely analogous land systems. At the broader scale of regional planning, it would be most useful to recognize land systems within a framework of land provinces, while for areas of a few kilometres or a few tens of kilometres, land facets, and in critical areas their subdivisions, are needed.

Further reading

Brink *et al.* (1966), Howard and Mitchell (1985), Mitchell (1971), Mitchell and Perrin (1966), Mitchell *et al.* (1979).

Part II

Terrain data requirements

6

The geomorphological framework

6.1 Types of variation in the landscape

Terrain results from the action of climate on rocks whose nature and distribution reflect their structure and lithology. Thus, differentiations between landforms are due to a combination of 'climatovariance', 'epeirovariance', and 'petrovariance'. Landform recognition and analysis have been greatly assisted in recent years through the availability of remotely sensed imagery and illustrated texts which use it (e.g. Verstappen, 1983) or are based entirely on it (e.g. Dresch *et al.*, 1985; van Zuidam, 1985; Short and Blair, 1986).

6.1.1 *Climatovariance*

Landforms strongly reflect the earth's climatic zonation. Those in arid and polar climates can be distinguished from those in humid by their greater angularity, marks of wind action, absence of graded stream profiles, and the dominance of mechanical over chemical weathering. The 'competence' (i.e. erosion resistance) of rocks in arid regions is determined by their physical coherence, in humid regions by their chemical resistance. Among humid climates, the tropical can be distinguished from the temperate by the more intense chemical and biotic weathering and the absence of frost action.

Weathering causes rapid differentiation of rocks and exaggerates relief differences. This leads to sharp contrasts between steep hillslopes and level plains. Semi-arid regions are intermediate in character but are distinguished by the intensity of fluvial erosion. This is because the rainfall, which is more frequent and intense than in arid regions, falls on land whose vegetal cover is too sparse to form a protective mat.

There have been two approaches to the classification of landforms on a zonal climatic basis (Stoddart, 1971). Peltier (1950), followed by Tanner (1961) and Leopold *et al.* (1964), sought to identify geomorphological regions theoretically by partitioning the space on temperature/rainfall graphs into categories relevant to specific processes such as frost shattering, chemical weathering, mass movement, and wind action. For instance, chemical weathering increases directly with rainfall and temperature (Fig. 6.1). Such analyses have an explanatory value, but are inevitably highly generalized.

On the other hand, Passarge (1926), followed by Büdel (1963) and Tricart and Cailleux (1972), sought inductive regionalizations on the basis of the assemblages of landforms of different climates. Tricart and Cailleux also recognized some analogy with the zonal concept in soil science. Landforms typical of specific climatic zones are called 'zonal', and those with features common to all climates such as beaches and bare rocks as 'azonal'. Those characteristics of one zone which occur in another are 'extrazonal'. These can take the form either of 'prolongations' such as of river valleys which continue the characteristics of upper catchments into climatically different ones downstream, or 'survivals' such as landforms on geological outliers which repeat the forms of parent outcrops from which they are separated. 'Polyzonal' landforms are those such as dunes or karst features which can occur in some, but not all, climates.

Climates change with time. The Tertiary and Quaternary oscillations have, in many areas, left the imprint of processes which have ceased. In particular, the apparent equatorward displacement of climatic zones during the Quaternary glaciations has left glacial and periglacial landforms in the temperate zone, pluvial in the arid zone, and aeolian in the semi-arid tropics. Climatovariance in landforms is thus a result both of present 'climatic' and past 'climatogenetic' forces. Butler (1959, 1967; Walker and Butler, 1983), from Australian experience, suggested systematizing the interpretation of polycyclic landforms in terms of 'K-cycles'. In fluvial landscapes, a K-cycle begins with an unstable phase of erosion and deposition (K_u) and ends with a stable cycle in which soils can form (K_s). The sediments and soils of each K-cycle, represented by its ground surface, are related to the modern cycle (K_0) by reverse chronological numbering from K_1 to Kn, with the latter representing the nth cycle before the present.

6.1.2 *Epeirovariance*

Landforms can also be classified according to tectonic structure, and this is usually the dominant factor in distinguishing them at the intermediate levels from land division to land region. Structure is governed by deep-seated pressures in the earth's crust which cause regional folding, faulting, and fracturing to form uplands and lowlands. The area over which a particular rock type appears at the surface is its 'outcrop'.

The varieties of rock are almost endless, but if classified according to mode of origin, they fall into three major classes: 'igneous', 'sedimentary', and 'metamorphic'. These can either be exposed at the surface or covered by a

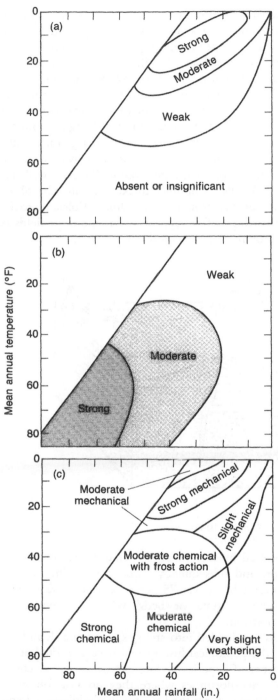

Fig. 6.1 Climatic control of (a) frost action, (b) chemical weathering, and (c) weathering regions (after Peltier, 1950)

59

mantle of unconsolidated materials constituting a fourth class which may be called 'drift'. Many examples of the appearance of rock types on satellite imagery and both vertical and oblique aerial photography are given in texts such as van Zuidam (1985) and Short and Blair (1986).

Igneous rocks are those formed at high temperatures by solidification from a molten state. They compose about 95 per cent of the earth's crust, but only about 25 per cent of its surface. They may be 'plutonic', 'hypabyssal', or 'volcanic'. Plutonic rocks are formed by the solidification of 'magma' (molten rock) deep within the earth's crust and occur at the surface as large masses like lopoliths, laccoliths, or batholiths. Hypabyssal rocks are formed by more rapid cooling at shallow depths in confined spaces and give rise to (1) small wall-like dikes, (2) cylindrical plugs, and (3) sills, sometimes thickening into phacolites, when in the crest of anticlines or the trough of synclines. Volcanic rocks, formed by solidification at the earth's surface, appear as volcanoes, or fields of cinder or ash. These types are shown diagrammatically in Fig. 6.2.

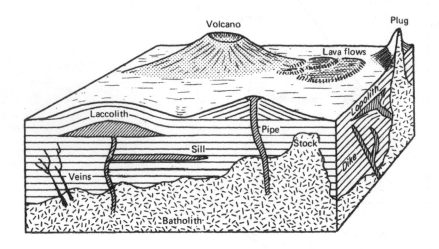

Fig. 6.2 Igneous landforms (Source: Strahler, 1969)

Sedimentary rocks are deposited by ice, water, or wind either on land, or more usually, on the sea floor, and subsequently consolidated. They occupy only 5 per cent of the earth's crust but cover 75 per cent of its surface. They generally occur in layers, known as 'strata' or 'beds', whose age sequence is shown by their order of superposition.

The land surface can broadly be divided into areas experiencing denudation or deposition. In any locality, the geological history will have consisted at one time of denudation and at another of deposition. This alternation is reflected in the strata being separated by unconformities where they do not follow the direct sequence. These unconformities form the basis for subdividing geological structure into the stratigraphic sequence from Pre-Cambrian to Holocene.

Most rocks undergo some tilting or folding after their deposition, so that bedding planes no longer lie horizontal, but have an inclination or 'dip'. This may be defined as the direction of steepest slope on a bedding plane, and its direction is shown on geological maps by an arrow. A line drawn at right angles to the dip is horizontal and thus a contour along the bedding plane. Its direction is known as the 'strike'.

The outcrop of a sedimentary rock depends on the slope of the ground and the dip and thickness of the beds. Where horizontal strata overlie beds of strongly contrasting lithology or inclination, they are known as 'overlapping'. Tectonic activity gives rise to landforms which are due to folding (Figs 6.3–6.11), single strike faults (Figs 6.12–6.17), single dip faults (Figs 6.18–6.20), and double faults (Figs 6.21 and 6.22).

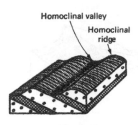

Fig 6.3 A monoclinal fold of simple strike ridges, cuestas, and vales

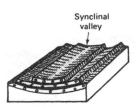

Fig. 6.4 A symmetrical syncline

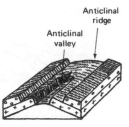

Fig. 6.5 A symmetrical anticline

Fig. 6.6 An asymmetrical syncline

Fig. 6.7 An asymmetrical anticline

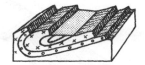

Fig. 6.8 Simple strike cuestas with related overfolds

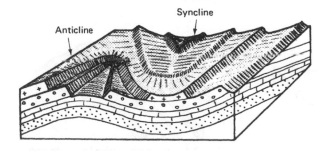

Fig. 6.9 A pitching syncline and anticline

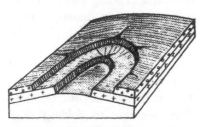

Fig. 6.10 A dome with denuded crest

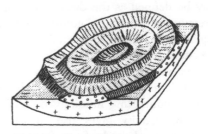

Fig. 6.11 A synclinal basin

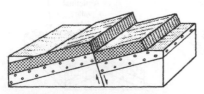

Fig. 6.12 Normal strike fault with hade (angular deviation of the fault plane from the vertical) opposite to dip

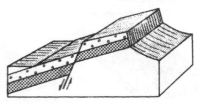

Fig. 6.13 Normal strike fault with hade with dip but steeper

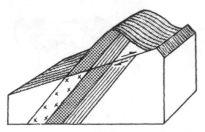

Fig. 6.14 Normal strike fault with hade with dip but less steep

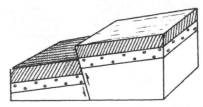

Fig. 6.15 Reversed strike fault with hade opposite to dip

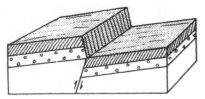

Fig. 6.16 Reversed strike fault with hade with dip but steeper

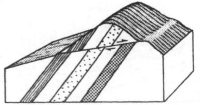

Fig. 6.17 Reversed strike fault with hade with dip but less steep

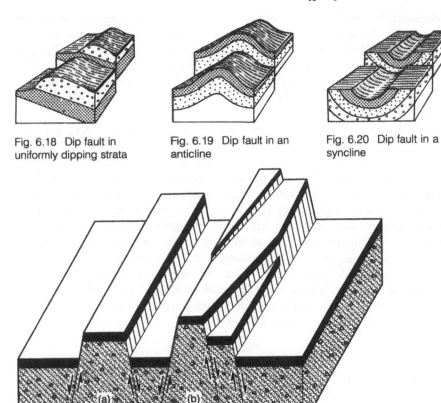

Fig. 6.18 Dip fault in uniformly dipping strata

Fig. 6.19 Dip fault in an anticline

Fig. 6.20 Dip fault in a syncline

Fig 6.21 (a) Horst and (b) compound horst

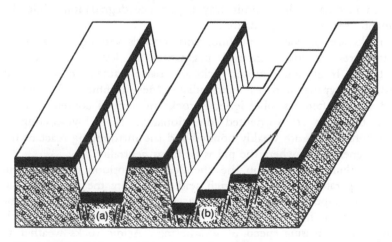

Fig. 6.22 (a) Graben and (b) compound graben

63

Metamorphic rocks are of igneous or sedimentary origin but have been altered by the pressures or temperatures which accompany mountain building. They are generally harder and more contorted than the original types. Regions of metamorphic rock are normally much fractured and faulted, but are distinguished by a strong 'grain' or 'lineation' in the topography. The resulting ridges and valleys tend to be shorter and less sharply marked than those due to sedimentary strata.

Drift deposits are formed by a combination of climatically induced processes and gravity. Ice, running water, wind, waves, and tides each create landforms reflecting their regimes of movement. Glacial deposits are mixtures of a wide range of particle sizes because of the viscosity of the carrying medium. Colluvium has a longitudinally concave surface and downslope fining of particles due to the dominance of gravity in its formation. Fluvioglacial and alluvial deposits show a similar but more complete longitudinal and lateral sorting by particle size. Dunes and loess deposits are composed of particles which are too small to withstand movement by the wind, but too large to cohere in immobile aggregates. Their accordance of summit levels and repetitive surface articulations reflect the even depth and harmonic variations of the air movements which form them. Beaches result mainly from the high energy attack of waves on the land–sea interface.

6.1.3 *Petrovariance*

Lithology is important in terrain formation because it governs weathering. Rocks are aggregates of minerals whose properties, relative abundance, and grain size determine their character. Therefore they tend to be more elevated in the landscape and more resistant to weathering and erosion than unconsolidated materials. They dominantly experience degradation while the latter dominantly experience aggradation.

Igneous rocks are generally crystalline, with crystals varying in size from several centimetres in coarse grained pegmatites to less than 1 mm in amorphous basalts. In general, plutonic rocks are coarse grained, volcanic rocks fine grained, and hypabyssal rocks intermediate. Metamorphic rocks such as gneisses and schists often resemble igneous rocks in mineral content but are distinguished by their stress-oriented directionality. Rock weathering is maximized where crystals are weakly bonded and the surrounding reactant has substantial free energy, i.e. where the pH is far from neutrality.

Rates of weathering depend on the chemical composition of constituent minerals. In temperate climates, silica–rich acidic rocks are the most resistant, while basic rocks, especially those of ferromagnesian type which tend to be dark in colour, are the most vulnerable. The order from left to right in Table 6.1 thus indicates increasing speed of weathering,. Acidic rocks weather to gravelly, sandy, and silty soils while basic rocks weather mainly to clays. In the humid tropics, this tendency is less marked and may even be reversed because the higher temperatures and lower soil acidity make for a greater relative mobility of silica in comparison with iron compounds.

Table 6.1 Approximate contents of silica (SiO_2) in some important minerals

65%	55–65%	45–55%	35–45%
Acid	Intermediate	Basic	Ultra basic

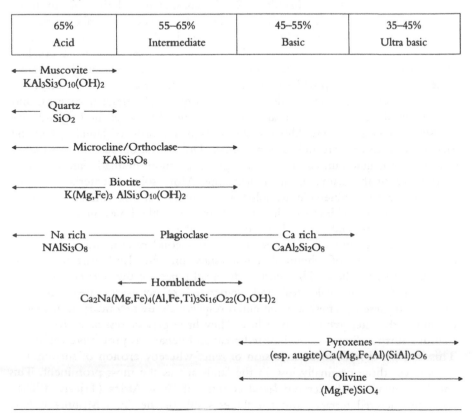

The subdivision of sedimentary rocks according to chemical composition indicates the types and sizes of product into which they will be converted. The commonest surface rocks are formed of clay-sized particles of silicate and aluminosilicate minerals, which form shales and mudstones. These tend to be soft, impermeable, and vulnerable to erosion. The result is that they form low ground between uplands and embayments between coastal headlands. They give gentle slopes, slump features, and a dense 'drainage texture'.

Next commonest are the siliceous sediments, formed of silica fragments ranging from stones to silt but not containing much clay. Their resistance to weathering and erosion increases generally with the size of the particles and the strength of the intergranular matrix. They range upwards from weakly consolidated sandrocks through malmstones and sandstones to hard quartzites and conglomerates. Because the dominant form of weathering is granular

65

disintegration along rectilinear fissures, they present generally rounded relief forms. Where the matrix is especially hard, they can have a bold castellated appearance as in the Zion National Park in the USA, the Petra area of Jordan, and the 'Tassili' of the western Sahara.

Calcareous rocks such as limestone, chalk, and dolomite are relatively competent and so form uplands, but this competence derives more from permeability than from hardness and results in an absence of surface drainage. The solubility of calcium carbonate causes the development of karst features such as gorges, sink holes, and subterranean channels, leaving a residual concentration of insoluble materials such as clay and flints at the surface. Where limestone is exposed to strong solution, one may find assemblages of spectacular landforms such as deep narrow gorges, subterranean caverns, and precipitous slopes, as in scenic areas such as Cheddar Gorge in England, the Carlsbad Caverns in New Mexico, the 'cockpit country' of Jamaica, and the 'stone forest' district around Guilin, China. Dolomites, because of the higher solubility of magnesium carbonate, can give even more dramatic landscapes, as in the area of this name in the Italian Alps. More soluble materials, notably gypsum, are too vulnerable to solution to survive at the surface in humid climates, but are widespread in arid. They resemble limestones and show karstic phenomena but have a less accidented relief.

Duricrusts, described by Goudie (1973), are secondary indurations caused by the concentration of chemical compounds on the land surface, whose topography they follow. The most widespread types are the 'ferricretes' of the humid tropics, the 'calcretes' and 'gypcretes' of semi-arid areas, and the 'silcretes' of deserts. These are cemented respectively by secondary iron oxides, calcium carbonate, gypsum, and silica. They have greater resistance to erosion than the surrounding non-duricrusted areas and behave as protective cap rocks. This sometimes leads to an inversion of relief whereby erosion of surrounding areas leaves those originally low in the landscape as the most prominent. This can be seen in the ferricreted *bowal* country of West Africa (Tricart, 1965a), and the calcreted terraces and fan slopes south of the Atlas Mountains (e.g. Joly, 1962).

The surface form of drift deposits reflects the materials of which they are composed. The steepest slopes are on coarse detritus such as talus, stony colluvium, river-bed boulders in mountain streams, and beach shingle. At the other extreme, the most level plains, lake beds, and deep water off coasts, are usually associated with clays. Intermediate slopes are associated with particle sizes between these extremes. Dunes reflect the angles of rest of their constituent particles. These are generally rounded, mainly quartzitic, and in the size range from about 0.05 to 1 mm, which are too large to cohere but too small to remain immobile in normal winds. Loess mantles extensive areas where it has been trapped after its journey from arid or periglacial source regions. It generally follows the contours of the underlying terrain, although the surface is usually modified by wind-formed undulations and intersected by valleys with steeper and more abrupt slopes than sand dunes. This distinctive character is due to its being composed of silt-sized particles from about 0.002 to 0.05 mm, of which quartz forms a lower proportion than in dune sands. The steep valley slopes result from the greater cohesion of such particles due to their larger aggregate surface area and the interlocking effect of their more platy shapes.

6.2 Surface geometry

The third main aspect of terrain is the geometrical form of the ground. Landforms, being three-dimensional, can be understood from block diagrams, but detailed consideration is helped by seeing them both in plan and in section. Reduced to bare essentials, landscape consists of hilltops, slopes, and valleys.

In plan view, hilltops can be seen either by stereoscopic observation of aerial photographs or by viewing 'hilltop envelopes' or crest lines on maps. Crests can be either circular or elongated, and either parallel or randomly distributed, giving rise to the four possible plan arrangements shown in Fig. 25.4. These permutations of shape can be defined quantitatively and the resulting values used to compare landscapes, and help in studying origins and processes. For instance, a statistical count of the number of hilltops in each altitude range may reveal a preponderant frequency at certain heights, indicating relict land surfaces from an earlier erosion cycle. Hills which are not elongated crests are either peaked or flat-topped, the former usually being the result of circumdenudation, the latter caused by the occurrence of a horizontal cap rock which may be a competent stratum or an indurated relic of a former erosion surface.

Quinones–Garza (1983) derives 12 topographic elements by adding planes, domes, and basins to Ruhe's 9 basic slope geometries (1975), shown in Fig. 6.23. This three-dimensional classification is able, for instance, to model the interrelations of changes in gradient or convergences of surface flow in order to discern their effects on surface runoff.

Valleys can be viewed in all three dimensions: plan, longitudinal section, and transverse section. In plan they appear as networks of branching streams which can be considered in terms of the hierarchy of 'orders' considered in Chapter 19. Individual channels can be straight, sinuous, or meandering, depending on the degree of curvature of their courses. They may be 'braided', i.e. interrupted by numerous islands, when the ratio of the stream's suspended load to its velocity exceeds a certain critical value. Although many classifications of drainage patterns have been made, summaries such as those by the USAEWES (1963a) and Haggett and Chorley (1969) have simplified them and reduced them to two main groups: 'integrated' and 'non-integrated', distinguished by the dominance of closed depressions in the latter. The main integrated patterns are 'dendritic', 'rectangular', 'trellis', 'pinnate', 'parallel', and 'reticulated' as illustrated in Fig. 6.24. Each can have number of different forms and occur in many combinations. Certain integrated patterns are associated with particular types of geomorphology: radial with isolated mountain masses, trellis with cuesta-and-vale topography, and dendritic with an arborescent erosion of uniform rocks.

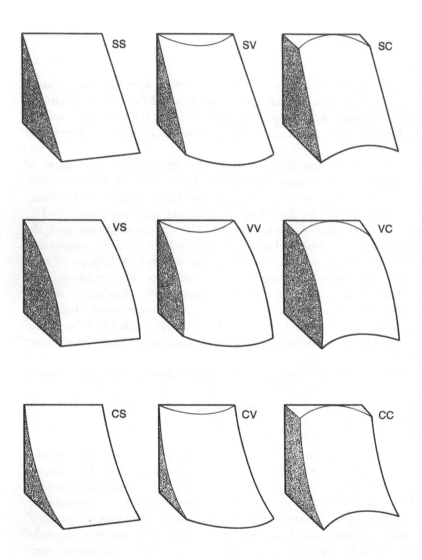

Fig. 6.23 The nine basic slope geometries: S = straight; V = convex; C = concave. The first letter indicates the character of the slope in a vertical direction; the second, its character in a horizontal direction. Each of the nine forms will have a specific pattern of surface drainage determining the loci of potential erosion and deposition. (Sources: Ruhe, 1975; Quinones-Garza, 1983)

Dendritic: the commonest pattern. Indicates uniform materials.

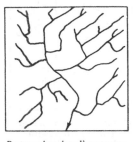

Rectangular: implies strong bedrock jointing and thin soil cover. The stronger the pattern, the thinner the soil.

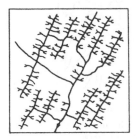

Trellis: implies strike ridge topography.

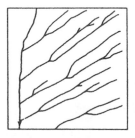

Parallel: characteristic of outwash areas of low topography, where main stream may indicate a fault.

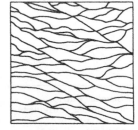

Anastomosing or braiding: in alluvial areas where sediment load exceeds carrying capacity of a stream.

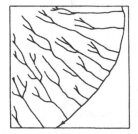

Radial (centrifugal): in isolated circular hill masses.

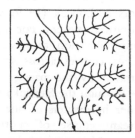

Pinnate: generally indicates high silt content as in loess or on flood plains.

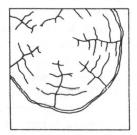

Annular: indicates igneous or sedimentary domes with concentric fractures or escarpments.

Fig. 6.24 Types of integrated drainage pattern (after Way, 1968)

Deranged: with many ponds, bogs, or lakes. Indicates flattish landscape often glaciated.

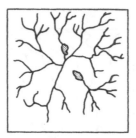

Centripetal: a variation of the radial pattern with drainage towards a central point, usually a sink or the centre of an eroded anticline or syncline.

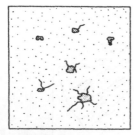

Internal: indicates highly porous level materials or karst conditions.

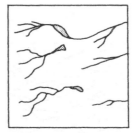

Dislocated: due to interruptions of drainage lines by faults or extrusions.

Fig 6.25 Types of non-integrated drainage pattern (after Way, 1968)

Non-integrated patterns usually occur on plains of glacial or river alluvium where drainage is poor, or on limestone or evaporites where it is partially subterranean. Some are shown in Fig 6.25.

These patterns, when interpreted by concepts originated by Davis (1909, republished 1954) and illustrated by such authors as Lobeck (1939) and de Martonne (1948), can indicate the history of a drainage system. For instance, 'superimposed drainage', developed at relatively high elevation, can give rise to 'antecedent' streams which preserve their ancient courses across structures which have since been exposed. This is illustrated by the passage of the Thames through the Goring Gap and of the Nile through the Sabaloka Gorge, north of Khartoum. 'Captured' streams are former headwaters of one stream which have been diverted into the downstream course of another, as evidenced by some of the former tributaries of the Meuse which have been captured on one side by the Seine and on the other by the Moselle drainage system. 'Consequent' streams indicate the original slope of the land, and 'subsequent' streams are their tributaries which have exploited faults or soft strata which have appeared during the process of denudation. A characteristic combination is a trunk consequent stream cutting across a series of parallel cuestas in the common direction of dip, fed laterally by subsequent streams along the strike vales. This is seen in southern England where the Ock, Thame and Kennet are strike subsequents joining the Cherwell–Thames between Banbury and Reading, or in the USA, where the Shenandoah joins the Potomac in the same way.

The profile along the lowest points in a valley is known as its 'thalweg'. This is paraboloid in longitudinal section with slopes declining asymptotically to the base level of erosion, as shown in Fig. 6.26. It most closely approaches the parabola in streams which have attained a state of 'grade'. This can be defined as a state in which there is an equilibrium between erosion and deposition and a tendency to absorb the effects of any displacement caused by changes in controlling factors (Mackin, 1948). Grade is interrupted by convexities or breaks of slope in the stream bed. These often give clues to the processes active in the valley. They may be due to an outcrop of resistant rocks, to glaciation leaving the upper parts of a valley 'hanging' above its lower reaches, or to rejuvenation at some intermediate point in its course. The Sabaloka Gorge, previously mentioned, separates the relatively level middle courses of the Blue and White Niles, where they have helped form the Sudan Plains, from the lower course of the main Nile whose gradient over the cataracts is steeper.

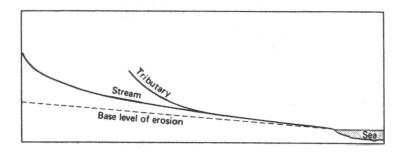

Fig. 6.26 Typical longitudinal profile of thalwegs of stream and tributary

Valley cross-sections often indicate geomorphic origins. Although most are complex, a certain number of basic types are indicative of origins. These are shown in diagrammatic form in Fig. 6.27.

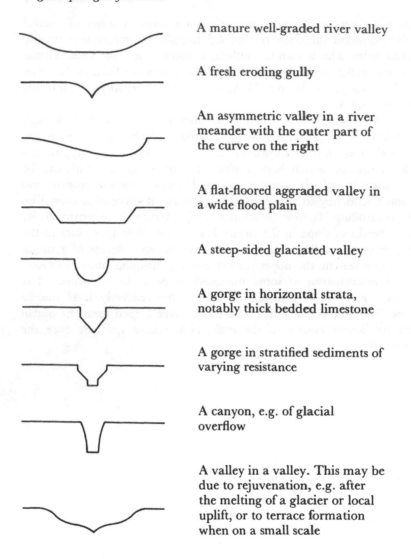

A mature well-graded river valley

A fresh eroding gully

An asymmetric valley in a river meander with the outer part of the curve on the right

A flat-floored aggraded valley in a wide flood plain

A steep-sided glaciated valley

A gorge in horizontal strata, notably thick bedded limestone

A gorge in stratified sediments of varying resistance

A canyon, e.g. of glacial overflow

A valley in a valley. This may be due to rejuvenation, e.g. after the melting of a glacier or local uplift, or to terrace formation when on a small scale

Fig. 6.27 Some representative valley cross-sections

Further reading

Büdel (1963), Butler (1959), Dresch *et al.* (1985), Goudie (1973), Haggett and Chorley (1969), Leopold *et al.* (1964), Quinones–Garza (1983), Short and Blair (1986), Stoddart (1971), Tricart (1965a, b), Tricart and Cailleux (1972), van Zuidam (1985), Young (1971).

7

Terrain processes affecting land uses

7.1 Types of process

A wide variety of processes damage land and soil, which can have natural or human causes or a combination of the two. These processes include water and wind erosion, physical, chemical, and biological degradation, salinization, alkalization, and air and noise pollution. Each has three aspects: state, rate, and risk. The state is the extent to which the land has already been degraded and is, for instance, visible in lands which are gullied, deflated, or salinized. The rate is the speed at which degradation is now occurring and can be measured in terms of annual change. Water and wind erosion can, for instance, be assessed in terms of weight of soil lost, salinization by the rise in soil electroconductivity and pollution by the increase in toxic substances per unit area. The risk is the rate of lowering of present or potential land productivity, measured in economic terms, caused by activities currently in progress such as the replacement of vegetation by bare fallow. Methods have been developed in order to quantify, grade, and map each of these aspects (e.g. Riquier, 1978; Morgan, 1979). Noise pollution is an increasing problem. Although it depresses property prices (Walters, 1975), its effect on land values still awaits detailed investigation.

7.2 Fluvial erosion

Erosion caused by running water has given rise to an extensive literature. This includes general surveys such as those by Hudson (1981) and Morgan (1979), reports of symposia (Morgan, 1981; de Boodt and Gabriels, 1980), the development of measurement formulae such as the universal soil loss equation (USLE) (Wischmeier and Smith, 1965, 1978) and the soil loss estimation method for southern Africa (SLEMSA) (Elwell, 1981), the applications of

73

remote sensing (Williams, 1981; Mitchell, 1981; Fenton, 1982), and control by land management (Greenland and Lal, 1977; Cannell and Weeks, 1979). Irving (1962) and Hails (1977) discuss the physical aspects of coast erosion.

The risks were first widely recognized in the USA. This resulted from the 'dustbowl' experience of the southern Great Plains in the 1920s and 1930s where pioneer deforestation in areas of marginal rainfall, followed by overgrazing and cereal monoculture, coincided with both drought and the world depression. Since that time, the global nature of the erosion problem has been widely recognized and countermeasures adopted in most countries. It is today most serious in the tropics and subtropics, especially in Africa. It is almost always associated with the destruction of vegetation either by natural causes, such as fire and drought, or by human activities such as deforestation, overgrazing, or overcultivation. The removal of vegetation exposes the surface to increased rain impact and accelerated runoff, and the lack of roots weakens the cohesion of the soil, making it more vulnerable to movement by water infiltrating from the surface.

Water erosion can be seen as a spectrum of processes ranging from those which are dominantly fluvial with a relatively high water content and low gradients at one extreme to those which are dominantly gravitational with less water and high gradients at the other. They form the sequence: (1) sheetwash, (2) soil creep, (3) stream flow, (4) mud flow, (5) land slump, and (6) landslide. In the FAO scheme for classifying the different forms of water-induced erosion (Riquier, 1978), the first two can be categorized as sheet-and-rill erosion, the third as gullying and the last three as mass movements. All cause damage both by the removal and by the deposition of materials. They may occur singly or in combination and there is clearly some overlap between them. Rills, for instance, are merely small gullies, and sheet erosion is only a shallow form of mass movement, but they merit distinction because of their different effects on the terrain and the different methods of assessment used.

Sheet-and-rill erosion is characteristic of many agricultural lands. It results from saturating the surface soil and lubricating its contact with the underlying rock, so that it slides down over it. This is especially evident when the latter is frozen, when the word 'solifluxion' is sometimes used to describe the process. The result is that soils tend to thin on upper slopes and thicken on lower. Various approaches have been made to categorize and quantify the causative factors.

A basic approach, used for instance by Higginson in the Hunter Valley of New South Wales (1973), has been to describe erosion risk qualitatively within the framework of land systems. In Zimbabwe, a scoring system was used for the factors causing erosion which could be quantitatively combined on a map (Stocking and Elwell, 1973, 1976). There has generally been a move towards the use of quantitative indices which combine the controlling factors into single equations for predicting erosion loss in any given area. One which has been shown to be significantly related to sediment yield in rivers is the ratio p^2/P where p is the highest mean monthly precipitation and P is the mean annual precipitation. From this, Fournier (1960) developed the following empirical relationship:

$$\log Q_s = 2.65 \log p^2/P + 0.46 \,(\log H)(\tan S) - 1.56 \qquad [7.1]$$

where Q_s = mean annual sediment yield (g m^{-2});
  H = mean altitude (m);
  S = mean slope of the basin (°).

This equation has been used by Low (1967) to evaluate regional variations of erosion risk in Peru.

The most widely used general formula is probably the universal soil loss equation (usually abbreviated to USLE). This was first developed for cropland east of the Rocky Mountains in the USA, but has since been applied, with modifications, in a number of other areas (Wischmeier and Smith, 1978; Wischmeier, 1976; Riquier, 1978; Soil Science Society of America, 1983). The basic formula is

$$A = RKLSCP \qquad [7.2]$$

where

A = the amount of soil lost in tonnes per hectare per year. Tolerable limits vary but are generally between 2 and 12 t ha^{-1} yr^{-1}.

R = the 'rainfall erosivity index', a dimensionless variable based on both the total rainfall of an area and its degree of concentration into short periods. The recommended measurement of this is the greatest average intensity of rainfall in any 30-minute period recorded in the area, expressed as millimetres per hour, and incorporating an allowance for a threshold factor whose importance in Africa has been emphasized by Hudson (1981) and in temperate regions by Morgan (1979). Table 7.1 shows a simple method of calculating this factor.

  Since the value for rainfall intensity is often difficult to obtain, approximations have been used. These have included the annual summation of monthly values for p^2/P where p is the monthly and P the annual rainfall in millimetres and, for West Africa, $P/2$.

K = the 'soil erodibility factor', a dimensionless variable based on soil texture, structure, and permeability reflecting its liability to erosion. It must be determined empirically between values of 0 for a totally resistant soil to 1 for a totally vulnerable one. In the USA values range from 0.03 for a gravelly loam in New Jersey to 0.69 for the Dunkirk silt loam in New York. Figure 7.1 is a nomograph which can be used for calculating it.

L = the 'slope length factor', is the ratio of the length of the given area to the arbitrary value of 22.1m (72.6 feet).

S = the 'slope angle factor', is the ratio of the slope in the given area to 9 per cent. S and L can be determined together as SL ('soil loss ratio') from the following empirical formula:

Table 7.1 Kinetic energy of rainfall (expressed in metric tonne-metres per hectare per centimetre of rain)*

Intensity (cm h^{-1})	0.0	0.1	0.2	0.3	0.4	0.5	0.6	0.7	0.8	0.9
0	0	121	148	163	175	184	191	197	202	206
1	210	214	217	220	223	226	228	231	233	235
2	237	239	241	242	244	246	247	249	250	251
3	253	254	255	256	258	259	260	261	262	263
4	264	265	266	267	268	268	269	270	271	272
5	273	273	274	275	275	276	277	278	278	279
6	280	280	281	281	282	283	283	284	284	285
7	286	286	287	287	288	288	289 †			

* Computed from the equation $E = 210 + 89 \log(10)\ I$

where E = kinetic energy in metric–tonne metres per hectare per centimetre of rain (t m ha^{-1} cm^{-1}); and

I = rainfall intensity in centimetres per hour (cm h^{-1}).

† The 289 value also applies for all intensities greater than 7.6 cm h^{-1}.

Source: Wischmeier and Smith (1978).

$$SL = L/30.5\ (0.76 + 0.53S + 0.076S^2) \tag{7.3}$$

where L is the slope length in metres and S is the slope percentage. Figure 7.2 gives a graph for determining SL.

$C =$ the 'crop management factor', approximating to the amount of vegetal cover on a scale from 0 for highly protective management to 1 for the most risky management. Experience from West Africa gives values of 0.001 for dense forest or culture with a thick straw mulch, 0.01 for ungrazed savanna or grassland, 0.1 for rice and crops giving good ground cover with or without tree crops, 0.2–0.8 for cassava, 0.3–0.9 for maize, sorghum, or millet, 0.4–0.8 for groundnuts, 0.5 for cotton and tobacco, and 1.0 for bare soil.

$P =$ the 'conservation practice factor'. This is a dimensionless variable ranging from 1 for a total absence of conservation practices to 0 for total protection. The following values have been used: in the USA 0.75 for contour ploughing, 0.5 for contour ploughing with contour trenches or 0.25 for contour ploughing with contour grass strips. In West Africa, grassland reduces it to <0.5, anti-erosion strips to <0.3, contour trenches to <0.2, and straw mulch to 0.01 (Elwell, 1981).

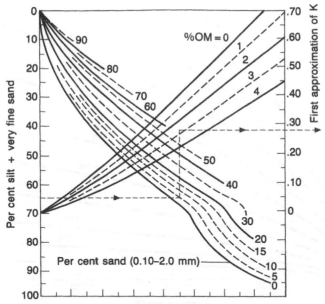

PROCEDURE: With appropriate data, enter scale at left and proceed to points representing the soil's % sand (0.10–2.0 mm). % organic matter, structure, and permeability, *in that sequence.* Interpolate between plotted curves. The dotted line illustrates procedure for a soil having silt + very fine sand, sand 5% 65%, OM 2.8%, structure 2, permeability 4. Solution *g* = 0.31.

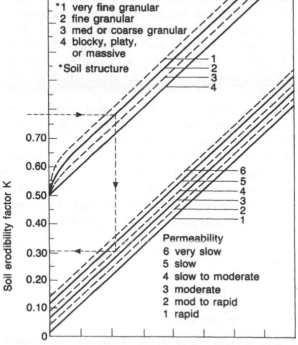

Fig 7.1 Soil erodibility nomograph (from Wischmeier *et al.*, 1969)

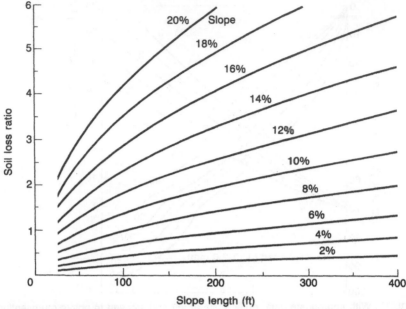

Fig 7.2 Chart for adjusting plot soil loss to length and degree of slope (from Wischmeier *et al.*, 1958)

The USLE has been modified for Africa by Elwell (1981) and an alternative approach more suited to European conditions has been suggested by Morgan (1979, 1980). This emphasizes the need for local testing and for caution in applying mathematical models of this sort, when applied outside the areas where they were developed. Roose (1975), for example, when applying the USLE to the Ivory Coast, defined *C* values for crops which are not grown in the USA.

Although the USLE is simple, easy to use, and widely accepted as an appropriate tool, having been based on over 10 000 plot-years of data, it contains both theoretical and practical weaknesses. An example of the former is that there is some interdependence between the variables so that some are counted twice. For instance, rainfall influences the *R* and *C* factors and terracing the *L* and *P* factors. Because there is no inclusion of a factor for runoff, Foster *et al.* (1973) suggested replacing *R* with an energy term, *W*, which is a function of rainfall and runoff energy, defined as

$$W = 0.5R + 5.9Q\left(\frac{q_p}{2.54}\right)^{1/3} \qquad\qquad [7.4]$$

where: R = rainfall erosivity factor;
$\quad\quad\;\; Q$ = storm runoff (cm);
$\quad\quad\;\; q_p$ = storm peak runoff rate (cm h^{-1}).

The following weaknesses also need to be borne in mind when using the USLE:

78

1. It is not accurate for a specific storm event, season, or year;
2. It does not include erosion by concentrated flow;
3. It does not estimate on-site deposition;
4. It does not accurately estimate sediment yield from fields using sediment delivery ratios which relate suspended load in a stream to gross erosion in a watershed;
5. It does not estimate sediment concentrated in runoff;
6. It does not enable an estimate to be made of chemicals lost.

Gullies begin where runoff concentrates into a flow which overcomes soil cohesion at a critical point. Their initiation is favoured by heavy rainfall, steep long slopes with converging drainage lines, soft impermeable soils, a rough surface, and a relative absence of vegetation cover. They are often enlarged rills but may also be initiated by the coalescence of small hillside nicks resulting from the localized weakening of vegetation cover, by the collapse of underground drainage pipes and channels, or by the exploitation of scars caused by landslips. The locations of future gullies can sometimes be predicted. Schumm (1977, 1979), for instance, showed that in small uniform semi-arid catchments gullying was likely where the ratio between the valley slope above a gully and its drainage area exceeded a certain threshold.

After initiation, gullies grow by headward erosion, downward incision, and lateral undersapping, those with the steepest gradients growing at the expense of the others by micropiracy. As the fingertip streams extend, the lower reaches of the network become simplified by concentration into fewer channels. At the same time the zone of maximum erosion migrates towards the head of the basin.

The general effect of gullying is to transfer sediment from the upper to the lower parts of a valley, accentuating relief in the former and smoothing it in the latter. Gullies deposit in three ways. Deltas extend the watercourse at its downstream extremity, riffles and bars form along its bed, and terraces build laterally on its meander plain. The combined result of erosion and deposition is that the thalweg tends towards a parabolic form in longitudinal section, both in reaches between topographic barriers such as waterfalls or cataracts, and along the trunk valley as a whole.

It is sometimes necessary to calculate erosion rates for whole catchments. A useful empirical solution is to multiply a mean USLE value by the number of hectares in the catchment and then multiply the product by a correction factor for the proportion of eroded material which is actually removed by the river. This factor is known as the 'sediment delivery ratio' (SDR). It is defined as the ratio of sediment delivery to gross erosion in the watershed and is usually expressed as a percentage. The larger the catchment the smaller the SDR in a fairly regular way, so that commonly used values drop from 53 per cent for a catchment of 0.1 km^2 to 4.9 per cent for one of 26 000 km^2 (the Potomac) (Robinson, 1983).

Numerous attempts have been made by both engineers and scientists for over a century to relate sediment transport rate to volume of river flow. Empirical calculations of both erosion and deposition rates have related field measurements such as the exposure or burial of stable markers over selected time periods to estimates of the volume and speed of flow of the rivers and a sampling of their suspended solids. Theoretical methods have attempted to relate erosion and deposition to the discharge, cross-sections, and sediment load of confined watercourses. Unfortunately, the empirical relationships are generally found to be inapplicable beyond the limited circumstances of their origin, while the theoretical ones often depend on invalid physical assumptions. The most valuable approaches would seem to be the derivation of formulae based on the energetics of fluid flow as suggested by Allen (1970), or the detailed analysis of the parameters of the catchment in question only within the range of conditions over which their validity has been tested.

Mass movement is the gravitational slippage of large masses of material where running water is not the main agent. Its nature depends on whether the hillslope materials are weathering-limited or transport-limited. Weathering-limited mass movements occur where the rock is bare because the removal of detritus is faster than its generation, transport-limited where the speed of weathering exceeds that of removal so that a soil mantle is able to form.

Erosion of weathering-limited slopes depends on the dominant scale and type of fissuring and jointing in the rock. Where macro-jointing is more important than micro-jointing, large slabs or boulders fall (Fig. 7.3). Where micro-jointing is dominant, smaller blocks form scree, dejection cones, or talus glaciers depending on the topography of their resting places. When the dominant form of weathering is granular disintegration, the fine fragments are readily removed by water or wind, leaving the rock slopes bare. In a landscape of weakly consolidated granular rocks such as soft sandstones, this may lead to an assemblage of steep rounded or castellated forms surrounded by mobile sandy detritus. To predict the likelihood of mass movement on weathering-limited slopes, it is necessary to infer the position of the true angle of rest of the poised material and the strength of the cohesive forces which are preventing it from falling.

Mass movement on transport-limited slopes tends to be along relatively deep zones of weakness, and is accelerated where the underlying layers are frozen and the plane of weakness is lubricated by water. It is often initiated by undersapping at the foot of the slope. The type of movement depends largely on the fluidity of the moving mass. Where the material remains solid and cohesive, it forms single or multiple landslips depending on the number of planes of weakness. Their amount of slippage may be measured by recording the site photographically over a period of time or by noting the amount of bending in probes inserted into the ground through the zone of weakness. The land surface of such areas often appears as a series of terracettes. Where movement is pronounced, these may also appear on the reverse side of the slipped block (Fig. 7.4).

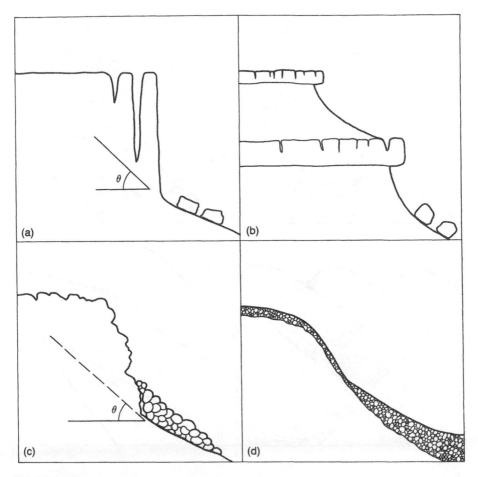

Fig. 7.3 Types of erosion in weathering-limited materials: (a) slab failure, occurring where macro-jointing dominates. θ = angle of rest towards which the slope will tend; (b) rockfall, a special case of (a) where the slope is undersapped; (c) rock avalanche, where micro-jointing dominates; (d) granular disintegration through surface loss of fine materials. (All forms sometimes occur together.)

Where the mass is somewhat more fluid as a result of soaking by interstitial water, the result is a slump, leaving a relatively steep exposed back wall and a hummocky surface on the slumped material resulting from multiple fissuring and internal flowage. Slumps are unstable and may experience further periodic movements, especially after heavy rains. They thus present hazards to every form of land use. When water permeates the whole soil mass to such an extent that it exceeds the liquid limit, it causes mud–flows, or where relatively rapid, mud spates or mud avalanches, known as *mures* in the Alps. These occur when four conditions are simultaneously fulfilled: a mass of materials rendered relatively porous by being both unconsolidated and having a mix of particle

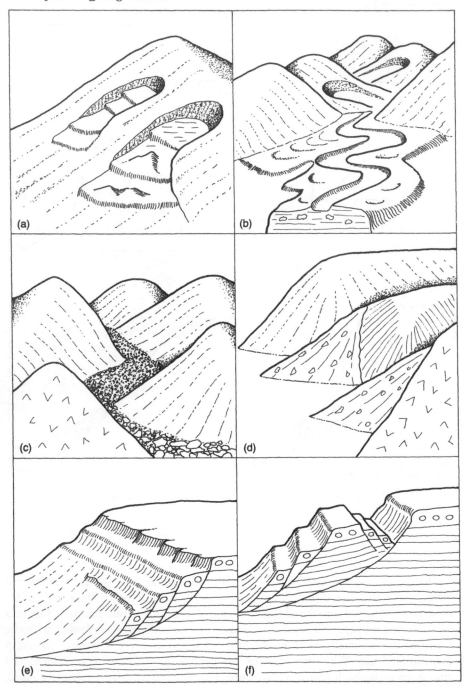

Fig 7.4 Types of mass movement in transport-limited materials: (a) normal slumping or landslip; (b) mud-flow; (c) talus glacier; (d) dejection cone; (e) terracettes (simple); (f) simple and reversed terracettes

sizes, an unstable poised position in the landscape, a relative absence of binding vegetation, and a period of torrential rain or rapid melting of snow. Mud-flows are thus commonest in spring in temperate latitudes and during monsoonal rains in the tropics. Their size and speed of occurrence in mountain areas have caused serious losses of life and destruction of property. The resulting surface is lobate, resembling the advance of liquid lava. Once movement has ceased, the surface of the flow remains unstable for agricultural or engineering uses both because of the surface unevenness and the risk of further movement.

Erosion control measures depend on an understanding of causes. They can be classified as physical or agronomic. Physical methods aim to reduce gradients by construction work such as slope terraces, check-dams in gullies, and on more level ground, cross-slope banks or bunds (French *banquettes*). Agronomic methods aim to bind the soil against creep and inhibit the concentration of surface runoff. They include strip cultivation, contour ploughing, the control of grazing, and the replanting of steep slopes, damaged areas, and gullies with appropriate vegetation. A detailed discussion of control measures is given by Hudson (1981).

7.3 Coastal erosion

Coastal processes differ from fluvial in their diurnal, rather than seasonal, character and in the intensity of energy concentration along the narrow littoral zone. Erosion, transport, and deposition are by waves, assisted by tides. The latter determine the level of wave action and sometimes are guided by the configuration of the coast to control its direction. As the shore is approached and the depth decreases, advancing waves suffer refraction and the crest lines tend to become parallel to the bottom contours. This necessarily concentrates wave energy on exposed headlands. Bays, by contrast, are low-energy zones where constructional forms predominate except when eroded by storm waves at high tides. Waves erode through the pounding of water hurled against the land margin, the abrasion by its sediment load, and the hydraulic pressure of air compressed into cracks. The long-term effect resembles the action of a horizontal saw and creates a sharp knick between the near-vertical retreating cliff and the near-horizontal wave-cut bench. With jointed or faulted rock, differential erosion forms features such as caves, arches, and stacks. There may also be a contribution from chemical erosion, especially on limestone coasts, and from seasonal freezing and thawing in polar areas.

Erosion rates on cliffed coasts depend on rock hardness as well as exposure, with drift materials being especially vulnerable to rapid recession. A bird's-eye view of any exposed coast shows the differential resistance of rocks. Around southeast England, for instance, headlands tend to be of relatively resistant chalk or sandstone, inlets of erodible clays. Where coastal retreat leads to the loss of important land, protective measures are adopted such as the dumping of large boulders along the cliff foot, the construction of sea walls, and the building of groynes to inhibit longshore drift and create a protective beach.

The materials eroded by marine action, supplemented by those brought down by rivers, are moved by waves and tides. Wave action within bays is largely normal to the beach and alternates between inward-moving 'swash' and outward-moving 'backwash'. This causes a to-and-fro movement of materials which varies in intensity and direction in accordance with the weather, but over a long period may show relatively little change.

Where coasts are less embayed and onshore winds blow relatively constantly from a long fetch, refraction is incomplete and the waves approach the coast obliquely. While the swash approaches from this angle, the backwash reponds to gravity and runs back into the sea in a direction normal to the beach. The sediment therefore follows a zigzag path, moving most effectively when the angle of wave incidence is approximately 30–45° to the beach line (Thompson *et al.*, 1986). If there are no interruptions it may go long distances, as in the easterly drift of beach materials between the major headlands on the south coast of England.

Where beach ridges build out to sea to continue the line of the coast, they are called 'spits'. These tend to be especially large near the mouths of rivers where extra sediment is available. Wave refraction around the points of growing spits produces rounded hooked forms. These may be very complex, and where breached, give rise to 'barrier islands'. Where there are two or more dominant directions of wave approach, the beach may build out into a 'cusp' or 'cuspate foreland', such as that of Dungeness where easterly drift along the English Channel coasts meets a westerly drift from the southern North Sea. Landward of these features, fine material accumulates and salt marshes can develop. Such coastal features can extend usable land but choke harbours and inlets. They increasingly separate docks and wharves from deep water, causing seaports to decline, as in the case of Ur in southern Iraq, Ostia in Italy, and Chester and Winchelsea in England. Such trends must be countered by dredging or the seaward relocation of port facilities.

7.4 Wind erosion

Wind erodes and deposits unconsolidated materials wherever its velocity exceeds a certain threshold. This is most frequent in the absence of vegetation as in arid lands. Particles coarser than about 2 mm are too heavy to be transported and those finer than about 0.002 mm are too cohesive. In between these values, three types of transport occur. The finest materials are removed in suspension and carried long distances to be deposited as loess. Those between *c.* 0.1 and *c.* 0.5 mm are moved mainly by saltation and those between 0.5 and 2 mm by creep, the degree of mobility increasing with sphericity (Williams, 1964). The overall result is to leave the finest and coarsest materials as lag deposits and to sort the intermediate ranges geographically by particle size.

Wind tunnel experiments have shown that the detachment capacity of wind varies with the square of its velocity, and the transporting capacity with its cube. Using the second of these relationships, Chepil (1945) obtained the following equation to describe the sediment discharge ($Q_s W$) per unit width ($kg\ m^{-1}\ h^{-1}$) for grains of 0.25 mm diameter:

$$Q_s W = 52(V - V_t)^3 \hspace{2cm} [7.5]$$

where V = wind velocity ($cm\ s^{-1}$) measured at a height of 1 m;
V_t = threshold wind velocity, usually taken as 400 cm s^{-1}, measured at 1 m height, required to initiate particle movement at the ground surface.

Bagnold (1941, 1966), working in the Libyan Desert, developed an equation for potential deflation rate which also included a factor for particle size and assumed an unlimited sand supply:

$$q = 1.4 \times 10^{-5}\ V*^3\ \sqrt{d} \hspace{2cm} [7.6]$$

where q is the weight of sand moved (g-s^{-1} along a lane of 1 cm width) and $V*$ is the wind shear velocity or drag velocity. This latter value is most conveniently measured from the ratio of wind velocity to the logarithm of the height above ground. Since these have a linear relationship, $V*$ is proportional to the slope of the line, i.e. y/x or $\tan \theta$ in Fig. 7.5.

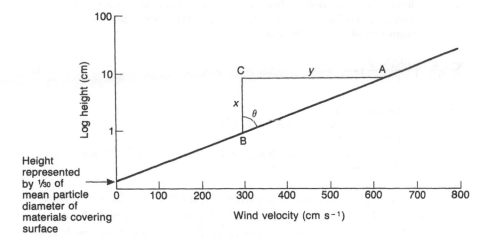

Fig. 7.5 A method for determining drag velocity ($V*$) (after Bagnold, 1941, 1965)

The proportionality constant has been found to approximate to 5.75, so that

$$V* = \frac{\tan \theta}{5.75} \tag{7.7}$$

The graph does not intersect the origin because near to the ground there is a small surface zone of dead air with a height K_0, which is sometimes assumed to be equal to 1/30 of the height of the surface irregularities (the latter taken as being equal to the grain diameter d, expressed in centimetres).

Equation 7.6 was reconfirmed experimentally (Williams, 1964) and by field observation (Bagnold, 1941), agrees well with natural dune migration rates (Bagnold, ibid.; Finkel, 1959) and with sandstorm records (Dubief, 1952), and formed the basis for Wilson's model for the development of ergs ('sand seas') (1971).

Because sand supply is seldom unlimited and other factors such as moisture and vegetation intervene, this rate is very seldom attained in practice and represents only a theoretical maximum. Various attempts have been made to derive empirical formulae with application to a wider range of environments. The FAO (Riquier, 1978) adds a consideration of rainfall, evapotranspiration, and land use to that of wind speed and particle size, using the formula

$$A = CEU \tag{7.8}$$

where
A = soil loss (t ha^{-1} yr^{-1});
C = climatic erosivity, derived from a summation of monthly indices relating windspeed, rainfall, and evaporation. The formula, involving the summation of 12 monthly indices, is

$$\sum_{1}^{12} \frac{V^3}{100} \left(\frac{PET - P}{PET} \right) n \tag{7.9}$$

where, for each month,
V = mean wind speed at 2 m height (m s^{-1});
P = rainfall (mm);
PET = potential evapotranspiration (mm) by the Thornthwaite (1948) method;
n = number of days in the month.

E = soil erodibility, based on multiplying empirical values for soil texture and structure. Texture values are: stony and gravelly 0.3; coarse 1.5; medium 1.0; fine 0.5. Values for structure range from 1.5 for those (such as peats or saline soils) which provide many loose deflatable aggregates to 0.5 for those which provide only large, firm, interlocking aggregates (such as chernozem, chestnut, or prairie soils).

U = land use, a dimensionless variable ranging from almost 0 for thick forest cover to 1 for bare soil. The values used are: forest 0.01; savanna 0.1; steppe 0.5; arable 0.7; dry farming 0.9; and bare soil 1.0.

The worst effect of wind deposition on terrain is its burial under moving sand. This is widely experienced on the windward side of desert settlements, such as the Nile and the oases of North Africa and the Middle East. In general such burial is relatively slow and is predictable from knowledge of the wind regime and the observed direction of dune movement. Because sand transport is proportional to the cube of wind speed, much, if not most, of the damage is caused during occasional violent storms.

Measures for defence against wind erosion can be classified as surface protection, windbreaks, and planting, and they are most effective when used in combination. The first is achieved by producing or bringing to the surface aggregates or clods which resist the wind force, reduce its velocity, and trap drifting soils. A method which has been used in Iran is to pour crude oil on to the surface of vulnerable areas to increase particle aggregation. Wind speeds may be reduced by walls, fences, or planted shelter belts of trees or shrubs. In Tunisia, for instance, palm fronds are driven in lines into the tops of dunes to arrest their movement. Because of the increase of wind velocity with altitude, it is more important that barriers contain a number of high points than that they are dense and continuous. A line of well-spaced tall trees will be a more effective windbreak than a dense low hedge. The best protection to the surface is vegetal cover. This can be maintained by preventing excessive grazing, maximizing the use of cover crops, strip cropping across the wind direction, leaving stubble in the ground or crop residues on the surface, or permanently vegetating especially vulnerable lands.

7.5 Physical degradation

The physical degradation of land is its loss of economic potential because of a decrease in porosity caused by compaction and the formation of pans. The latter may be due either to the concentration of clay at certain depths to form 'fragipans', or to the chemical cementation of layers to form duricrusts or 'duripans' (subsoil pans formed by silica cementation). Crusts and pans form slowly but cause an irreversible decline of land productivity which may locally become total, as on the laterite shields in tropical Africa and some calcreted terraces around the Mediterranean. Crust and pan formation is normally an end-point in the process of compaction and diminishing porosity and so physical degradation is usefully measured in terms of the rate of increase of soil density. The FAO proposes the indices shown in Table 7.2 (Riquier, 1978).

The improvement of soil structure is best achieved by minimizing the compaction of the surface by heavy vehicles and by stimulating the development of organic matter to improve aggregation and reduce the blockage of pores by moving particles. It is very difficult to remove pans once they are formed. The main methods are to destroy them by deep ploughing and digging and the encouragement of plants whose roots will penetrate them.

Table 7.2 Soil physical degradation classes (0–60 cm layer)

	Increase in density $(g\ cm^{-3}\ yr^{-1})$	*Decrease in porosity* $(\%\ yr^{-1})$	*Decrease in permeability* $(cm\ h^{-1}\ yr^{-1})$
None to slight	<0.1	<1	<0.5
Moderate	0.1–0.2	1–3	0.5–5
High	0.2–0.3	3–5	5–20
Very high	>0.3	>5	>20

7.6 Chemical degradation

Chemical degradation of land can take many forms, but can be broadly defined as chemical changes in the soil which inhibit the production of economic plants. These changes can be due to natural causes, human activity, or a combination of these. The commonest manifestation in humid regions is acidification associated with reduced levels and lowered availability of plant nutrients, and in arid regions, salinization and alkalization. Chemical pollution is the type of chemical degradation which results from acid rain or the addition to the land of noxious chemicals from fertilizers, insecticides, industrial wastes etc.

In humid climates where rainfall exceeds evapotranspiration there is a net downward movement of water in the soil. This causes leaching of bases, acidification, and the replacement of normal vegetation by less useful acid-loving plants. The loss of calcium or 'decalcification' is the most important part of this process because of that element's role in maintaining soil structure, organic activity, and nutrient availability to plant roots. Such availability generally decreases markedly when the pH falls below about 6. It is therefore usually adequate in humid climates to measure degradation in terms of decreases in pH and base saturation over time. Table 7.3 indicates standards used by the FAO in making such estimates (Riquier, 1978). Countermeasures are based on the neutralization of the acidity by the addition of calcium compounds in conjunction with a policy of improving organic activity in the soil by encouraging a more lime-loving vegetation.

Table 7.3 Chemical degradation: Standards for assessment (decreases in 0–30 cm soil layer per year).

	pH (units)	*Base saturation (%)*
None to slight	<0.05	<2.5
Moderate	0.05–0.2	2.5–5
High	0.2–0.5	5–10
Very high	>0.5	>10

It is also sometimes important to measure chemical degradation in terms of the rates of decrease of specific plant nutrients. Among the major nutrients, nitrogen and potassium are mainly lost through leaching, phosphorus through fixation in complex insoluble compounds unavailable to plant roots, especially at high and low pH values. The most suitable indices for measuring rates of decline are therefore the decreases in total nitrogen and potassium and in available phosphorus per unit time. These conditions can be reversed by additions of compounds containing the elements or by altering the pH to render the existing phosphorus more available.

Degradation from loss or excess of micronutrients is often difficult to determine from chemical analyses. This is because the uptake of a particular element by plants depends not only on the level of its active form in the soil, but also on the availability of elements with which it interacts, both in the soil and in plant tissues. The main micronutrients are iron, manganese, copper, zinc, boron, molybdenum, and, for leguminous plants, cobalt. Silicon and chlorine are required in low concentrations but are never deficient in soils. Some others such as selenium, iodine, and chromium are not needed by plants but are by grazing animals. Too high a level of micronutrients in the soil can sometimes be as serious as too little. Plants commonly have their growth affected by excess manganese and aluminium in acid soils, and by heavy metals such as nickel, cobalt, and chromium in acid soils derived from some ultrabasic igneous rocks. Molybdenum can be beneficial to plants if added to agricultural land at rates as small as 35–70 g ha^{-1}, but applications of over 1.5–2 kg ha^{-1} may be toxic to most (Russell, 1978; Wild, 1988).

All, with the exception of molybdenum, become less available with increasing alkalinity, and improved drainage usually decreases the solubility of cobalt, copper, manganese, molybdenum, and zinc. Therefore, although generally beneficial for the main nutrients, these changes are undesirable for trace elements.

7.7 Salinization and alkalization

In arid regions and after marine inundations the characteristic forms of land degradation are salinization and alkalization. Soil salinity threatens land uses because over certain critical levels it inhibits the osmotic pressures whereby roots absorb moisture. Its effect on crops is thus akin to that of drought. Where salinity levels are high, it will rot bricks and cement and rust metal.

Salt builds up in soil as a result of evaporation from mineralized water. Although measurable in terms of the weight of dissolved solids per unit weight of soil, it is more conveniently assessed in terms of the electrical conductivity (EC) of extracted soil water to which this is linearly related. The units of measurement are the milliSiemen (mS) and the microSiemen (µS) per centimetre at 25 °C (1 mS=1000 µS). The rate of salinization is expressed in terms of the increase of EC per unit time. The FAO standards for this in the 0–60 cm layer are (in mS yr^{-1}): <2 none to slight; 2–3 moderate; 3–5 high; and >5 very high.

Salinization can only be controlled by decreasing the salt content of the soil water. This is achieved by ensuring that water added to the surface either in rain or by irrigation is sufficiently in excess of evapotranspiration to maintain a net downward movement of water in the soil over the year even if this requires special underdrainage. This excess is calculated as a percentage of the total water added and is known as the 'leaching factor' or the 'leaching requirement'. It depends on the salt tolerance of the crop and the salinity of the irrigation water, as shown in Fig. 7.6 (Hoffman and van Genuchten, 1983).

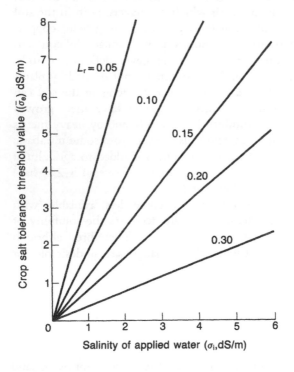

Fig. 7.6 Graphical solution for the leaching requirement (L_r), the minimum leaching fraction that prevents yield reduction, as a function of the salinity of the applied water and the salt tolerance threshold value for the crop. (Source: Hoffman and van Genuchten, 1983)

Alkalinity results from a relatively high ratio of sodium to other ions in the soil solution whether or not it is also significantly saline. The soil solution normally contains a mixture of common salts: mainly the chlorides, sulphates, nitrates, carbonates and bicarbonates of sodium, potassium, calcium, and magnesium. These salts ionize and the cations migrate towards and are adsorbed on to the negatively charged surfaces of clay and organic particles, known as the 'exchange complex', in proportions related to their concentration in the soil solution and to the order of preferential attraction: H>Ca>Mg>K>Na. Where sodium is sufficiently preponderant to dominate the exchange complex, the soil clays become deflocculated and much less permeable and friable. This inhibits plant uptake of water, as well as that of needed K and Ca, and when serious, renders the use of such soils uneconomic. Land degradation by alkalization can thus be measured in terms of the rate of increase of the percentage of sodium ions on the exchange complex (ESP) per unit time. The FAO standards for such increases, in ESP per year in the 0–60 cm soil layer, are: <1 none to slight; 1–2 moderate; 2–3 high; and >3 very high. Countermeasures involve the replacement of sodium by calcium. They are based on leaching with water accompanied by the addition to the surface of a suitable chemical. Where the soil is rich in $CaCO_3$, it is desirable to add sulphur or sulphuric acid, where not, gypsum. Methods are given by the USDA (1954) and the FAO (1978b, 1985).

7.8 Land pollution

The environmental threat from pollution has grown rapidly in recent years because of population growth and the increased volume and changing patterns of consumption made possible by new technology. It ranges in seriousness from the defacement of local amenities by litter to serious damage to public health through the poisoning of water supplies or vegetation over considerable areas. The four main sources of such damage to terrain are solid, liquid, and aerial wastes, and noise.

Solid wastes are of three types: rubbish dumps, toxic agents in the food chain, and fertilizers. Everyone is familiar with the paper, plastic, scrap metal and other rubbish that disfigure our environment, while the large amounts of toxic wastes generated by industry are less conspicuous. Even where concentrated into tips these occupy considerable space around settlements. The best modern systems of disposal include incineration, injection between rock layers, and landfills (which may have to be lined to prevent leakage into groundwater). Many of the toxic agents used in the human food chain – pesticides, hormones, antibiotics, drugs, preservatives, and food additives – remain in the environment and may be concentrated as they progress through the trophic sequence. A study in California, for instance, showed that a plankton–fish–bird chain concentrated DDT by 100 000 times (Brubaker, 1972). Agricultural fertilizers are widely employed, especially in advanced countries, and a considerable increase in their use is forecast for developing countries in the next decades. Although important in raising land productivity, some fertilizers, notably those containing nitrogen, increase soil acidity. On the other hand, excess calcium or potassium may reduce the availability of other cations. The main long-term problem resulting from the excess of chemical fertilizers is the addition of toxic amounts of nitrogen and phosphorus to the drainage water which harms aquatic plant growth by increasing the draft on the limited dissolved oxygen. The best ways to minimize the amounts and control the dispersal of their undesirable effects are to maximize the use of organic fertilizers and to make careful selection and restrained applications of artificials.

Water pollution comes from much the same sources as the solid refuse, and includes solid objects, industrial chemicals, and domestic wastes. These can be classified into the non-degradable, which are persistent, and the degradable, which are less so. The former include chemicals such as cyanide, mercury, cadmium, synthetic organic chemicals, and radioactive wastes. Complex molecules are often very slow to break down and are toxic to fish and thus to the animals and humans who consume them. Organic pollution is generally non-persistent. Its main sources are sewage and agricultural wastes. Untreated sewage consists largely of compounds in a chemically reduced state. When discharged into a river, it often progresses through a series of stages represented by zones on its downstream course. The first zone is characterized by the mixing of the unchanged waste with the water; the second by microbial decomposition, using up the available oxygen. The third zone is septic through the exhaustion of the free oxygen and may include the emission of reduced gases such as methane and hydrogen sulphide. The fourth is a zone of recovery due to the entry of further oxygen via bacteria and algae, so that the fifth and final zone is again one of clean water.

The most harmful liquid agricultural wastes are the same as the most harmful solids – compounds of nitrogen and phosphorus which find their way via drainage into water supplies. They are primarily responsible for the eutrophication of streams and lakes, and have had a very harmful effect on the seas, notably off the eastern coast of North America and in the North Sea. In these areas, the overload of nitrogen feeds marine algae which bloom into large growths that block sunlight and deplete the oxygen supply, smothering fish and crustaceans. This damages the whole marine food chain. Nitrate in drinking water is harmful because it is reduced to nitrite in the stomachs of animals and humans, which combines with haemoglobin to reduce the oxygen capacity of the blood. Where its concentration is over about $38 \text{ mg} \, l^{-1}$, it can be fatal to infants and cattle. The only effective control is to keep the amount of fertilizers added to the land to levels which will ensure their total use by plants and avoid overcharging drainage waters.

Some radioactivity affecting land comes from normal background radiation. There is also some from the manufacturing and testing of nuclear weapons. Radioactive wastes, for instance, are still causing serious problems not only in Japan, but also in towns such as Canonsburg, Pennsylvania where early manufacture of nuclear weapons took place. Today, however, other than major disasters such as Chernobyl, the most serious threat is probably the coolant water from nuclear power stations which has become radioactive through contact with the reactor core. This water not only brings waste heat which kills aquatic life, but also introduces radioactive isotopes into the environment. Even when washed out to sea they may be returned to shore by incoming tides. Radioactive materials may cause leukaemia or cancer both from external exposure and from being consumed in food or water. Perhaps the most serious aspect of the danger is our relative ignorance both of the total amounts of radionucleides being released into the environment and of their long-term effects on health.

7.9 Biological degradation

Biological processes making for the improvement or degradation of soils can most usefully be assessed in terms of changes in their organic matter content. The value of organic matter as an index is because its influence on soil properties is far out of proportion to the small quantities present. It commonly accounts for at least half of the cation exchange capacity of the soil and strongly influences nutrient availability. It is responsible more than any other factor for the stability of aggregates which enable soils to combine permeability with water-holding capacity. Furthermore, it supplies both energy and body-building constituents for the microorganisms which recycle plant residues, churn the soil, and help to maintain its structure. Levels of soil organic matter depend on climate, vegetation and human uses of the land. In humid regions, natural forest or grassland generate organic matter through leaf fall, root decay, or animal wastes. The amount thus generated decreases under controlled forestry, agriculture, or pastoralism due to the periodic removal of those plants and animals which renew the supply of organic materials. Organic matter may also be removed from the soil by erosion, especially after the destruction of vegetation and the drying of the surface. The drainage of organic soils in the English Fenlands has, for instance, led to their shrinkage and deflation to a depth of several metres, in places exposing the underlying clays. The FAO have suggested that the annual percentage decreases of the percentage of soil organic matter from the 0–30 cm layer should be characterized as follows: <1 none to slight; 1–10 moderate; 10–20 high; and >20 very high; (Riquier, 1978).

Continuous grassland and forest, especially when fertilized, will in time correct the shortage of organic matter resulting from their temporary removal. In cultivated soils the most important sources are crop stubble and residues, and the amounts are greatest when the crops are grown with adequate lime and fertilizers. Additions of farmyard manure and the ploughing-in of succulent immature crops known as 'green manuring' are also useful. There is, however, an optimal level of organic matter for the soils of any given area, dictated by climate and land uses, above which it is both difficult and uneconomic to go.

7.10 Air pollution

Two factors control aerial pollution: the nature of emissions and the state of the atmosphere. Table 7.4 gives a general listing of the most important types of air pollutants.

Table 7.4 Types and sources of atmospheric pollutants (after Oke, 1978)

Type	*Source*	
	Natural	*Anthropogenic*
Particulates	Volcanoes Meteors Sea spray Forest fires	Combustion Industrial processing
Sulphur compounds	Bacteria Volcanoes Sea spray	Burning fossil fuels Industrial processing
Carbon oxides	Volcanoes Forest fires Animals (CO_2 only)	Combustion engines Burning fossil fuels
Hydrocarbons	Bacteria Plants	Combustion engines
Nitrogen compounds	Bacteria	Combustion

About 90 per cent of all atmospheric particles are derived from natural sources. The main anthropogenic sources are from combustion (domestic and power station coal and oil burning, car exhaust, and refuse incineration), industrial processing (cement and brick works, iron foundries, metal processing mills), and surface disturbances due to building works. They most commonly consist of carbon and silica, but may also include iron, lead, manganese, mercury, cadmium, chromium, copper, nickel, beryllium, aluminium, and asbestos.

The particles can be either liquid or solid and generally range in size from greater than 100 μm to less than 0.1 μm in diameter. Those over about 10 μm, because of their greater size, hover mainly over the city or industrial plant which spawned them, creating clouds which settle on the local landscape either as dry deposition or water-borne in raindrops and snowflakes. Some of these cause immediate damage to vegetation, wildlife, and buildings, while the rest enter runoff and soil water.

The finest pollutants are gaseous combustion products and aerosols. The former are mainly the oxides of sulphur, nitrogen, and carbon. Some of these climb skywards, especially when emitted from tall chimneys. Here they unite with the weather system and become the main contributors to acid precipitation. Although 100 per cent of the materials may be deposited within a few kilometres of the source during a rainstorm, generally only 19–20 per cent falls within 50 km of it. Their flight may last for days and take them hundreds, even thousands, of kilometres. *En route* the pollutant molecules interact chemically with sunlight, moisture, oxidants, and catalysts, to become other compounds of carbon, nitrogen, and sulphur. Ultimately they are precipitated as acid rain or snow, which are in reality merely dilute solutions of carbonic, nitric, and sulphuric acids.

Downwind effects depend also upon the source configuration, including its shape, duration, and height. They can be classified into point sources, line sources, and area sources. The most important point source is the chimney stack which can be the originator of concentrated toxic materials. A busy highway is the commonest form of line source, where exhaust emissions from many vehicles constitute a continuous output along its length. Cities are the paramount area sources because they bring together a multiplicity of small sources which are hard to separate. Scale affects this classification. For instance, when considering a whole continent, cities can appear as point sources.

The movement of pollutants occurs within the 'atmospheric boundary layer', more simply referred to as the 'boundary layer'. This can be defined as the lower part of the troposphere where the frictional drag of the atmosphere moving across the roughness of the earth's surface causes turbulence and resulting winds and where air parcels from the heated surface 'bubble up' (Oke, 1978). The dispersion of pollutants is greatest where such winds and bubbling up are at a maximum, as in sunny daytime conditions, especially in summer and where relief locally increases wind speeds. Conversely, it is at a minimum and pollutants are deposited nearest to their source when the boundary layer is stable and there is a temperature inversion near to ground level.

Atmospheric motion serves both to diffuse (dilute) and to transport air pollutants. The further the wind travels the more dilute it becomes. It may therefore become so weak to be of little consequence. The greatest risk therefore often comes from weak winds because both horizontal transport and turbulent diffusion are curtailed. Wind direction determines which lands receive the contamination. This can clearly make the difference between serious contamination and total escape. A particular wind direction may also result in multiple pollutant inputs due to the coincident alignment of sources. In addition to the cumulative loading, this might cause reactions between chemicals to develop secondary pollutants downwind.

Aerial pollution is most commonly deposited as acid rain. The worst affected regions today are those around and downwind of the main industrial areas of the Northern Hemisphere, notably southern Scandinavia, northeastern United States, eastern Canada, and neighbouring seas. Some Scottish lakes have experienced a decline of 0.5 pH units since 1850 (Battarbee *et al.*, 1989). The pollution tends to be worst where precipitation is highest, as in mountainous areas, which also have the thinnest soils and the most vulnerable vegetation cover. The acidity concentrates in closed water bodies, and is only neutralized where soils are alkaline, as where the parent material is limestone.

The total amounts of pollutants emitted annually are staggering. In 1980, the USA and Canada between them emitted over 32 million tonnes of sulphur dioxide and over 24 million tons of nitrogen oxides, and when the European figure is added, the total for sulphur dioxide alone is almost 100 million tonnes. Measurements of rain and snow have shown that in virtually all of the eastern USA and much of southeastern Canada, the precipitation has pH values of 4.1–4.6, while occasional storms have recorded values as low as 1.5.

The effects are most obvious in closed lakes where the pH falls below critical values. This destroys fish stocks, partly as a result of increased aluminium concentration. The rate of fish-kill is often at a maximum when polluted snow melts in the spring. Few fish survive in water with a pH below 5 and none below 4.5. Their destruction allows some insects to multiply, but the lowered pH brings an overall reduction in the activity of the aquatic ecosystem and a change to more acid-tolerant fauna and flora. Neighbouring lakes can vary widely according to the chemical buffering capacity of their catchments, so that it is not unusual to find one where all fish have been destroyed near another which is unaffected. As previously discussed, the effects are multiplied by the presence of nitrates originating as agricultural fertilizer.

Unlike the dramatic effects on aquatic life, the influence of acid rain on land is harder to measure. This is partly because nitrogen and sulphur, even in the form of mild acids, serve as plant nutrients, and the areas previously noted as the main recipients of acid rain are largely those with nitrogen-poor podzolic soils. On the other hand, the acid rain has the adverse effects of delaying the decomposition of plant litter and causing the build-up of toxic metals in the soil over long time spans. Wildlife suffers from the pollution of the trees and plants, as evidenced, for instance, by the poisoning of song birds in Lapland through feeding on insects contaminated by aluminium (LaBastille, 1981). The acidification also extends to groundwater and causes corrosion and an increased heavy metal content of water supplies.

The cure is to control the emissions at source and to counter the acidification in the recipient areas by measures to raise pH.

Further reading

FAO (1971, 1978b, 1985), Hudson (1981), Morgan (1979), Oke (1978), Thompson *et al.* (1986), Wischmeier and Smith (1978).

8

Vegetation in terrain evaluation

8.1 The place of vegetation in terrain evaluation

There is no part of the earth's land surface, except permanent snow and ice fields and the most extreme deserts, which does not support some vegetation. In humid climates, the cover is generally thick enough completely to obscure the surface of the ground. There are three ways in which vegetation is important in the evaluation of terrain: as an index for the recognition of terrain types, as an attribute in their definition, and as a natural resource physically attached to them.

8.2 Landform recognition from vegetation

Vegetation cover is usually an important guide in recognizing terrain types, particularly when they are viewed from the air. This is not only true in undeveloped areas of low population density where the ecological climax closely reflects site characteristics, but also in developed areas such as western Europe where there is a close relationship between agricultural land use and site. The most important single factor in this relationship is probably the effect of ground configuration on soil moisture conditions. For this reason, vegetation is a valuable index of terrain in areas where soil moisture levels fall below thresholds which are critical to plant growth, as in semi-arid lands. Other terrain factors such as slope, aspect, soil depth, and nutrient status are also important, and are each determinants of plant cover in particular localities

A good example of the use of vegetation as an index was the survey of the range potential of land in Jordan carried out by Hunting Technical Services, Ltd (1954–) (Fig. 8.1). The Australian CSIRO has also made wide use of vegetation in recognizing terrain types in surveys of, for instance, semi-arid Australia (Mabbutt and Stewart, 1965) and Papua New Guinea (Robbins, 1976). Geobotanical studies have used the relations between vegetation and terrain to indicate the characteristics of underlying ore bodies. They have been of particular value in reconnaissance surveys of remote terrain, both where individual vegetation associations give spectral responses whose distribution patterns disclose structural and lithological features on satellite imagery, and where anomalous plant communities associated with mineralization appear on aerial photographs and on the ground. Stratiform copper and lead–zinc deposits in southern Africa, Australia, and the UK; iron ore bodies in South America and Australia; phosphate deposits in Australia, Sweden, and Finland; and carbonatites and kimberlites in Europe and southern Africa have been so located (Cole, 1980; Brooks, 1983).

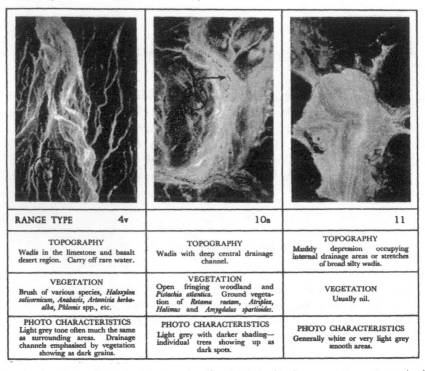

RANGE TYPE 4v	10a	11
TOPOGRAPHY Wadis in the limestone and basalt desert region. Carry off rare water.	TOPOGRAPHY Wadis with deep central drainage channel.	TOPOGRAPHY Muddy depression occupying internal drainage areas or stretches of broad silty wadis.
VEGETATION Brush of various species, *Haloxylon salicornicum, Anabasis, Artemisia herba-alba, Phlomis* spp., etc.	VEGETATION Open fringing woodland and *Pistachia atlantica.* Ground vegetation of *Retama raetam, Atriplex, Halimus* and *Amygdalus spartioides.*	VEGETATION Usually nil.
PHOTO CHARACTERISTICS Light grey tone often much the same as surrounding areas. Drainage channels emphasised by vegetation showing as dark grains.	PHOTO CHARACTERISTICS Light grey with darker shading—individual trees showing up as dark spots.	PHOTO CHARACTERISTICS Generally white or very light grey smooth areas.

Fig. 8.1 Sample sheet and range classification key showing some range types in Jordan. (Source: Hunting Technical Services Limited, 1956)

8.3 Vegetation as definitive of land units

There have been two contrasting views about the use of vegetation in the definition of terrain units. Some workers, especially those with a background in the earth sciences, have tended to define landscape units in terms of landform and soil characteristics, using vegetation merely as a useful key to their recognition both in the field and on remotely sensed imagery. This has been for two chief reasons. First, the plant population of an area, because it is organic, has been regarded as an intrinsically different phenomenon from its terrain. The two can vary independently, so that changes or natural boundaries in the one are not necessarily reflected in the other. Second, as emphasized by Zimmermann and Thom (1982), vegetation is an ephemeral rather than a permanent characteristic of landscape. It is in a continuous state of change through seasonal variations, natural colonization, the activities of wildlife, and above all through the impact of human activities, such as grazing, burning, and new plant introductions.

Other workers, especially those with a background in the life sciences, have adopted a more holistic approach, according vegetation a definitive, rather than merely a diagnostic, function in the identification of terrain units. The earliest expressions of this view seem to have been from Bourne (1931) and Veatch (1933) who distinguished 'sites' and 'natural land types' on ecological as well as physical grounds. Hare (1959) relied even more strongly on vegetation units in the reconnaissance survey of Labrador–Ungava, and the 'ecoregion' concept incorporating both vegetation and landforms has been used fairly extensively in both the USA and Canada since this time (Bailey *et al.* 1985). The Land Research and Regional Survey Division of CSIRO also used vegetation as well as landforms and soils as a definitive criterion for land systems (Mabbutt and Stewart, 1965), although later Australian work has tended to move towards a purer earth science base (e.g. Gibbons, 1983).

There are substantial reasons for using flora and sometimes fauna to define units. Some features, such as coral reefs, termitaria, and mangrove swamps are wholly or largely biotic in origin. Elsewhere, by providing the organic components of the soil, vegetation is closely integrated with the land surface and influences its character. In the undeveloped areas covered by most land resource surveys, it changes so slowly that it constitutes a relatively permanent land attribute. Practically, it is often the key factor in the evaluation of unimproved land. The case for using it as definitive is thus strongest in forest and range areas which are least subject to changes in land use. It is included in land unit definitions for these purposes, for instance, by the FAO (1976a) and Vink (1983).

The differences between these two viewpoints partly derive from the nature of the landscapes and the scale of survey of the areas considered by each. In general, vegetation is most useful in defining terrain units where it is (a) natural and undisturbed, (b) important in forming the ground surface as in marshes or humid forests, and (c) considered over either very small or very large areas. In small areas, vegetation differences form those variations in the landscape which are clearly visible on the ground, and both articulate and magnify the effect of soil nutrient or drainage contrasts, such as between 'grykes' and 'clints' in the limestone country of northern England, or the alternating tussocks and hollows characteristic of raised bogs. Over large areas, on the other hand, the variations of terrain are largely climatically determined and result from the secondary effects of vegetation in, for instance, weathering rock outcrops, holding slopes against erosion, and transforming basin sites into bogs or fens. Howard and Mitchell (1985) have shown the parallelism between geomorphological and ecological classifications of the landscape and have attempted to resolve this question by advocating a synthesis of the two into 'phytogeomorphic' units whose definitions give a greater importance to vegetation at extreme scales.

8.4 Vegetation description and mapping

The classification and mapping of vegetation use different principles from those used for landforms. Subsequent chapters will show how the two systems can be coordinated for different purposes, but it is first necessary to consider the approaches that have been developed for vegetation alone. Three are discernible, although there is considerable overlap between them. These are the floristic, the physiognomic, and the ecological, the last being to some extent intermediate between the first two.

8.4.1 *The floristic approach*

The floristic approach views plants in terms of their position in the classical Linnaean groups, recording the species composition of each plant community (defined as the smallest recognizable floristic unit of vegetation) either qualitatively or quantitatively and characterizing their geographical distributions. The commonest floristic classification of vegetation is based on the dominant species of the tallest stratum. More refined classifications may use the combination of dominants in each stratum, but this may result in an inconveniently large number of small units. A compromise is to use groups of characteristic species, a method which has been given a quantitative basis by Goodall (1952).

The approach has been applied where economic properties of soil or rock are associated with particular plant species or genera, for instance in geobotanical studies or in identifying sites with particular soil properties.

Two practical studies which relate floristic classifications of vegetation to site value are those by Smith (1949) in the Sudan and Beckett and Webster (1965d) in Britain. The main limitation on natural plant growth in the Sudan is almost invariably aridity. Smith showed that the natural occurrences of certain tree species could be accurately predicted if only three variables were known: rainfall, soil texture, and site type. Given the same clay soil and site type, a succession of different *Acacia* species occurred over the 1500 km transect from the river Atbara at about 16° N to the Sobat near Nasir at 9° N. At the same time, a single species, such as *Acacia senegal*, occurred on sands with under 400 mm of rainfall and on clays with under 600 mm. So general was this relationship that it was possible to say that a tree species in the Sudan required 1.5 times as much rain on clay soils as it did on sands. It was further found that, with no recorded exceptions, all species progressed through their rainfall span via the same sequence of site types, and that, given the same rainfall, the site types could always be arranged in the same order of provision of soil moisture for tree growth. The 'site transect' showing equivalent rainfall–site combinations is represented diagrammatically in Fig. 8.2. It makes it possible to relate plant species to terrain types in the Sudan so that rough predictions of each can be made from the other. It is probable that similar relationships occur in areas of natural vegetation elsewhere, either where rainfall varies evenly within large tracts of relatively homogeneous terrain, or where terrain varies predictably over an extensive area with constant rainfall.

In a study of similar type in the widely different Oxford area, Beckett and Webster (1965d) found statistically significant correlations between land facets and certain tree, shrub, and climbing species in hedgerows. On the whole, shrubs showed the greatest degree of association. For example, willow and guelder rose were found where there was shallow groundwater, crab-apple on well-drained sites, and blackthorn in the areas of highest fertility.

8.4.2 *The physiognomic approach*

The second method of vegetation classification is the physiognomic, which gives units which are generally the easiest to quantify and to relate to land classes. The definition of classes is based on the three main characteristics of the vegetation which can be accurately measured and mathematically defined: height, ground cover, and life form. Height, or more precisely 'stand height', is the mean height of the dominant or codominant trees. Ground cover is the proportion of the ground covered by the vegetal canopy, and life form is the character of the dominant plants as expressed by such terms as conifer (gymnosperm), deciduous hardwood, sclerophyllous forest, etc. Because they are not concerned with the detail of plant floristics, physiognomic schemes emphasize dominant plants and are relatively simple to quantify and to relate to land classes. They are thus especially useful to non-specialists concerned with vegetation in the mass, i.e. as an amenity, a source of cover, or in relation to cross-country mobility.

Classifications of vegetation physiognomy have been approached from three points of view: from the ground, from the air, and from the aspect of practical applications.

The pioneering scheme for the ground description of plant physiognomy was by Kuchler (1949), but this was considerably developed by Dansereau (1958) whose structural classification has formed the basis of later work, and is described in connection with mapping in section 16.4. The plant component of the biogeocoenose is the 'phytocoenose', which is approximately synonymous

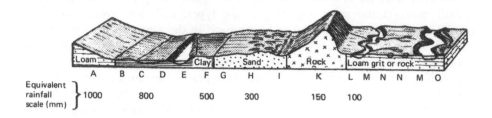

Fig. 8.2 Schematic transect of sites in order of relative abundance of moisture for plant growth in the Sudan: (A) hard-surface slopes subject to sheet flow; (B) high flood plain (flooded only for days); (C) low flood plain (flooded for weeks at a time); (D) mounds or banks in swamp or beside river; (E) beds of landlocked pools; (F) flat clay plains without runoff or standing water; (G) sand plains; (H) sand dunes and hills; (I) pockets, hollows, or valley beds in sand country; (K) rocky hill slopes; (L) seasonal watercourses flushing for an hour or two after rains; (M) hard plains of grit or rock; (N) seasonal runnels flushing for an hour or two after rains; (O) banks of permanent rivers or streams. (after Smith, 1949)

with 'plant community'. Its physiognomic subdivision is the 'synusia', which can be defined as 'a group of plants of one or several related life forms, growing under similar environmental conditions'. Synusiae may correspond to the horizontal layers of the community, i.e. 'tree', 'shrub', etc. or else to lateral subdivisions such as 'grassland', 'forest', etc. Each synusia is made up of individual plants, which have been classified physiognomically in a number of ways (Kuchler, 1967). One of the most widely adopted systems is that of Raunkiaer's 'life-form classes', based on the way plants react to the critical season of drought or frost limiting their growth (1934). It is simply expressed in terms of the location of the perennating buds in relation to the ground surface. The main classes are:

1. Phanerophyta   Buds >25–30 cm above ground.
2. Chamaephyta   Buds above ground but <25–30 cm
3. Hemicryptophyta Buds at ground surface.
4. Geophyta   Buds below ground surface.
5. Therophyta   Buds in the seed annuals.

In essence, the phanerophytes are the trees and shrubs, the chamaephytes are grasses and high-seeding herbs, the hemicryptophytes the low-seeding herbs, the geophytes the plants which grow from tubers, corms, and bulbs, and the therophytes the annual plants.

The study of vegetation in plan view has been greatly stimulated by the availability of aerial and satellite photography. Their synoptic viewpoint favours the use of physiognomic classifications because it preferentially reveals the physical properties of vegetation such as size, shape, and life form. This has encouraged the use of such classifications in forest and range assessments and

102

for military purposes. The latter especially favour them because the properties visible from the air are those which govern cross-country mobility and concealment from aerial observation. The Canadian army, for instance, in a study of Camp Petawawa, found that cross-country mobility depended on the diameter and spacing of tree stems, concealment on plant height, canopy closure, and seasonal changes. All these except stem diameter could be determined from the air (Parry *et al.*, 1968).

The classification of patterns of vegetation physiognomy from the air depends on scale. The land cover classification of Anderson *et al.*, (1976) for use with remote sensor data, discussed in section 23.5, is suitable for use at a wide variety of scales. An example of a semi-quantitative local scheme was that for the survey of the Aitape–Ambunti area of northwest Papua New Guinea, using 1:50 000 black and white aerial photography (Heyligers, 1968). There were three sorts of grassland: mid-height (2–6 feet, 60 cm to 1.8 m), tall (over 6 feet, 1.8 m), and scrub. Palm vegetation was subdivided into low, mid-height, and tall, with limiting values at 50 and 100 feet (15 and 30 m). The upper canopy was classed as open, irregular, or closed, and the lower storey was classified in accordance with crown size, regularity of cover, and presence of a particular species (sago palms). This and other classifications from aerial photographs, such as that of Howard (1970a), are only partially quantitative, and cannot become more so except to the degree that plant species and structures, rather than mere photo tone and texture, can be recorded. Howard (1970b) made a step in this direction with the design of a 'stereoprofiler' by which land and vegetation profiles can be derived automatically from aerial photographs.

8.4.3 *The ecological approach*

The third way of classifying vegetation in relation to terrain is by associative groups of plants, or ecologically. This is the normal method employed in atlases, textbooks, on small-scale maps, and for regional studies of plant distribution such as that by Tansley of the British Isles (1953). Mapping units, summarized by Howard and Mitchell (1985), range in size from the simple 'plant associations' representing the smallest units of floristic classification up to the 'plant panformations' extending over whole climatic zones. The names of individual units are a mixture of the botanical and the physiognomic, using such terms as 'gallery forest', 'grassland', 'Acacia tall grass scrub' etc. Notable developments of the method are: (a) mapping systems using plant associations with their grouping and components in the French Swiss school of Braun-Blanquet (1964); and (b) detailed vegetation maps of local communities tied to more widely recognized associations of Tuxen and his school, and combinations of these two systems (Vink, 1983).

Hierarchies of mappable units have been developed, such as that by Long

(1974) which to some extent parallels that for terrain units and is shown in Table 5.1. Another such scheme which has been widely used in Latin America is that by Holdridge and Toshi (1972). It is based on three hierarchical levels. The 'life zone' is level I, determined by ranges of rainfall, 'biotemperature', and potential evapotranspiration. Examples include 'cool temperate steppe' or 'subpolar dry tundra'. Each life zone contains a number of examples of level II, the 'association'. This is a phytogeomorphic rather than a purely vegetal class, because it may be based on climatic, geomorphic, or soil criteria wherever these are appropriate in a particular region. Associations may in turn contain a number of examples of level III, the 'non-climax successional stage' or the 'man-made cover type', which is the practical planning unit analogous to the land facet.

8.5 Practical applications

Various approaches have been adopted to predict the practical effect of vegetal cover on human activities. Research has been devoted to environmental conservation and development, and also, particularly for military purposes, to the prediction of cross-country mobility.

Man's colonization of the biosphere has fundamentally changed its patterns of species abundance and distribution, often in ways that have been harmful to their productivity. It is thus important to ensure that the ecological results of present or proposed changes of this type are fully considered in planning development projects. Since the biosphere consists of ecosystems, the approach must be through their control and management.

Ecosystem productivity results from the transfer and transformation of solar energy within the biosphere. Its study is highly complex, involving inventory and description of the variables, analysis of their changes over time, and study of various states of different types of ecosystems, usually concentrating on a few examples. To approach these problems, the Unesco 'Man and Biosphere' programme (1971), which coordinates national approaches, has developed 'ecosystem modelling'. This is based on 'systems ecology' which assumes that the nature of an ecosystem and the processes within it can be simplified in such a way that, using information from ground survey and remotely sensed imagery, the behaviour of the system as a whole in time can be simulated. It further assumes that its state at any particular time can be expressed quantitatively, and that changes in it can be described by mathematical expressions, commonly in the form of differential equations. These are built into a computer program, which gives a numerical solution of the equations over a specified time interval.

Such modelling allows potential modifications to the ecosystem to be tested without repercussions so that plans to manage it to obtain some specified goal can be incorporated into a computer program. Such a program can test all possible combinations of management practices and hence optimize the results in any sense specified.

Grabau and Rushing (1968) described a computerized system for answering questions about vehicle trafficability in forested areas using a physiognomic vegetation classification of the Dansereau type. They determined sample area sizes in terms of the 'structural cell' which, simply stated, was the minimum area showing all the internal variation in an assemblage. In practice this meant a circular site enclosing 20 plants of the type with which the survey was concerned. Only physiognomic attributes which were mutually independent were selected and these were recorded in such a way as to minimize transcription errors. Field observations were made on a form which could subsequently be punched out and used in a computer. This approach was related to cross-country mobility problems by the US Army, who fed information on tree stem girths and spacings, representing the vegetal obstacles, into a mathematical model for predicting vehicle speeds. The method was as follows: the computer read numerical values characterizing the tree height, crown radius, and stem diameter from punched sample cards and automatically calculated the stem sizes and spacings to be expected in the area from which the sample was drawn. The data could also be printed out in visual form as a cumulative graph relating stem diameter classes to numbers of plants in a given area.

A system of this type is relevant to forestry and ecology where information is required on the spatial geometry of plant assemblages. It could, for example, be used to determine such facts as the degree of association of two species or the spatial arrangement of stem diameter classes. An easy way to achieve such objectives would be to print out annotated maps of the sample areas, which should each be equivalent to structural cells, using data on stem size, compass bearing, and the distance of individual plants or plant groups from the centre of the sample area.

Further reading

Howard (1970a, b), Howard and Mitchell (1985), Kuchler (1967), Vink (1983).

Part III

Collection and analysis of terrain data

9
Project planning

9.1 The organization of terrain surveys

The word 'planning' comes from the Latin *planus*, meaning a plain or an area which is all visible. This well expresses the objective of the planning stage of a terrain survey. The aim is usually to obtain the environmental data for some form of development, such as road building, irrigation, or landscaping, within acceptable cost constraints. The plan must include the outline of all the work to be completed before it is actually implemented. The normal sequence of operations in the survey is:

1. Preliminary definition and programming of specialized skills, critical path analysis, and literature orientation;
2. Assessment of land by remote sensing;
3. Field sampling;
4. Laboratory analysis of soil, water, and vegetation samples;
5. Data processing;
6. Presentation of results.

These topics are dealt with in this and the following chapters. There is considerable overlap between them. In particular, stages 2 and 3 must largely run concurrently.

Surveys begin with a plan that states the main steps and the costs of achieving the overall objectives. Since they usually require the collaboration of a number of specialists, the first task is to divide the work into missions to be handled by individuals, and to arrange a framework for their coordination, both formal and informal, to avoid duplication and unused time. This may involve the formation of groups, which are best when kept small. The success of the survey will depend on the degree of 'lateral' integration between the specialist disciplines within each stage and on the overall 'linear' integration between stages. This requires ensuring liaison between groups within a team by arranging regular communications and opportunities for discussion. It is especially important when they are physically separated, i.e. between headquarters and field sites.

9.2 Detail and scale of survey

The choice of scale depends on the purpose of the survey, the time available for its completion, and the amount and nature of prior knowledge available. In surveys of large areas there are three clearly recognizable stages (Robertson *et al.*, 1968):

1. *First stage regional surveys*. In countries where knowledge of natural resources is insufficient to indicate development possibilities, the initial requirement is for rapid general surveys to show the range of such possibilities and where more detailed investigations will be most rewarding in accordance with an overall view of national development priorities. Such surveys are generally based on the land systems approach and the interpretation of aerial and satellite imagery at scales of 1:40 000 or smaller.
2. *Second stage surveys* analyse the development possibilities in a project area or group of areas which have been identified in the first stage. This involves the acquisition of sufficient information to indicate whether specific development projects are feasible, both from the physical point of view and from the general cost/benefit aspect. This means that the scale of survey has to be enlarged to give fairly detailed information on the various resources involved. This can again be based on land systems, but also requires the inclusion of soil surveys to series level, the quantification of vegetation resources, the identification of water storage sites, the measurement of groundwater supplies and quality, and an assessment of present land uses. Work at this stage will generally be based on maps and aerial photographs at scales of 1:25 000 or larger.
3. *Third stage* or *detailed surveys* are analyses of an area to formulate a specific project plan, and to show in detail the cost–benefit ratios of the activities considered. Studies are made of the crop, mineral, or manufactured item to be produced, its management in terms of likely costs and yields, and the optimization of the environmental changes involved. The scale of observation need not be much, if any, larger than in second stage surveys but there will inevitably be a heavier emphasis on detailed economic assessment.

9.3 Flow diagrams, bar charts, and critical path analysis

Planning a survey is aided and more easily envisaged by preparing 'flow diagrams', 'bar charts', and 'critical path analyses'. Flow diagrams are simple graphic illustrations of the sequence and interactions of different activities in a project. An example, illustrating the stages of a terrain survey, is shown in Fig.

9.1. Bar charts, such as that shown in Fig. 9.2, are timetables showing the phased inputs of a number of specialists in a project. They are used for programming working and leave periods, planning travel arrangements, and identifying the times when specialized equipment or facilities are required.

Critical path analysis assists project planning by dividing the major activities into their components and depicting them in the form of a chronological network. This can be used for optimizing the use of resources, progress reporting, and control.

The procedure, as outlined by Dale and Michelon (1966) and Baboulene (1969), is to construct a diagram in which 'events', represented by circles or diamonds, are linked by 'activities', represented by lines or arrows. Figure 9.3 illustrates the main points of the method. Each event marks the completion of one or more activities and must occur before other dependent activities can begin. Events requiring decisions are represented as diamond-shaped 'decision boxes', those which do not, as circles. Event 2 is a decision box, the others not. Durations can be given for activities or dates for events. If the latter are imposed, they are written in a 'flag'. Data are not complete (event 3) until both the field survey and the archive search have been done. It is thus a 'merge point'. On the other hand, the completion of the data analysis must precede both the report writing and the production of maps and illustrations. Event 4 is thus a 'burst point'. Event 1 is both a merge point and a burst point.

The critical path is the longest time sequence through the network because this is the path that controls the time of final completion. Events and activities along this path are known as 'critical events' and 'critical activities'. The critical path is marked by doubling the activity lines along it as shown in Fig. 9.3. It is usually desirable to draw the diagram so as to make this line horizontal and central. The 'float' is the time available for an activity in excess of its necessary duration, which is, for example, 20 days for activity 2 on Fig. 9.3. The 'slack' is the difference between the earliest and latest times for an event, both of which can be written inside the event circle. Where there are maximum and minimum times as well as average times for an activity, these can be written in after the activity as shown for activities 1 and 2. Critical path networks can be expanded to include a number of sub-networks, such as subdividing activity 7 into components for arranging transport, planning routes, field traversing, soil sample analysis, and data compilation.

The advantage of the method is that planning is comprehensive and the risk of overlooking needed inputs is reduced. It permits calculation of whether individual jobs can be completed in time and focuses attention on potential bottlenecks and where the application of resources can be most effective. It shows which activities can be carried out simultaneously and where there is likely to be spare time. Finally, it makes it easier to monitor progress, especially in detecting slippage along the critical path. Where the critical path analysis is complex, it is advantageous to use a computer to generate and update diagrams and to calculate dates and times.

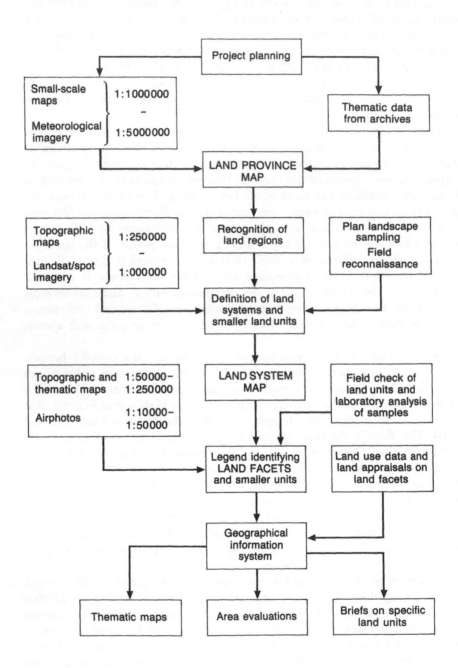

Fig. 9.1 A flow chart for a land system survey

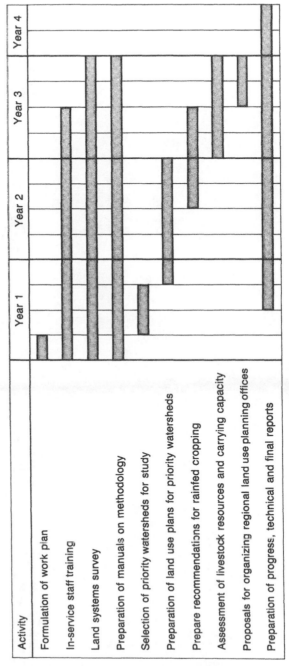

Fig. 9.2 Bar chart for formulating a regional land use plan

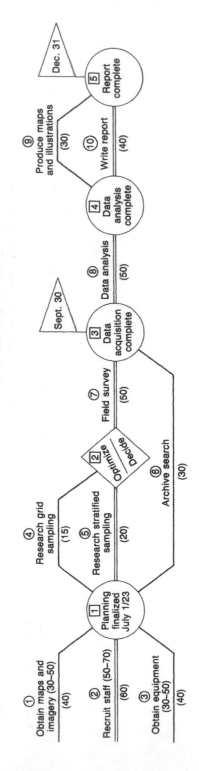

Fig. 9.3 Example of a critical path analysis for a terrain survey. Numbers in parentheses represent days.

9.4 Archive work

An early step in evaluating an area is to undertake a systematic review of all the available information about it in order to arrive at a preliminary assessment. Such information is usually in the form of books, articles, maps, unpublished reports, and remotely sensed imagery.

Libraries are of many sorts, some rather unexpected. They include major national collections such as the Library of Congress and the British Library, those of international organizations such as FAO in Rome and Unesco in Paris, the archives of government departments, and the libraries of universities, colleges, research institutes, commercial firms, and cultural agencies such as the British Council or the United States Information Service. Most towns and some villages and local institutions have libraries with information about local history, and for comprehensive coverage, it is desirable to refer to these.

Books and articles are relatively easy to obtain. The key background is that on geology and soils derived from such bibliographies as the American Geological Institute *Bibliography and Index of Geology*, originally published annually by the Geological Society of America separately for North America and elsewhere, but now monthly and combined for the whole world. *Elsevier Geoabstracts*, published monthly by Elsevier Science Publishers, covers geomorphological and pedological literature but has the disadvantage of only being indexed thematically and not by geographical area. The Commonwealth Agricultural Bureau at Harpenden provides bibliographies of any agricultural topic on request, and the Land Resources Development Centre at Tolworth has produced bibliographies of the land resources of a number of countries.

Topographic maps exist for all the land surface of the earth, but the best scales available vary widely. While for developed countries much coverage exists at scales larger than 1:50 000, 1:100 000 is more common in developing countries, and most of their more remote parts are mapped at only 1:250 000. Geological and other thematic maps are rarer and at smaller scales. Although almost all Britain, for instance, is covered by geological maps at 1:63 360, some is still at only 1:250 000. Most of Asia and Africa is represented at only 1:2 000 000 or smaller. Soil and other thematic maps are still more infrequent and at even smaller scales. For example, a complete 1:250 000 soil map of England and Wales was not published until 1983.

Much data about terrain can be learned from topographic maps Settlements and routeways reflect the relief and drainage. Place names including terms such as 'cliff', 'ford', 'highway', 'holloway', 'marl', 'nab', 'clough', 'haugh', 'chine', and 'winterbourne' carry indications of local terrain, and vegetal terms such as 'heath', 'moor', and 'forest' often allow it to be inferred. Archaeological remains such as abandoned routeways, seaports, castles, and forts indicate terrain conditions significant in the past. For instance, mountain areas in North Africa and the Middle East often show traces of settlements based on water supplies which are no longer exploited, and saline flats in Iraq and Pakistan indicate irrigable tracts abandoned because of salinization.

Further reading

Baboulene (1969), Dale and Michelon (1966), Dent and Young (1981), FAO (1976a), Howard and Mitchell (1985), Robertson *et al.* (1968).

10

Remote sensing: principles

10.1 Platforms and sensors

Remote sensing is today one of the basic preliminaries to land assessment. It can be defined as any perception and recording of phenomena by devices not in contact with them. Applied to the terrestrial environment it is the identification and analysis of earth surface phenomena using instruments carried in aircraft or spacecraft. These sample the radiant energy within the electromagnetic spectrum, which moves at the wide range of wavelengths and frequencies summarized in Fig. 10.1.

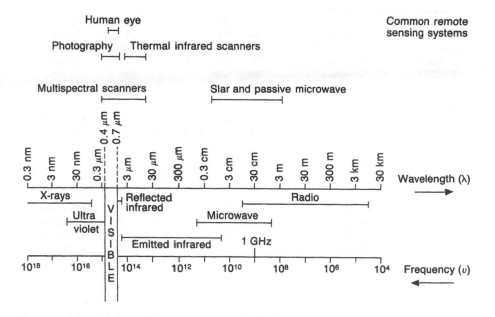

Fig. 10.1 Electromagnetic spectrum (after Curran, 1985)

The principal source of energy is the sun and most sensing devices are 'passive' in that they depend on solar radiation which has been received at or near the earth's surface and radiated back into the atmosphere. On the other hand, some sensors, such as radar, are 'active' in that they generate their own energy and record it as it returns to source.

Another group of sensors are those used by geophysicists to penetrate deeper into the earth's crust. These include magnetometers to record anomalies in the earth's magnetic field, gravity meters to record variation in its gravitational field and hence the type of rock mantle, and scintillation counters to indicate radioactive areas. These can be used either on the ground or from aircraft, but not as yet, except experimentally, from satellites.

The most important platforms are shown in Table 10.1.

Table 10.1 Summary of the main remote sensing platforms used since 1971 (in approximate order of altitude)

Platform	Orbit	Altitude (km)	Main Sensors	Best resolution (m)	Return period
Meteorological satellites					
SMS (GOES, METEOSAT)	Geostationary	35 900	Rad	2400	48 per day
NIMBUS	Sun-synch.	995	CZCS	800	6 days
TIROS/NOAA	Sun-synch.	c. 850	AVHRR	1100	2 per day
Unmanned earth resources					
Landsats 1–3	Sun-synch.	920	MSS	80	18 days
SPOT	Sun-synch.	832	HRV	10	26 days
Seasat	Circle 72° N–S	790	SAR	25	152 days
Landsats 4–5	Sun-synch.	705	MSS, TM	30	16 days
ERS (ESA)	Sun-synch.	700	SAR	10	3 days
HCMM	Sun-synch.	620	Rad	500	16 days
Manned earth resources					
Skylab	Low	435	C	c. 10	No return
Shuttle	latitudes	c. 300	C, SAR, MOMS	c. 20	1 day
Mercury Gemini Apollo	Circle, low latitudes	c. 250	C	c. 100	No return
Non-orbital					
Skylark	Rocket	c. 270	C	c. 100	No return
Balloons	Airborne	<40	C	5	No return
Aircraft	Airborne	<32	C	<1	c. 10 years
Helicopters	Airborne	<3	C	<1	c. 10 years

Notes:

1. The platforms vary widely in length of life, some operating for years, some for only a few orbits.

2. Abbreviations: AVHRR advanced very high resolution radiometer; C camera; CZCS coastal zone colour scanner; GOES geostationary operational environmental satellite; ERS earth resources satellite; ESA European Space Agency; HCMM heat capacity mapping mission; HRV high resolution visible scanner; MOMS modular opto-electronic multispectral scanner; MSS multispectral scanner; NOAA National Oceanographic and Atmospheric Administration; Rad radiometer; SAR synthetic aperture radar; SMS synchronous meteorological satellite; SPOT satellite probatoire pour l'observation de la terre; synch. synchronous; TIROS television infra-red observation satellite; TM thematic mapper.

3. A comprehensive list and full details about current platforms and sensors are given by Curran (1985) and United Kingdom National Remote Sensing Centre (1987).

4. Aircraft and orbiting satellites are the most important because, unlike balloons or rockets, they can be controlled both in altitude and direction. Their differences of motion, however, govern their differences of application. Aircraft, restricted to the earth's atmosphere, have an absolute maximum altitude of about 32 km and normally do not fly above 20 km. These altitudes are appropriate for obtaining imagery suitable for the use of stereoscopy. Satellites, on the other hand, give imagery which is small scale, generalized, and covers large areas, but is hardly ever appropriate for stereoscopy.

10.2 Aircraft imagery

10.2.1 *Passive direct systems*

Description

'Passive direct systems' of remote sensing from aircraft include photographic or scanner systems which employ the visible and near-infra-red (IR) wavebands of the electromagnetic spectrum, usually reproduced as photographic prints or transparent diapositives. Such photographs are still by far the most important tools in the study of terrain.

Aerial photographs can be oblique or vertical. Examples of these are given in Figs. 10.2 and 10.3. Obliques are classed as 'low-level' if the view angle is near to the horizontal, or 'high level' if it is near to the vertical. The former are more useful photogrammetrically because they normally show the horizon, and this permits their rectification on to a rectangular grid.

Vertical photographs are of wider usefulness. They are taken in runs looking vertically downwards from an aircraft, which normally ensures that consecutive prints in the same run have a 60 per cent overlap ('fore and aft lap') and adjacent runs a 20 per cent overlap ('side lap'). This means that every spot on the ground appears on at least two, usually three, and often as many as six prints, giving complete stereoscopic coverage. Because of 'tilt' (the angular divergence of the aircraft from a horizontal flight path), no photograph is ever rigorously vertical, but normal survey photography requires that tilt be under 2°, and it is in fact almost invariably under 1°. This source of error as well as the scale distortion away from the centre of the photograph must be borne in mind in making measurements.

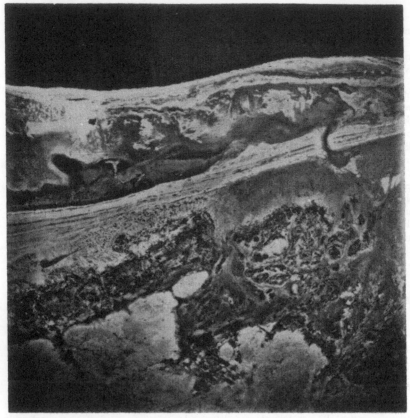

Fig. 10.2 Example of a vertical aerial photograph: coastal marsh area south of Misurata, Libya. Note that this photograph reveals enough ground detail to be almost usable as a geomorphological map as it stands. (Ministry of Defence, Air Force Department, photograph V:533/RAF/2684 no.0015 of 6 February 1964. Original scale 1:80 000 approx. British Crown Copyright Reserved)

The quality of aerial photography depends upon three factors: contrast, resolution, and parallax. The first two depend essentially on the type and grain of the film, the lighting conditions, and the quality and focal length of the camera lens. A consideration of the detailed methods used to quantify these is beyond the scope of this book, but they are described by the American Society of Photogrammetry (1980) and summarized by Curran (1985).

Scale

The scale of aerial photography is derived from the ratio between the focal length of the camera and the flying height (see Fig. 11.1). Since the former in most survey cameras is about 15 cm, the horizontal scale of contact prints is often about 6.7 × the flying height in metres. There are four commonly recognized ranges of scale:

120

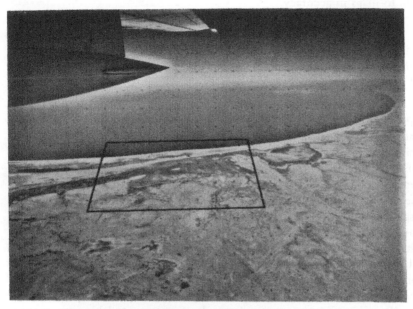

Fig. 10.3 Example of an oblique aerial photograph of the same general area as can be seen in Fig. 10.2, whose approximate boundary is shown by the trapezoidal line. Note the wider spread of country visible and the more graphic impression. It would be hard, however, to use this photograph as a map. (Ministry of Defence, Air Force Department, photograph TRI S:543/2129 no. 0106 of 15 March 1963. Original scale 1:60 000 approx. British Crown Copyright Reserved)

Large   >1:20 000
Medium   1:20 000–1:50 000
Small   1:50 000–1:85 000
Very small   <1:85 000

Very small-scale imagery is today taken almost entirely from satellites, which today also convey the advantages of repetitive and multispectral coverage in both analogue and digital form. Technological advances are increasingly making high–resolution satellite imagery competitive with aerial photography.

Mosaics

To cover larger regions, aerial photographs are combined into 'photomosaics'. These are of two main types: uncontrolled and controlled. 'Uncontrolled mosaics' or 'print laydowns' are composed of photographs which have not been accurately fitted into a surveyed grid. They are relatively cheap and quick to produce and are useful as rough location charts, and as a plot of the component photographs whose numbers they show. 'Controlled mosaics' are

121

based on a number of geodetically surveyed points and can be treated as equivalent to maps in accuracy.

Since the mid 1970s, photomosaics have been largely replaced by satellite imagery from multispectral scanners. This can vary in planimetric accuracy from being 'scene–corrected', when the image is enhanced by processing and approximately grid–referenced, to 'precision', when internal detail is accurately adjusted to imposed coordinates.

Film types

There are four main types of film used in normal aerial photography: black and white (panchromatic), IR monochrome, true colour, and false colour. Black and white panchromatic film still accounts for over 90 per cent of all film used, due to its relative cheapness and availability. The other types of film have specialized uses depending on their sensitivity to certain terrain features whose spectral responses cannot easily be distinguished on monochrome or are most conspicuous in the near-IR part of the electromagnetic spectrum.

'Multiband sensing' is a method of recording simultaneously signals from several bands of the spectrum, using a 'multiband camera system'. This consists of an array of simultaneously recording cameras (usually nine) with matched lenses containing filters and films representing different portions of the visible and near-IR parts of the spectrum. It can be supplemented by extra cameras holding true colour and false colour film, and other remote sensing devices for longer wavebands.

10.2.2 *Passive indirect systems*

Passive indirect systems sense the earth's radiation by translating it into electrical signals which can be transformed into visible light via a cathode ray tube and then recorded on film or magnetic tape. They can be used to record in the visible, near-IR or thermal IR wavelengths. They do the last by sensing through 'windows' in the IR spectrum that allow terrestrial radiation to escape through the atmosphere with least loss of absorption. The most important of these windows are those between approximately 2 and 5 μm and between 8 and 14 μm.

The sensitivity of detectors, and the wavelengths to which they respond, vary widely, but most important for terrain evaluation is the technique known as 'linescan', or 'infra-red linescan' (IRLS) when recording in the IR wavelengths. The technique involves scanning a succession of parallel strips across the track of an aircraft or satellite with a scanning spot a few metres across. The spot goes back and forth in such a way that no gaps are left between consecutive passes. As it records only an average radiation, the limits of resolution depend on the size of the spot. Its diameter is normally of the order of 2–3 milliradians, i.e. if the plane is flying at 1000 m, the spot will be 2–3 m across. The radiation received in the aircraft is reflected by a rotating parabolic mirror on to a detector which generates signals which vary in intensity with the amounts of incoming radiation. The signals modulate an electron beam which can be recorded in digital form on magnetic tape or projected on to a cathode ray tube to generate a black and white photographic print of the ground. Imagery obtained in this way is distorted laterally in proportion to its deviation from the line of flight, limiting the total useful angle of scan to about 60° from the vertical. A diagram of a simple line scanner is shown in Fig. 10.4 and of its method of information management in Fig. 10.5.

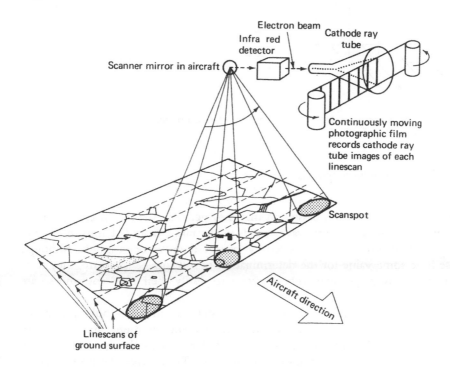

Fig. 10.4 The principles of obtaining IR linescan imagery. (Source: Cooke and Harris, 1970)

Multispectral linescanners are a refinement of this method, differing from it in one important respect. Instead of having a single detector mounted at the focus of a parabolic mirror, they employ a dispersing system which produces a focused spectrum of energy. By correct placement of an array of detectors, the dispersed signal can be recorded in narrow, discrete, spectral bands or channels. Scanners with 12, 18, and 24 channels are in use. This method opens the possibility of selection, allowing non-relevant data to be eliminated and discarded before producing an image, something which is not possible with camera systems (Olson, 1970).

123

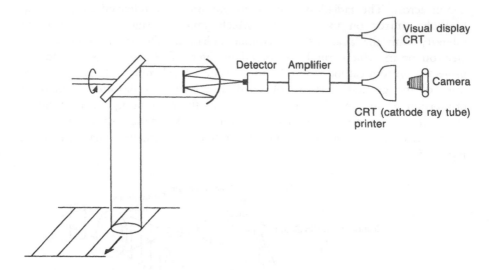

Fig. 10.5 The system for the management of information from a simple IR linescanner. (Source: Olson, 1971)

10.2.3 *Active indirect systems*

Active sensors differ from passive in that they generate their own energy. They operate at the longer 'microwave' bands in the range from 0.1 mm to beyond 1 m. These bands are also recorded by 'passive microwave' scanners. Although these have some value for the determination of soil surface moisture, they have insufficient advantages over other systems to justify their high cost and so are not discussed here.

The most important active microwave system is 'sideways–looking airborne radar' (SLAR) and its higher-resolution refinement 'synthetic aperture radar' (SAR). Short pulses of energy are directed sideways to the ground on both sides of an aircraft from a transmitting antenna. They strike the ground along successive range lines and are reflected back again at time intervals related to the distance of the ground from the aircraft. The returning signals are transformed into black and white photographs via a cathode ray tube using a technique similar to that used in IRLS. One limitation is that return pulses cannot be accepted from any point within 45° from the vertical, so that there is a blank space under the aircraft along its line of flight, and increasing distortion of the scale towards the track of the aircraft, the reverse of that which affects IRLS. A diagram of the method is shown in Fig. 10.6

As with camera and linescanning systems, recent developments in radar have been towards multiband systems which use a number of different wavelengths in combination rather than a single one. Although such systems have not been used extensively, the differences in soil and vegetation penetration capabilities in different bands have led to some valuable applications.

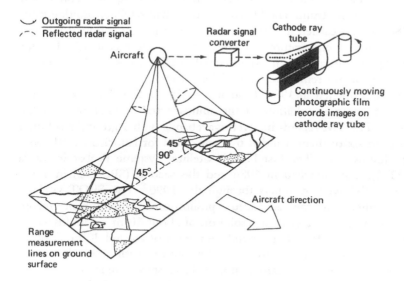

Fig. 10.6 The principles of radar scanning. (Source: Cooke and Harris, 1970)

Promising developments include 'multiple polarization' and 'coherent' radar systems. The former have the extra capacity of distinguishing objects by their different degrees of polarization, while the latter improve resolution of images of objects to the point at which they are essentially independent of their distance away. Stereoscopic SLAR obtained by combining horizontally and vertically polarized imagery, and colour recording to improve radiometric resolution have also been used experimentally. Recent advances have also made it possible to determine the height of certain objects on Seasat imagery to within a few centimetres (MacDonald Dettwiler, 1987).

Another development has been the use of 'lasers'. Airborne profile recorders, used in conjunction with lasers, can give an accuracy of less than 0.1 per cent of flying height in recording microrelief.

10.3 Satellite imagery

Satellites use essentially the same sensors as do aircraft, but with a greater emphasis on scanners and radar. The first photographs of the earth from space were taken from the USA's National Aeronautics and Space Administration (NASA) Mercury, Gemini, and Apollo satellites, and by the end of 1966 over

2400 photographs had been taken (Curran, 1985). The 1970s saw rapid growth with the launch by the USA of Skylab, Landsats 1–3, the Heat Capacity Mapping Mission (HCMM), and Seasat. Most of the imagery from the Landsats was taken with multispectral scanners covering the visible and near-IR parts of the spectrum. The HCMM was a small satellite, orbiting from 1978 to 1980, which carried a scanning radiometer for recording maximum and minimum surface temperatures for conversion to thermal inertia data for geological mapping. Seasat was an experimental satellite to establish the utility of microwave sensors for remote sensing of the oceans.

Landsat 4 was launched in 1982, Landsat 5 in 1984. They carry a thematic mapper (TM) Scanner in addition to the multispectral scanner used in the earlier Landsats. The TM records in seven wavebands with a ground resolution of about 30 m in six of them. Landsat 6 is scheduled for launch in 1991, with Landsat 7 to follow later. The first French satellite, Système Probatoire de la Terre (SPOT-1), was launched in 1986 and the second (SPOT-2) in 1990. They are to be followed by others through the 1990s. The SPOT satellites carry two scanners of the advanced 'pushbroom' type which record panchromatic images with a ground resolution of about 10 m on multispectral images in three wavebands with a ground resolution of around 30 m. A novel feature is the inclusion of a mirror in the optical path. This has three advantages for the study of the earth's surface. First, stereoscopic coverage can be obtained by viewing one area from successive orbits; second, the sensor can be pointed to cloud-free areas; and third, it can obtain repeated coverage of one area at a number of look angles, thus decreasing the repeat time to only 1 or 2 days. The European Space Agency (ESA) is launching the earth resource satellite ERS-1, in 1990, and Japan has a similar programme. Canada is preparing a high-resolution radar satellite, Radarsat, aimed largely at monitoring sea ice. Less information is available on the Soviet satellites, but the manned Voshkod and Soyuz and the unmanned Cosmos and Meteor satellites include similar capabilities to Landsat and SPOT.

The Space Shuttle is a series of four NASA spacecraft that are designed to shuttle backwards and forwards between earth and space. They started in 1981 and will probably continue well into the 1990s. The orbiter vehicle can launch earth-sensing satellites into space and has itself carried a battery of sensors which include a survey-type 'metric' camera and high-resolution SAR sensors. The latter, named shuttle imaging radar (SIR), records in the L (23.5 cm) waveband and produces imagery with a nominal ground resolution of 25 m over a 100 km-wide swath. The first, SIR-A, recorded in 1981; SIR-B, beginning in 1984, is the first space-borne radar mission providing quantitative calibrated imagery with multiple incidence angles. Early results have shown the potential utility of data from the Shuttle sensors, especially for geological applications, but it is still early fully to evaluate their potential.

Meteorological satellites have operated since 1960 and are better developed than earth resource satellites. Their prime use is for the study and prediction of short-term weather changes, but they have also proved valuable for land applications. They carry high-resolution radiometers recording in the visible, near-IR, and thermal IR parts of the spectrum. They can be polar-orbiting or geostationary. Polar orbiting types such as NIMBUS, the television infra-red observation satellite (TIROS), and National Oceanic and Atmospheric Administration (NOAA) satellites occupy sun-synchronous orbits between about 500 and 1500 km high and typically cover the same spot on the earth's surface twice a day. The geostationary satellites are placed in orbit at over 35,000 km altitude above the Equator and advance in the same direction as the earth's rotation. They are thus able to record images of the same wide portion of the earth's surface many times a day. Both types have proved valuable for land applications requiring low spatial resolution, rapid repeat time, and large areal coverage, such as the monitoring of seasonal changes in vegetation cover at continental scale.

Satellite imagery has thus become an important resource in studies of the earth's surface, and is wellnigh irreplaceable for a number of uses. It is free from many of the administrative and technical constraints to flying survey aircraft and producing aerial photographs. Its advantages include the provision of scene-corrected images with marginal coordinates. The repetitive coverage ensures that some cloud-free imagery is available virtually everywhere and allows monitoring seasonal and annual changes of vegetation and other ground features. The wide synoptic viewpoint allows interpretations not easily appreciated on the multiplicity of aerial photographs required to cover the same area. Its importance appears likely to increase relative to aerial photography in the future as it is increasingly able to match the latter's two main advantages: high ground resolution and stereoscopic capability.

10.4 Image processing: general

The products of remote sensing are either in photographic ('analogue') or numerical ('digital') mode, and are mutually convertible. The first is pictorial, the second a two-dimensional array of 'pixels' (the component 'cells' or picture elements from which an image is made up), each with a quantifiable reflectance value in one or more spectral bands.

Analogue imagery can be transformed into digital by scanning with a 'microdensitometer'. This records and quantifies the reflectance value of each pixel, converting the picture into a stream of numerical data. Conversely, digital information can be represented pictorially by translating the quantified reflectance of every pixel into a grey shade or colour tone and projecting it on to a cathode ray tube from which it can be photographed to obtain hard copy. This capability, which is now generally available, of changing with relative ease between analogue and digital formats is known as 'image processing'. It can be subdivided into 'continuous' and 'discrete' forms of processing according to whether it handles analogue or digital data respectively.

10.5 Continuous image processing

Continuous image processing corrects, enhances, and classifies analogue imagery, using essentially photographic techniques. The most widely used of these are selective enlargement to facilitate the study of detail in a particular areas, contrast modification to optimize the range of grey tones, tonal edge enhancement to heighten the contrast around features such as geological faults, breaks of slope, and soil textural boundaries, the production of colour composites to combine several spectral bands on to one image, and directional and spatial filtering to remove image blemishes such as scan lines and clouds.

A number of instruments are used for these tasks. Enlargers can be used to emphasize tonal edges by printing a positive image while it is in register with a blurred negative image of itself. This enhances detail in complex areas and reduces it in simpler areas. Colour composites are produced by illuminating sets of imagery from different spectral bands with different coloured lights and combining the results on to a single photograph.

Contrasts within the imagery can be enhanced by a number of methods. The 'electronic dodger' decreases image contrast as a means of exposing the detail in areas of shadow and highlight. The method is, when producing an image from a negative, to use an illuminating source which varies in relation to the density of each area of the image. It sacrifices detail in the middle density range in order to enhance it by high illumination in areas of high density and by low illumination in areas of low density.

Analogue image processing is essentially enhancement by using a television camera on either black and white or colour imagery. Black and white 'image analysers' can enlarge, alter the contrast, density slice, and measure the areas of all or part of an image. The technique known as 'video image analysis' uses a computer to convert imagery from analogue to digital format. This permits an enhanced discrimination of grey scales (Reybold and Petersen, 1987). Colour image analysers have these capabilities and can also edge enhance and density slice in colour.

'Additive viewers' are instruments which project a colour composite image on to a ground glass screen. They consist of a set (usually four) of transparency projectors, each with adjustable illumination and a choice of colour filters, focused on a single viewing screen. Identical black and white transparencies, such as Landsat frames from different multispectural scanner (MSS) channels, are placed in each projector, so as to obtain a single multicolour picture. The advantage of the method is the interactive control by the operator of the hue and brightness of the colour composite, but the disadvantage is the time required to secure exact registration between the different projectors. Hard copy can be photographed from the screen. Their use has declined with the increase in digital processing.

The 'Diazo colour printer' is simple to operate and much cheaper to buy than the additive viewer. To produce a colour composite image, each transparency is copied in the machine on different coloured film, and these are then sandwiched and viewed over a light table. The procedure, using Landsat MSS diapositives, is shown in Fig. 10.7. The printer is simple to operate and much cheaper to buy than the additive viewer and provides comparable image quality with the additional advantage of transparent hard copy. However, it is less spectrally flexible since there is no visual control over the hue, value, and chroma of the colour composite except by repeated processing.

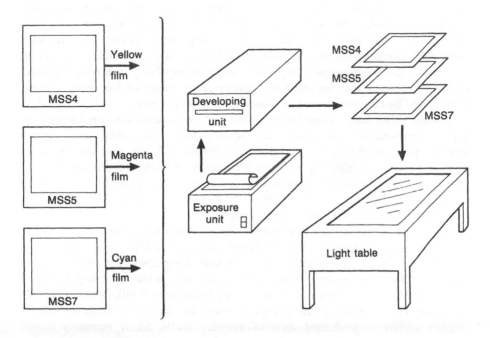

Fig. 10.7 The Diazo system for printing and overlaying Landsat MSS imagery

A final method of continuous image processing is by 'optical processing'. This is concerned mainly with enhancing directionality in an image in order to emphasize linear features such as geological faults and strike lines. It is based on the fact that a laser passing through a film transparency is scattered in a manner which is dependent on the spatial distribution of tones within the transparency. The light passing through the transparency is spread in a way called a 'diffraction pattern', which shows the relative intensities of the different directionalities within the original transparency. These effects can be further enhanced by the use of filters, which emphasize particular directions in the diffraction pattern by removing others. For example, they can remove unwanted directions caused by scan lines, aligned clouds, or unwanted geological lineations.

129

10.6 Digital image processing

10.6.1 *Stages and equipment*

Digital techniques have several advantages over photographic. Both must rely ultimately on the skill of the interpreter, but digital analysis eases this task, because of its speed in execution, suitability to large amounts of data, and elimination of subjective judgements in differentiating colour tones. It is also less labour intensive because, being computer assisted, it handles multidimensional spatial data without being limited to three primary colours or to two- or three-dimensional displays. It performs especially well when spectral landscape patterns are dominant and the radiance values of the scene can be represented by discrete numerical classes of individual pixels.

On the other hand, digital analysis not only represents a major increase in capital investment compared to photographic processing and visual interpretation methods, but also involves considerable increase in operating costs and technical skills. Against this, it should be said that in general the total cost of surveys involving remote sensing is a very small part of the costs of land development. It is important to choose the remote sensing technology for a project on the basis of a cost–benefit analysis.

Digital image processing is based on the manipulation of myriads of pixels, each with an 'address' and a 'digital number' (DN). The address in two-dimensional image space is measured first along rows, and second down columns. On the ordinary Landsat MSS scene, for instance, the top left-hand pixel would be 1, 1, and the bottom right-hand pixel 2340, 3240. The DN quantifies the electromagnetic energy received by the sensor from the earth's surface within its graduated scale of intensity levels. Many remotely sensed images (including Landsat MSS) are given in an 8-bit (2^8) intensity range which stretches from 0 for minimum to 255 for maximum radiance.

There are three stages in processing a digital image, which normally comes on computer-compatible magnetic tape. First, it is read in from the tape drive, then manipulated by computer, and finally displayed on a visual display unit (VDU). The computers used are mainframe, microcomputers with graphics, or purpose-built digital image processors. The last are today replacing the others for routine use because of their speed and flexibility of operation and decreasing cost. Curran (1985) reports that their rate of sale in the UK alone increased fivefold over the years 1980–1983.

The operator can manage the computer by typing commands on to a keyboard, and controlling the display from either a pen-guided graphics tablet or a finger-guided control panel. Some image processors further simplify the task by offering instruction cards which appear either on the VDU or the keyboard screen.

10.6.2 *Procedures*

The normal sequence of operations in analysing digital remotely sensed imagery is given in Fig. 10.8. The input is in the form of 'computer-compatible tapes', but it is also possible to use analogue imagery if it is first put into digital form by a 'digitizer'. The most commonly used processing techniques can be categorized as 'image restoration' and 'image classification'.

Image restoration has two purposes: scene correction and scene enhancement. First, it is necessary to correct distortions due to variations in the spacecraft's altitude, velocity, attitude, and the earth's curvature. The computer-based procedures to make the necessary radiometric and geometric corrections are increasingly being undertaken by the primary data handling organizations (e.g. NOAA, ESA). The imagery analyst may nevertheless also need software programs to correct the scene to a commonly used map projection (such as Lambert or horizontal Mercator), to harmonize the radiance values of adjoining scenes when making a mosaic map, or to reduce radiation distortion prior to comparing pixels of the same ground area.

Second, 'software enhancement programs' are used to improve contrasts within the imagery associated with the landscape features being studied. They include 'contrast stretching', 'two–band ratioing', and 'pixel transformation'. In contrast stretching, the pixels are displayed over a wider range of grey-scale values, and in 'histogram–equalized stretching' more grey-scale values are assigned to the most frequently occurring parts of the histogram. In 'two–band image ratioing', the spectral contrasts within images are enhanced by using the ratio between the grey scales of pixels in two separate bands, rather than the absolute values of either alone. Three- or multiple-band ratioing is also possible.

The techniques which extract the maximum contrast in a set of multispectral images of a scene, however, are statistical 'transformations'. The most popular of these are 'principal components analysis' (PCA) and 'canonical analysis'.

Principal components analysis replaces statistical axes based on pixel grey scales from two or more wavebands with orthogonal axes which are designed to maximize the separation of clusters of points in multidimensional space. To take an example, Landsat data on a single pixel come from four MSS bands. A hypothetical scatter plot of two contrasting bands, say MSS5 and MSS7, illustrates that while these two are correlated and thus hold some redundant information, neither alone adequately separates the data. To overcome this, PCA creates new axes, called 'band axes', along the lines of maximum variance within the data. Therefore, once the pixels have been located in the new coordinate system, a band axis A image would contain more information (i.e. provide better spectral separations), than any other, and what it missed would be largely covered by band axis B.

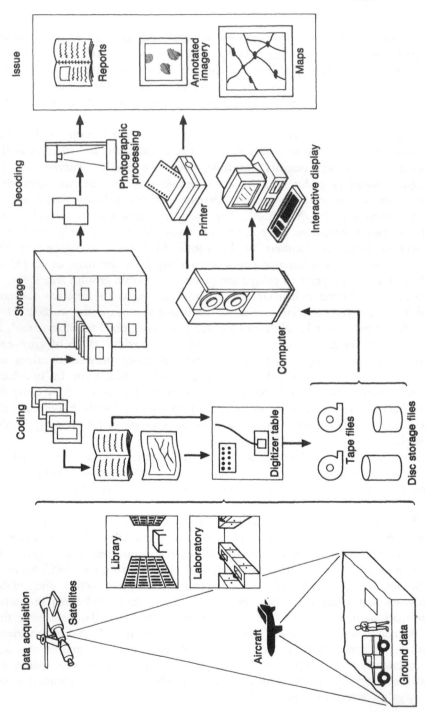

Fig. 10.8 Processing sequence for remotely sensed imagery

Canonical analysis, like PCA, compresses the information component of several images into one by inserting new axes into the multidimensional space to maximize the difference in DN between the major scene components. In practice, the difference from PCA is that the new axes are located by mathematically partitioning pixel values along the new axis directions.

Both PCA and canonical analysis, however, suffer from the disadvantage of requiring considerable computing time and and destroying the relationship between object radiance and image DN. Nevertheless, it is likely that their use will increase because of their capacity to integrate information from multiple-channel sensors such as the Landsat TM.

Image classification techniques may be used either to help the analyst interact with the computer display or to obtain automatic interpretations of the imagery. The information from each pixel, derived from one or more spectral bands, is evaluated and assigned to an information class, notionally representing a certain assemblage of ground conditions. The classes may be 'unsupervised' or 'supervised'. In unsupervised classifications, the computer divides the pixels into natural groupings based on their grey-scale values. Such groupings can be achieved using several readily available statistical packages of which 'cluster analysis' has proved the most popular. This works by locating the desired number of cluster centres in the waveband-to-waveband measurement space, and continues to move the cluster centres until the clusters have maximum statistical separability. The results of this classification depend on the number of classes initially chosen. Generally, the greater the number of classes, the narrower will be inclusiveness of each and the greater the possibilities of error in delimiting ground features.

Supervised classifications select pixel values (in one or more wavebands) which are the known spectral responses of particular ground features and use these as the basis of classification. The procedure is first to correlate pixel values with a 'training area' for each feature to be identified, and then to extrapolate these over the whole scene. The three most popular means of performing this extrapolation are the 'minimum distance to means classifier', the 'parallelepiped classifier' and the 'maximum likelihood classifier' (Lillesand and Kiefer, 1979). The minimum distance to means classifier calculates the mean DN of each class in all wavebands. This is called the 'mean vector'. Then the pixels are all assigned to their nearest mean vector, and finally a boundary is located around each class. This method suffers from being insensitive to the variance in the properties of each class, and can lead to misclassification of pixels with one aberrant spectral value. The parallelepiped or 'box' classifier is essentially a simultaneous density slice in all wavebands, and is currently the most popular for remote sensing applications as it is both fast and efficient. The maximum likelihood classifier is the most expensive and accurate of the three, and unlike the other methods, has the advantage of leaving no pixels unclassified. It first calculates the mean vector, variance, and correlation for each land type in the training data, on the usually valid assumption that the data for each class are normally distributed. It then uses this information to describe the spread of pixels around each mean vector using a probability function. All pixels are then allocated to the class in which they have the highest probability of membership.

Further reading

American Society of Photogrammetry (1983), Curran (1985), Lillesand and Kiefer (1979), Sabins (1978), Short and Blair (1986), Townshend (1981).

11

Remote sensing: applications

11.1 Landform interpretation using remotely sensed imagery

Remotely sensed imagery is an integral part of terrain survey because it reduces costs, saves time, and contributes information obtainable in no other way. It permits extrapolation from field observation, which in turn checks and improves imagery analysis. Nevertheless, because it cannot be more than an instantaneous and partial record of the immediate surface, it remains essentially complementary to, and cannot replace, ground information.

Remote sensing reveals the way certain parts of the electromagnetic spectrum are reflected from the earth's surface. It cannot, in general, penetrate through the vegetation cover or beneath the ground surface. The main exception to this is the capability of radar to penetrate a little superficial foliage or some depth into soil. The latter is highly variable and ranges from almost nil when the soil is moist and compact to as much as 20 m when it is completely dry and loose. In general, however, terrain characteristics must be interpreted from visible surface indications, using all available collateral information. As previously discussed, there is a considerable difference between the methods used for airborne and satellite imagery.

11.2 Indications on airborne imagery

11.2.1 *Distance and altitude interpretations*

As previously mentioned, aerial photographs can be oblique or vertical. Obliques have occasionally been used for survey purposes, but their scale distortion from foreground to background is so serious that they are ill-adapted for this purpose. Their main value is for graphic illustration.

Vertical photographs can to some extent serve as maps and the sciences of photogrammetry and aerial photographic interpretation have developed from this. Distance measurements can be derived directly from a photograph by multiplying by the 'scale factor'. This can be defined as the flying height divided by the focal length of the camera. Figure 11.1 makes this clear. CAO and aOd are similar triangles, therefore the distance CA on the photographs bears the same relation to the ground distance, ad as does *f to H*.

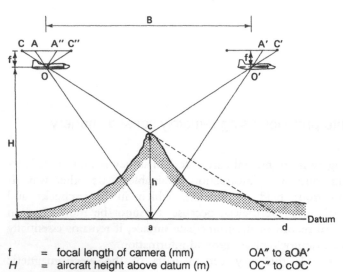

f	= focal length of camera (mm)	OA″ to aOA′
H	= aircraft height above datum (m)	OC″ to cOC′
a	= a point at datum	
B	= air base (m)	
pa	= absolute parallax of a = AA″ = the air bas *B* as measured on the photograph in mm	
pc	= the absolute parallax of c = CC″	
Δpac	= CC″ – AA″ = the difference of parallax between c and a	

Fig. 11.1 The geometrical relations between aerial photographs and the ground surface

Much more information can be derived from photographs if they are viewed in pairs in stereo fashion rather than singly. This is because of the quality of 'parallax'. It can be most simply defined as the relative movement of objects when seen from different viewpoints and is the basis of depth perception. Everyone is familiar with the necessity of using both eyes to perceive the relative distance of objects while with one eye they appear to lie in a common plane. In the same way, when consecutive views of the same piece of ground are taken from a moving aircraft, an observer who puts an eye at the position at which each was taken, obtains an awareness of depth.

The tool used for viewing photographs three-dimensionally is the 'stereoscope', shown in Fig. 11.2. The area covered by both photographs appears in relief to an observer viewing each through a separate eyepiece. Subject to certain errors consequent upon the imperfect verticality of all photographs and image distortion away from their centres, it enables heights to be determined and form lines to be drawn.

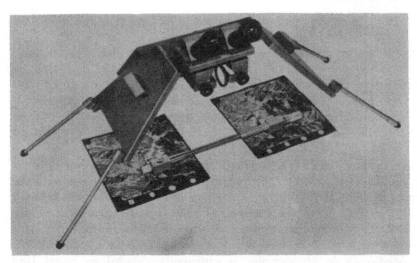

Fig. 11.2 A stereoscope. This example is an ordinary mirror type, of Topcon make, with a choice of ×1, ×1.8, and ×3 magnifications. A parallax bar is resting on the aerial photographs. (Photo: H. Walkland)

The determination of heights is based on the fact that, as in Fig. 11.1, the height of c above datum is equal to the flying height of the aircraft above datum (H) minus the airbase (B) times the camera focal length (f) divided by the absolute parallax of c (p_c). Expressed mathematically in the terms given in Fig. 11.1:

$$h = H - fB/p_c \qquad [11.1]$$

h, H, and B must be in the same units (usually metres or feet) and f and p_c must be in the same units (generally millimetres).

Normally H and f are printed on the margin of aerial photographs; B is determined by multiplying the separation of the 'principal points' of adjacent photographs (known as the absolute parallax of a, or p_a) by the scale factor. The principal point is the centre of the photograph determined by intersecting its marginal 'collimating marks'. The separation of the principal points of two adjacent photographs is measured with a ruler after marking the location of each principal point on to the other photograph. p_c is derived by adding to p_a the difference of parallax between a and c (Δp_{ac}), thus:

$$\Delta p_{ac} + p_a = p_c \qquad [11.2]$$

Δp_{ac} is measured with an instrument known as a 'parallax bar'. The essentials of this are a pair of glass plates, each having a black mark, whose separation can be adjusted by means of a micrometer screw. When viewed through the stereoscope the marks can be seen to fuse together into an apparently floating object which can be moved up and down by means of the screw. The height of objects can be determined by comparing micrometer readings when the floating mark is level with their tops and bases. In Fig. 11.1, this would be used to give the height difference between a and c.

The higher the aircraft flies, the smaller will be the parallax effect because, in formula [11.1] when h, f, and B are constant, an increase in H will be accompanied by a decrease in p_c. On the other hand, when the aircraft flies very low the parallax effect will be so exaggerated as to make stereoscopy difficult. Thus the intermediate scales between about 1:20 000 and 1:60 000 are generally the most useful, especially as they accord with the most commonly used map scales.

Different types of terrain interpretation are appropriate to the four different scale ranges for aerial photography noted in Chapter 10. Large-scale aerial photography is useful for detailed interpretation of small features such as beach ridges, river terraces, periglacial deposits, or vegetation patterns. Because this scale requires a large number of prints to cover a relatively small area, it is expensive and so tends to be used mainly in settled or economically valuable regions. Medium-scale photography is sufficiently detailed to reveal land elements and yet can show units as large as land systems without an excessive number of photographs. Small-scale photography has the advantages of a wider and more synoptic view of country and of being able to cover large areas with relatively few photographs. It is useful for topographic mapping and interpretation of low value terrain such as mountains and deserts. Very small scale imagery, today taken mainly from satellites, is from altitudes an order of magnitude higher than aircraft, so that the resolution is poorer and it must be enlarged and enhanced before use. It is generally inappropriate for mapping at scales larger than about 1:250 000.

11.2.2 *Lithological interpretations*

Photographic film

Lithological differences between surface materials can best be interpreted using a combination of different sensors. Black and white panchromatic film forms a useful high-resolution basis for most purposes, and has covered by far the greater part of the earth's land surface at least once, and most of it several times.

Infra-red black and white film makes use of three facts: (a) that near-IR film penetrates haze better than does conventional photography, because it senses at the slightly longer wavelengths which are less subject to scattering by atmospheric particles; (b) that near-IR radiation is strongly absorbed by water, and (c) that the gross structure and condition of the mesophyll tissue of plants are most clearly revealed in this spectral range. It is better than panchromatic film in distinguishing between land and water, and is thus especially useful for charting shorelines, showing the depth of features under water bodies, and

138

revealing the presence of water at shallow depth underground. It is also valuable in distinguishing between different types of vegetation and identifying areas of plant disease.

True colour film has the advantage over panchromatic that it displays variations of hue, value, and chroma rather than of tone only. Cooke and Harris (1970) demonstrated in a study of the Isle of Man that this facilitated the identification of small individual features. It does, however, suffer from certain technical limitations when compared with panchromatic. Apart from greater cost, these include the narrower range of exposure that will provide an acceptable image, the more critical film processing requirements, and the greater difficulty in duplicating colours in different parts of the same picture and on different prints from the same film roll. It is thus sometimes difficult to recognize features from tonal indications alone.

False colour is perhaps the most versatile and potentially valuable type of film for terrain interpretation because it combines the advantages of IR and colour, making it especially valuable in water and vegetation interpretation. It also provides a more sensitive means of identifying exposures of bare rock. Thermal IR imagery has proved its value in discriminating rock types from their thermal inertia and for revealing subterranean features associated with different soil moisture contents (Sabins, 1978).

Radar

Ground-penetrating radar can be used in the field to chart the extent and depth of diagnostic soil horizons (Reybold and Petersen, 1987), but most radar scanning is from aircraft and satellites. The first successful airborne radar project covering an extensive ground area (20 000 km^2) was the survey and mapping of Darien Province (Panama) in 1968 using K-band radar. The province, because of continual cloud cover, had not been successfully photographed for more than 30 years despite aircraft being available most of that time. Since then, projects in the tropics at scales of 1:125 000 or smaller have covered extensive cloud-prone areas in Andean South and Central America, Liberia, and parts of Indonesia and Papua New Guinea. All Nigeria was imaged in the late 1970s to prepare national small-scale coastal zone land use maps, and only old incomplete aerial photography at a mixture of scales had been available before that time. By the end of 1980, about 8 500 000 km^2 of Brazil had been covered by SLAR, including all of Amazonia, leading to landform studies and mineral exploration.

Black and white SLAR and more recently SAR have proved economically valuable in providing small-scale planimetric and thematic maps and thus in providing the basis for land systems surveys. Because of its relatively low resolution, most imagery is used at a scale of 1:250 000 or smaller, though it can occasionally be usefully enlarged to 1:100 000 or even 1:50 000 to provide a base for the compilation of landform and vegetation information.

In general maximum signals are obtained from slopes facing the aircraft, surfaces which are rough and diffuse, such as vegetation, and those with a high moisture content. Minimum signals are returned from 'specular reflectors' such as calm water, and relatively smooth surfaces such as beaches or dry lake beds and wet surface soil. Radar's low glancing angle of emission is better than aerial photography in highlighting small topographic contrasts in areas of low relief. This is especially valuable when an area is scanned from two look directions. SAR also tends to emphasize square or linear features especially when they are parallel to the line of flight. X-band SAR imagery, because of its partial penetration of vegetation, records ground features such as drainage patterns and geological lineations especially clearly.

Appearance of different rock types

In studying the lithology of a particular area, aerial imagery is of vital assistance in showing the boundaries between 'solid' rock and 'drift', and in subdividing each into the genetic classes discussed in earlier chapters. The indications are clearest in arid areas and where vegetation is relatively sparse. Seasonal differences are much less important in landform than in vegetation interpretation.

The main features in landform recognition were given in earlier chapters, but certain tonal characteristics emerge clearly on remotely sensed imagery. Igneous landforms reflect their modes of origin but volcanic rocks generally give darker photo tones than plutonic, and basic than acidic.

It is sometimes possible to identify geothermal phenomena on thermal IR imagery and to distinguish different types of igneous rock on the basis of their thermal inertia. Fig. 11.3 is an aerial photograph showing volcanic and granitic outcrops and Fig. 11.4 a ground photograph of an area where basaltic dikes intrude a lighter coloured and softer crystalline mass.

Among sedimentary rocks, calcareous materials are distinguished from siliceous by their usually lighter tones and the presence of karstic phenomena. Sandstones are often darkened by iron impregnation, especially in the highly oxidized and dehydrated states in which it occurs in deserts.

Karstic phenomena include solution features such as dolines and steep slopes along valleys due to solutional collapse. Chalk has somewhat less abrupt slopes but a low 'texture of dissection' and absence of surface drainage. Sandstones and other siliceous rocks vary in competence with grain size and matrix strength. They range from soft undulations to abrupt castellated hills, but all tend to be marked by an arborescent network of surface drainage. Figures 11.5 and 11.6 of the Grand Canyon give examples of the appearance of horizontal sediments of both types.

Aluminosilicate sediments such as shales, clays, and mudstones are recognizable from their generally dark hues, the 'fine texture' of the relatively abundant surface drainage, and that, because of their incompetence, they tend to occupy the low ground and to be the locus of greatest marine incursion.

Fig. 11.3 The Mohawk Mountains, Arizona. Note the contrasting appearance of the granite rocks and the isolated lava hill in the northeast (US Government photograph)

Fig. 11.4 Basalt dikes traversing granitic hills on Abdul Kuri Island, Indian Ocean

141

Fig. 11.5 Aerial photograph showing rock contrasts in the Grand Canyon, Arizona (US Government photograph)

Fig. 11.6 Ground view across the Grand Canyon showing the difference in competence of different strata

Figure 11.7 shows the way aerial photography can reveal differences in drainage texture associated with differences in material in the same climate.

The surface form of drift generally indicates its origin. This in turn gives a clue to the sense of its internal lithological variations. Gravity deposits are usually radially oriented from their place of origin which they resemble in tone. Parallel lines along a coast indicate beach ridges. These are most often siliceous and the fining seawards from coarse shingle to mud is associated with a darkening of photo tone. Wind-formed dunes are clearly recognizable. They reflect the nature of upwind materials, but their tendency to concentrate siliceous particles gives them a light tone. The proximity of arctic climates or mountains reaching the snow-line indicates the likelihood of glacial deposits. These can be recognized from their distinctive forms and alignments. Their lithology and particle size composition can be interpreted by reference to their source areas. They can be broadly subdivided into three types, depending on their origin from ice sheets, glaciers, or meltwater. Till plains, whose material is boulder clay, can be recognized from their lateral extent, directional eminences such as drumlins, and from the way they are 'plastered over' and mask the underlying topography. Moraines, formed of mixed unstratified materials, can be recognized from the indications of former valley glaciers. Fluvioglacial deposits resemble those of strongly seasonal rivers, and can best be interpreted from their location in relation to former ice sheets and glaciers.

11.3 Interpretation of satellite imagery

Satellite imagery has greatly extended the types of interpretation possible, although conventional aerial photography is still needed for ground detail (Cochrane and Brown, 1981). In spite of the differences in scale the methods of interpretation of satellite imagery do not differ greatly. The observer uses the same combination of image pattern recognition and reference to relevant collateral information.

The most useful format for image interpretation is usually the multispectral colour composite. Structure, relief, and drainage networks can be recognized from tones and patterns. This leads to the recognition of lithology and geomorphic processes. Approximate heights and gradients in mountain areas can be inferred from the patterns of snow cover and from contrasts of light and shade on slopes exposed to, or shaded from, the southeasterly solar illumination at the time of Landsat overpasses.

Lithological differences are often observable, allowing the separation of outcrops into the same types as those recognized on aerial photographs, and HCMM imagery has been useful in this regard (Kahle *et al.*, 1981; NASA, 1982). The pattern, direction, and density of drainage channels allow deductions to be made about valley gradients and the roughness and composition of different materials. It is also often possible to distinguish

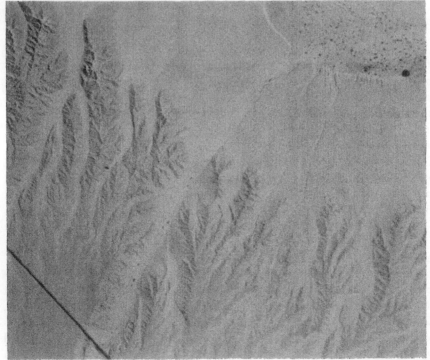

Fig. 11.7 A sequence of 'erosional textures' in southern California, reflecting increasing permeability and coarseness of soil textures from (a) to (d) (US Government photographs)

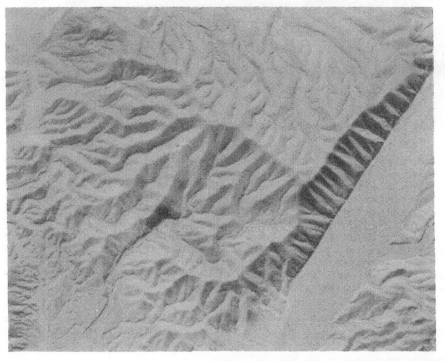

Fig. 11.7 continued

145

different lithologies. Figure 11.8 shows an area of southern Morocco with three major land regions, each of which reflects a distinctive pattern of internal processes and can be readily subdivided into land systems.

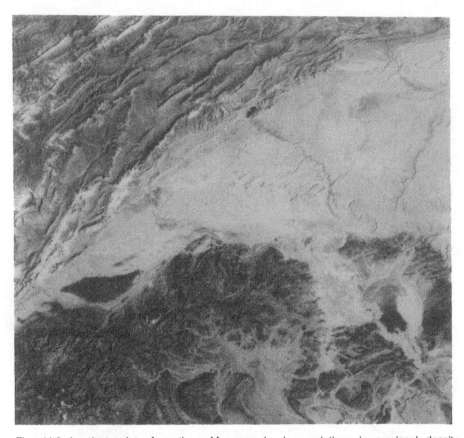

Fig. 11.8 Landsat print of southern Morocco showing variations in erosional density, landforms associated with limestone and sandstone, and three major land regions. The top left (northwestern) part of the photo is the High Atlas: sub-parallel Liassic and Jurassic limestone fold ranges with some snow-capped peaks. The bottom third of the photo is occupied by the anti-Atlas Massif, mainly composed of folded and metamorphosed Palaeozoic limestones and sandstones with associated igneous rocks. The light-coloured re-entrant within this at right of centre is the alluvial Tafilalt Oasis, with the multi-peaked sand sea called the Erg Chebbi between it and the right margin. The light-coloured triangle in the centre of the photo with its apex on the left margin consists of somewhat dissected low-dip Cretaceous and Tertiary limestone plateaux-hamadas. The area is described by Joly (1962). Each of the three land regions can be subdivided into land systems with distinctive geomorphological processes

Satellite imagery shows the approximate density of plant cover and to some extent its nature but cannot in general distinguish between particular plant

146

formations or associations. It sometimes also shows, on a single frame, geographical or altitudinal thresholds where drought or cold become limiting for plant growth. These may extend from arctic to temperate conditions or, for instance, near the southern or eastern coasts of the Mediterranean, from subhumid to arid. Local concentrations or unusual sparseness of vegetation can also indicate approaches to these thresholds. Low-lying plains and valleys tend to be enriched in vegetation everywhere. Basin sites usually hold lakes in humid regions or salt flats in arid regions, but support concentrations of vegetation elsewhere. Hilltops are generally impoverished in vegetation in humid areas, enriched in arid.

11.4 Interpretation of soils

Soil classification is based on profile morphology and soils are mapped on the basis of similarities in the nature and arrangement of their horizons. These result from the actions of climate and biota on terrain. Where the soil is bare, its interpretation on aerial and satellite imagery depends mainly on visible geomorphology and inferences about its past vegetal cover, but also on the direct physical appearance of some attributes. Where the soil is vegetation-covered, interpretation is based mainly on landform–plant relationships.

A number of soil properties can be seen or inferred by direct observation of the bare surface. Radar penetration of dry soils and thermal sensing of soil moistures can give indications of depth. In particular, satellite-borne radar penetrated several metres of sandy overburden to reveal underlying rock structures, drainage systems, and archaeological sites in the Sahara (Canby, 1983; Short and Blair, 1986). Soil colour is darkened by moisture, organic matter, and in the extremes of coarse and fine texture. Water is highly absorbent in IR wavelengths and more reflectant in the visible wavebands. Its presence brings the reflectances of all soils into greater similarity, so that moist soils are less easily distinguished from each other than dry. The best correlation between reflectance and moisture content is in the 2.08–2.32 μm waveband range. Adding water to a soil increases its thermal inertia, i.e. making it more sluggish in responding to changes in ambient temperature. Thus time-lapse thermal measurements can sometimes be correlated with moisture content. Soil moisture differences were distinguishable from orbital altitudes on HCMM imagery (Heilman and Moore, 1981).

Organic matter darkens soil, especially when in highly decomposed form such as peat when it has very low reflectance both in the visible and near-IR, notably in the 0.52–0.62 μm waveband. There is some correlation of reflectance with the degree of decomposition in the near-IR because of the cell structures surviving from the plant litter. Specifically, C/N ratios can be correlated with reflectance up to 1.89 μm (American Society of Photogrammetry, 1980).

Soil textures can likewise sometimes be inferred from photo tone. In general, the coarser the soil materials and the larger their aggregates, the rougher is the surface and the lower the reflectance. This reflectance, however, increases with the fineness of sand grains but decreases again in silts and clays, usually because of the larger amount of moisture and organic matter they contain. Increasing clay content is also associated with decreasing thermal conductivity.

Chemical compounds such as carbonate, sulphate, nitrate, and silicate can sometimes be detected from an analysis of soil spectral signatures. Carbonate, for instance, has peaks of reflectance in the 1.9, 2.0, 2.16, 2.35, and 2.55 µm wavebands. Surface salinity in arid areas is sometimes recognizable from light tones associated with irrigated or seasonally inundated basin sites in climates with high evaporation. Iron and manganese are perhaps the elements most amenable to recognition from their spectral reflectance characteristics. This is because of the distinctive colours of their oxides in the visible spectrum: ferric oxide is yellow when hydrated (limonite) and red when unhydrated (haematite) and ferrous oxide bluish grey. Manganese oxide is black, and contributes, for instance, to the dark colour of some desert surfaces.

But more widely useful than direct tonal interpretation of soils are those based on their geomorphic and vegetal, i.e. 'phytogeomorphic' environments. At global scale, the main soil orders can be recognized on satellite photography mainly from the broad climatic and vegetation zones with which they are associated, such as tundra soils with the subarctic, podzols the coniferous forests, and oxisols with tropical rain forests.

As the scale is enlarged, more and more articulation of the phytogeomorphic patterns become visible to the interpreter. Normal large-scale soil mapping based on aerial photography is at scales between 1:50 000 and 1:250 000. At these scales the main key to soil distributions is the variation of site conditions in relation to geomorphology and vegetation and the soil moisture and fertility conditions which can be interpreted from them. Specifically, soil depths can be inferred from the distribution of rock outcrops and soil chemistry from their lithologies. The presence of moisture can be deduced from geomorphic situations in valleys and basins, and from a greater vegetation density, especially in arid areas. Daily and seasonal soil temperature variations are reduced under vegetation. Textures can be inferred directly from certain landforms such as sands from the presence of dunes and beaches, and clays from claypans. Textures of transported soils tend to be finest in areas of fine drainage texture and transported soils fine with distance from their source areas, e.g. downstream in rivers and outwards on alluvial fans. Certain chemical characteristics of soils can be inferred from their source rocks, such as high pH and base status from the presence of basalts and calcareous sediments, and lower pH from acid igneous rocks and siliceous sediments.

Further reading

American Society of Photogrammetry (1980, 1983), Curran (1985), Kahle *et al*. (1981), NASA (1982), Sabins (1978), Short and Blair (1986), Townshend (1981), US Department of the Army (1979).

12

Landscape sampling and statistical analysis

12.1 Need for sampling

Terrain is continuous but variable, and if we are to discover its properties, our observations must be restricted to a small part only, that is, to a sample. Sampling and the associated field work considered in this and the following chapter provides the link between the project planning and remote sensing interpretation discussed in Chapters 9–11 and the data processing considered in Chapters 14 and 15.

Although by classifying the terrain we aim to reduce the variation with which we have to deal at any one time to manageable proportions, the variation remaining in each class is still much greater than is introduced by observational error. Samples differ therefore from one another, and worthwhile information can be obtained only when several, and possibly many, sampling sites have been observed. Replication alone, however, is not sufficient to ensure reliable information. We must also avoid bias, because information from biased sampling is often misleading. To avoid bias, samples should be chosen so that initially every site has an equal chance of inclusion.

12.2 Size and shape of sampling units

Many ground observations are of areas rather than sites. These will be hereinafter called 'sample plots' and 'sample points' respectively. Decisions need to be taken about the size and shape of both. The size depends on the scale being considered, the complexity of the landscape, and the nature of surface processes. The sample plots or points must be practically uniform or at least have a predictable sense of internal variation. In general they can be related to the appropriate level in the terrain hierarchy so that, for instance, at 1:250 000, one is considering land systems, at 1:50 000, land facets and so on. Sample plots or points must be smaller than these units but big enough to represent the point or area being sampled. They are thus comparable to the unit cells of a crystal. The lateral dimensions should be adequate to permit the study of soil profile layering and to include at least half a cycle of cyclic variations. The shape should be roughly hexagonal and one lateral dimension should not differ appreciably from any other.

The sample area should be large enough to be accurately located on remotely sensed imagery. This presents no problems on conventional aerial photography, but can do so at the relatively small scales of satellite imagery. The problem can best be understood by reference to Landsat linescanner imagery since the pixel provides a clear basic unit. It is not normally possible to locate a ground site in terms of an external coordinate reference system with sufficient accuracy to assign a particular site to a particular pixel. It is thus necessary to enlarge the sample area to ensure such locational accuracy. To achieve this, the minimum dimensions of a sample area, A, can be estimated as follows:

$$A = P(1+2L)$$ [12.1]

where P = the pixel dimensions;
   L = the required accuracy of location in terms of number of pixels.

Thus with an accuracy of ±0.5 pixel, the length and breadth of the sample unit using Landsat MSS data should be at least 158 m and for ±1 pixel it should be at least 237 m (Justice and Townshend, 1981).

12.3 Density and depth of observation sites

Standards for the density of site observations derive from those first established for soil surveys by the US Department of the Interior (1951) and land development consultants such as Hunting Technical Services Limited (1954–). Reconnaissances require a minimum of one site per 1250 ha, although one site per 250 ha should be regarded as normal. Semi-detailed surveys concerned with the feasibility of development projects such as irrigation require about one site per 100 ha, and detailed surveys require between one site per 25 ha and one site per 10 ha, depending on the complexity of the landscape and the amount of financial investment involved. These standards are applicable to most parts of the world, especially developing countries, and can also be used for surveys of other land resources, for which there are no established levels of survey intensity.

A larger number of samples will, however, be required at any scale to represent the variance when the landscape is complex, or to comprehend variations related to processes such as fluvial erosion or wind action. For these, the sampling should be arranged to include the whole area in which the process operates or a logically selected part of this area. For instance, to study

151

fluvial erosion one must sample a catchment or subcatchment, and to study wind erosion one must sample the area covered by the particular wind regime, or some recognizable part of this caused by, for instance, the arrangement of some major obstacles.

Soil observations are made to a depth adequate to penetrate any horizons present. This depth varies, but for areas considered for agricultural development, especially where irrigation is envisaged, 1.5 m is normal. This approximates to the maximum rooting depth of most annual crops and to the depth to water-table above which, in most soils, there is potentially a continuous capillary film reaching to the ground surface. A certain proportion of observations, however, are made to 5 m in order to determine the likelihood of development problems arising from the occurrence of underground rock or bands of gravel, clay, or gypsum. At least one such deep boring is done per 1250–2500 ha in all types of survey, but especially in areas where subsoil conditions cannot be inferred from the geology, e.g. alluvial plains.

12.4 Design of sampling schemes

Decisions about where to sample ground data are important since the validity of extrapolations depends on it. Obtaining a sample from a larger population is an inherent component of all field-work and its efficient execution is particularly important in reconnaissance surveys where the proportion of the area with ground data is likely to be very small. It is important to design the sampling scheme and data collection system in a way that will be suitable for later statistical analysis.

The two main types of sampling are 'purposive' and 'probability'. Purposive sampling is where 'typical' sites are chosen on scientific grounds, probability sampling where they are chosen in such a way that 'each individual or sampling unit has a known chance of appearing in the sample' (Justice and Townshend, 1981).

Purposive sampling exploits existing knowledge about the landscape to the utmost, subdividing it into component areas that are apparently homogeneous. One sample is taken from each of these areas. More often, however, 'stratification' is used. In this, the individual areas are grouped into relatively homogeneous types called 'strata'. In simplest form, a single sample is taken from each area or stratum, but it is normal to sample each internally by a random method to give a measure of statistical variance. The advantage of purposive sampling is that the samples are very representative since they are based on the skills and the local knowledge of the investigator. It also reduces the time spent on field survey by permitting the selection of accessible sites, and the reduction of sampling density in homogeneous or unimportant areas. The main disadvantage is that it lacks the randomness essential to statistical procedures (Cochran, 1963). Therefore, a formal statement of the representativeness of the sample is unobtainable and consequently the reliability of estimates for the whole survey area is unknown.

Probability sampling overcomes this problem. There are three types. When the chances are made equal over the whole area it is known as 'simple random sampling'. In the simplest scheme, points are chosen using pairs of random numbers to select grid intersections and observations are then made at these points (Fig. 12.1(a)). If the emphasis is on discrete areas, a similar scheme can be used to select sample areas (Bagwell *et al.*, 1976) (Fig. 12.1(b)). In order to secure a more even areal coverage 'systematic sampling', using the intersections of a regular grid, is preferred (Fig. 12.1(c)). This is sometimes called 'grid sampling'. The random element can be increased by aligning the grid randomly (Fig. 12.1(d)), the number of degrees of freedom being limited to the number of 'random starts' and not to the total number of sample points or plots. Grid sampling can lead to distorted results if the natural variations in the landscape are 'harmonic' with the grid. Sampling randomly within the grid squares combines the representativeness of the random sample with the comprehensiveness of the systematic sample (Fig. 12.1(e)). This is known as 'systematic random sampling'.

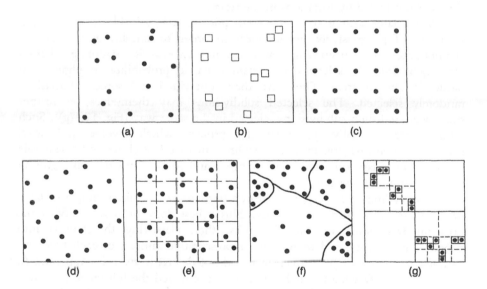

Fig. 12.1 Types of sampling schemes: (a) random point sample; (b) random areal sample; (c) systematic sample; (d) randomly aligned systematic sample; (e) unaligned systematic random sample; (f) stratified random sample; (g) random nested sample. (Source: Townshend, 1981)

153

At present the consensus of opinion is that systematic sampling will provide as good or sometimes even a better estimate of the mean for a specified number of samples than random sampling. However, as probability theory is based on random selection, the theory and techniques of random sampling no longer apply; and the calculation of standard error of the mean, variance ratio, fiducial limits, and other statistical parameters are no longer valid and therefore remain unknown. Nevertheless, in practice as it is often expensive and time-consuming to locate random plots in the field, purposive or systematic sampling may be preferred. Rosayro (1959) compared random sampling and samples taken along lines drawn between prominent map features (selected line sampling) and found there were no important differences.

The choice of sampling scheme depends on the scale of the survey, the complexity of the landscape, the amount of information already available, and time and cost constraints. The commonest compromise is to stratify the sampling by subdividing the area into the smallest terrain units recognizable by remote sensing, and then to use random sampling within these, with the highest density of observations in economically valuable and internally complex units. Field sampling is restricted by the difficulties of access and of accurate ground location, especially in remote areas and where maps and aerial photographs are inadequate. One can achieve an acceptable result by sampling random or equidistant points along roads or access lines, provided that their choice is not biased by variations in the terrain.

When a large area is to be covered quickly, it is valuable to use 'nested sampling' of progressively smaller areas in order to simplify and accelerate ground access by concentrating sampling sites. In stratified sampling schemes this begins with the selection of certain strata; in probability sampling, with randomly chosen areas. These are then subdivided and some subdivisions randomly selected. The selected subdivisions may themselves be further subdivided and their subdivisions selected in the same way (Fig. 12.1(g)). Such schemes are especially appropriate for projects which involve multi-level remotely sensed imagery. Primary sampling units are first chosen on small-scale imagery, secondary on larger-scale imagery and so on until the final ground sampling units are selected.

It is sometimes necessary, for research purposes, to obtain a sample of large areas of the earth's surface for study. This is best done by first stratifying the area involved into large terrain units, then classifying these units into morphological types, and then taking representative samples of each type. An example of such a scheme covering world hot deserts is given in Appendix A. In order to verify whether such a scheme has covered the full range of terrain variety, and that no gaps have been left, one may use a method such as the one illustrated in Fig. 12.2. This involves the construction of a three-dimensional diagram whose axes represent the main types of natural variety being sought: in this case lithology, texture of surface materials, and geomorphic form. The known examples are then plotted on to the diagram. This then reveals the presence of clusters where data are abundant, and gaps

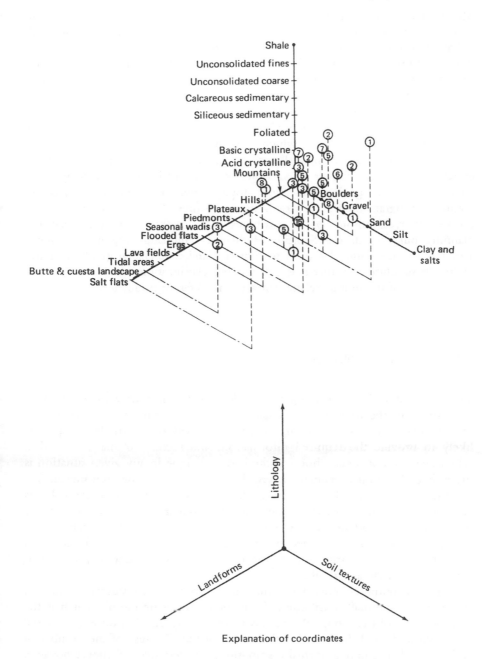

Fig. 12.2 Three-dimensional graph used to assess how far all combinations of the three variables: surface material, lithology, and geomorphic form, had been represented by a number of observed examples. (Source: Perrin and Mitchell, 1970 © British Crown Copyright 1970/MOD)

155

where more is needed. Figure 12.2, for instance, shows an abundance of examples of mountains and plateaux on acid crystalline and consolidated sedimentary rocks, the dominantly coarse texture of surface materials and the absence of examples of lava fields and tidal areas. Multidimensional schemes of analysis, using more than three such variables, would be possible with the use of a computer.

A stratification system for ecological sampling, devised by the UK Institute of Terrestrial Ecology, subdivided Great Britain into 1×1 km grid squares and described about 600 of these which is a sufficient sample (*c.*6 per cent) to characterize the whole country (Bunce *et al.*, 1983). Each square is quantified in terms of a large number of variables into 76 'attributes' identifying it in terms of climate, geology, location, topography, and human artefacts. A computer program then assigns each square to one of 32 'land classes', based on the interrelationships of its attributes. The method has been used for planning purposes in the Scottish Highlands (Highland Regional Council, 1985) and in the national project to monitor landscape change (Ball and Barr, 1986). Its significance is that it is a parametric classification which incorporates a calculation of the mutual relationships of the different parameters.

12.5 Statistical methods

The range of statistical techniques is wide and it is necessary to select those appropriate to the particular investigation. They are important at two stages: first, to ensure that the sampling scheme discussed earlier in this chapter is likely to provide the required information, and second to analyse the data when they are obtained. Choosing the best technique in any given situation is often difficult, and it is generally advisable to consult a qualified statistician, first to ensure that the sampling scheme is likely to provide the information that is required, and second to choose methods of analysing the data when they are obtained. It is good practice to write down algebraically every step in any analysis that is to be performed before the survey is undertaken. Unforeseen difficulties can arise during survey and the investigator should be prepared to modify the analysis as a result.

For most purposes individual values in a group are considered to have a 'Gaussian' or 'normal' distribution about their 'arithmetic mean', which is the average value of the group. When this is not so, the distribution is considered to be 'skewed', and the 'skew' is the extent of this. Because of the number of techniques that require a normal distribution, it is desirable, whenever possible, to transform the individual items of data so as to secure a normal distribution.

The numerical values of these quantities properly refer only to the sample from which they were derived. If we wish to use them to describe all the terrain from which the sample was drawn then we must realize that they are subject to 'sampling error'. We need to know what 'confidence' may be placed in the sample values as estimates of the true values for all the terrain studied. In order to calculate the 'confidence limits' an investigator must make certain assumptions. In most simple applications he assumes that the variation within classes is random, i.e. the sampling error associated with one observation is quite independent of other observations. He also assumes that such sampling errors are 'distributed' in a 'binomial' way if they are of the 'presence' or 'absence' of certain attributes, or normally if they are of measured attributes.

Statistics are most powerful in these circumstances, and it is always worth examining measurement data to see whether they are normally distributed before proceeding to calculate confidence levels. Data that are not normally distributed can sometimes be transformed so that they become normal.

Statistics are required for at least three purposes in terrain evaluation. First, they provide a numerical description of the population of sample plots or points and their characteristics. For some purposes the 'median', the middle value of those observed, or the 'mode', the most frequently occurring value, may be preferred to the mean. To indicate a range of values, the 'variance' or its square root, the 'standard deviation', is employed. When the standard deviation is expressed as a percentage of the mean, the result is known as the 'coefficient of variation'. When observations are of the presence or absence of some attribute, then the 'proportion' of those that possess the attribute is needed.

The second major requirement is to determine 'confidence limits' for predictions of terrain properties. These allow a worker, on the evidence of a sample, to predict the likelihood that the value of a given attribute, at a site randomly selected from the whole population, will fall within specified limits of deviation from the mean. This is obtainable from recognized statistical formulae relating sample size to its mean and standard deviation (see e.g. Cole and King, 1968). Statements of confidence take the form: 'on the sample evidence there is a k per cent probability that the mean value of the property in question lies in the range $x_1 - x_2$' or, 'the mean value of this property is $\bar{x} \pm c$, where $+ c$ and $- c$ are the symmetrical $(100+k)/2$ per cent confidence limits'. Confidence limits are usually established at the 95, 99, and 99.9 per cent probability levels. This means that one can state with confidence that the determined limits will only be exceeded by one value in 20, one value in 100, or one value in 1000 respectively. For example, it might be necessary to predict soil strength at an unvisited site to ensure that a vehicle with a given axle loading will pass without sinking. The best estimate will be a mean value from comparable sites, and it should be accompanied by a statement of confidence. This has the form: 'there is a 95 per cent chance that the soil strength will exceed y and that the vehicle will pass.

The third important use of statistics is for comparing two or more classes of terrain. Comparisons may be needed to decide which kind of terrain would be best to develop for agriculture or which would be most suitable to provide road-making materials. Alternatively, a surveyor might wish to know whether it was practically important to distinguish two neighbouring and similar types of terrain in terms of agricultural potential. We assume that the sampling evidence shows that some differences do exist in both cases. The question is whether they are large enough to justify separations in classification and mapping. Such comparisons involve statistical 'testing'.

To make a test the statistician acknowledges that there is sampling error, and that he cannot be sure whether any differences he observes between classes of terrain are more than sampling variations. He therefore adopts the following position: if there are no real differences between the classes being considered, what are the chances that the differences observed could have arisen by chance? The situation being tested is often referred to as the 'null hypothesis'. Conventionally the null hypothesis is accepted in a single test if the chance of observed differences arising is greater than 5 per cent and rejected if it is less. If the null hypothesis is rejected, the differences between classes are regarded as significant. If the investigator is unhappy with this degree of certainty, he can apply a more stringent 1 per cent, 0.1 per cent or other level. Alternatively, he can take the view that the evidence suggests the differences are real but would like to be more sure. To increase his confidence he can carry out further sampling. If there are real differences between classes then increased sampling will increase the chance of finding them; if there are no real differences, increased sampling will decrease the chance of differences appearing in the sample evidence.

Testing for differences of proportions in samples is usually done with the 'chi square test' (χ^2). Testing for differences between means of measured properties is usually done by classical methods such as 'analysis of variance', or, if there are only two classes, 'Student's *t*'. Both of these assume normal distributions of the data, but are robust against departures from this, i.e. they are unlikely to mislead over a wide range. Several methods have been developed to compare sample means without making any assumptions about data distribution. They are known by the names of their originators, and among the most important are the 'Mann–Whitney *U*-test', the 'Kolmogorov–Smirnov test', and the 'Kruskal–Wallis' analysis. They have less power than the classical methods for normally distributed data, but can be used when the classical methods cannot.

Statistics can also be used to generate their own terrain classifications as an alternative to the study and testing of predetermined schemes. The basis is numerical taxonomy leading to multivariate classification. This assumes that the site samples are drawn from a continuously varying population or continuum and that the classes into which they should be assigned are best determined by ordination on the basis of maximizing degrees of affinity between groups of attributes. This ordination by affinity can be developed by using principal components analysis (PCA). This technique places every site in multidimensional space with each dimension representing a measured attribute. Classes are determined by the clustering of sites within this space, and calculations are made to analyse this clustering to give a minimum number of composite factors (eigenvectors) to explain the observed inter-site variations.

12.6   Statistical references

The following statistical texts will enable the reader to follow up the brief survey in the preceding section. Two lucid and entertaining introductory texts are by Moroney (1965) and Campbell (1967). Yule and Kendall's (1950) classic covers a broad field in detail but still at an introductory level, while Siegal (1956) can be recommended for distribution-free statistics. Snedecor and Cochran (1978) give thorough treatment, with special emphasis on agricultural problems. King (1969) and Gregory (1978) write with reference to geography; Miller and Kahn (1962) and Krumbein and Graybill (1965) to geology; Greig-Smith (1964) and Pielou (1977) to ecology; and Webster (1977) to soil science. Standard works on sampling are by Yates (1960), Sampford (1962), Cochran (1963) and on fuzzy sets by Kandel (1986). Computing methods for geographers are described by Cole and King (1968) and Dawson and Unwin (1976).

Applications of statistics to terrain evaluation can be found in the testing of the land system method in the Oxford area (Beckett and Webster, 1965b, c, and d) and over the arid zone (Perrin and Mitchell, 1970, Mitchell 1971; Beckett *et al.*, 1972; Mitchell *et al.*, 1979). There have also been a number of approaches to describing landforms in predominantly mathematical terms. These include methods of manipulating spatially distributed data, including the PCA of a number of physical attributes (Cadigan *et al.*, 1972) and the spectral analysis of longitudinal profiles (Pike and Rozema, 1975). Digital terrain models have been reviewed by Doyle (1978).

Further reading

Beckett *et al.* (1972), Beckett and Webster (1965 b, c, and d), Cadigan *et al.* (1972), Cochran (1963), Dawson and Unwin (1976), Doyle (1978), Gregory (1978), Howard and Mitchell (1985), Maguire (1989), Mitchell (1971), Mitchell *et al.* (1979), Moroney (1965), Perrin and Mitchell (1970), Pike and Rozema (1975), Snedecor and Cochran (1978), Townshend (1981), Webster (1977).

13

Field techniques

13.1 The programme of field observations

The success of the contemporary field-worker depends on combining three skills with his basic professional training. These are an 'eye for country' obtained by field experience over a range of environments, practice in the management and interpretation of remotely sensed imagery, and familiarity with modern data handling techniques.

Field-work in terrain surveys differs from that in specialist surveys in the wider range of data collected. It involves observations of basic geology, geomorphology, pedology, hydrology, ecology, and both actual and potential land uses. These are normally done by a team of specialists travelling in company. This inevitably makes for somewhat slow progress because joint observations take longer than individual ones and it is not always possible to select observation sites of equal significance to each scientist. In responsible teamwork, these difficulties can be partly overcome by allowing some freedom of individual movement, provided that all scientists catch up with each other at each joint observation (Haantjens, 1968). The alternative is for all the work to be done by a single 'integrated surveyor', though this gives a less detailed result. However many scientists are involved, speed and economy necessitate emphasis on aspects of terrain related to immediate land uses, though this should not lead to a neglect of other aspects which may have a long-term significance.

Field observations are made at purposively or randomly selected sampling points or plots, the distribution of which has been predetermined according to the principles outlined in Chapter 12. Once the study area has been defined, it is scanned on the maps and aerial photographs to find routes and plan the daily field traverses. These aim to cover the area in blocks which do not leave isolated gaps requiring special visits to complete.

The activities at each site are best limited to those which are relatively quick and accurate under field conditions and which do not require equipment which is either complicated, heavy, or cumbersome to transport, easily damaged by shocks or dirt, or needing too high a standard of skill, sharp

160

eyesight, or manual dexterity to operate. More detailed observations usually require bringing samples back to a laboratory.

Although each sample is of a point or plot, it can be regarded as representing all the land up to the midpoint to the next sample in any direction within the same sampling stratum. Care is needed in extrapolating data observed at a single site to the surrounding area, since they vary in areal coverage. Some, such as slope angle or vegetation cover, refer to a tract of land; others, such as soil strength, to a point only. In practice, it is usual to assume that field observations are based on an 'envelope' with a relatively arbitrarily chosen radius, usually about 100 m, centred on the point or plot. If statistical analyses are planned, adjacent envelopes should not overlap.

It is useful to have comprehensive proformas for recording information. These may be in the form of cards to be punched at the study sites, provided they are not too difficult to use or vulnerable to damage under field conditions. But it is more common to tabulate field notes on printed forms and then abstract the data on to a different format for storage purposes. Current trends are towards transferring all such data on to computer software, and ultimately towards eliminating the necessity of recopying data by carrying portable computers into the field.

The features to be recorded at each site fall into a number of distinct categories, commonly carried out in the following order:

1. Landscape and mesorelief, with associated broad-scale processes;
2. Ground surface and microrelief, with associated small-scale processes;
3. Vegetation, sampling when necessary;
4. Land uses: past and present;
5. Soil profile and groundwater, sampling when necessary;
6. Field tests.

Each of these aspects has its own techniques and most involve the use of special tools.

13.2 Landscape and mesorelief observations

Landscape and mesorelief observations may begin with a statement about microclimatic factors if these differ significantly from the situation of the study area as a whole. For instance, a site may have experienced exceptional rain or frost affecting the soils or vegetation. Much of the terrain data can be derived from aerial photographs and maps before visiting the site, so that the field visit can concentrate on final checking. There are three major aspects: relief, landscape aesthetics, and erosion risk.

Observations of relief must interpret the basic geology and landscape evolution and specifically include angle and aspect of maximum gradient, relief amplitude, and the proportion of the total area occupied by unusable land such as rock or gullies. Landscape assessment is most simply done by a combination of the 'isovist' method (see Ch. 26) and a scaled estimate of landscape quality within a 360° circuit of the site with high values for vertical contrast, pleasing colour patterns, and the presence of water. Because of the natural complexity of landscape, it is valuable to accompany observations with field sketches and ground photographs.

The assessment of erosion risk includes all aspects of land conservation. Observations are made of the important geomorphic processes influencing the landscape within an appreciable time scale, such as mass movements, sheet and rill wash, gullying, and wind and wave action.

13.3 Ground surface and microrelief observations

Ground surface observations include microrelief and the nature of surface materials. Microrelief includes the shape and distribution of features less than 1 m in height and not more than a few metres in diameter, which are too small to appear on normal maps. It is hard to assess quantitatively and must generally be based on a verbal description of the ground surface unless either a morphometric map or a detailed study of a sample area is being undertaken. Descriptions are expressed in terms of genesis and morphology and should include the size, shape, and areal distribution of such features as rills, hummocks, pits, sink holes, and *nebkas* (small dunes). Stone and Dugundji (1965) describe a measuring device, called the 'continuous ride geometry vehicle' which can record a microrelief profile. As the vehicle traverses the terrain, it gathers, edits, amplifies, and digitizes data on x and y coordinates yielding a magnetic tape suitable for use in a computer. Such tools are, however, expensive, and not always justified by the information they yield.

The character of surface materials includes both their composition and their appearance from above. They constitute the part of the regolith which occurs at depths accessible for quarrying and opencast extraction, and include ore bodies, building stones, constructional gravel or aggregates, lime, china, brick clays, etc. The character of the ground surface is recorded in detail. It is especially important in assessing the spectral response of different types of surface material on remotely sensed imagery. Where not covered by vegetation, it should be subdivided into proportions occupied by stones, sand, silt, and clay. Assessment is then made of the lithology, alignments, and the maximum and average size of stones, aeolian features in the sand, and the crusting and cracking of finer materials.

13.4 Vegetation observations

The type of vegetal data collected will depend on the objectives and intensity of the survey (exploratory, reconnaissance, extensive, or intensive) and whether it is concerned primarily with the collection of physiognomic or floristic data.

The worker should become familiar with the flora of the geographic region, the vernacular and botanical names of the species (especially the trees), and be able to identify dominant species when not in flower, to know how to use botanical field keys, and how to collect, dry, and preserve floristic samples in a botanical press for herbarium identification. It is often useful or even essential to establish a temporary herbarium against which newly collected samples of plant species can be checked.

An early step is the visual assessment of vegetation types as portrayed on the satellite and aerial imagery. A comparison of old with new photography will provide information on vegetation succession and changes in land use. This includes the identification of the major boundaries of the vegetation types, and with the aid of stereoscopy, the plant cover and stand characteristics.

Field-work for intensive surveys will include both a physiognomic classification of the vegetation according to the methods developed by Dansereau (1958) and Kuchler (1967), considered in Chapter 8, and a species classification with an estimate of their relative abundance. Exploratory surveys will concentrate on acquiring basic information such as percentage ground cover and the height and spacings of dominant plants.

13.5 Land use observations

Information about land uses will overlap with that about vegetation, specifically where the latter is induced rather than natural. They should be related to the local socio-economic conditions, by translating the field observations, which are essentially area-based, into economic evaluations, which are essentially activity-based.

The field data can be broadly subdivided into those concerned with agricultural and engineering types of use. For both it is important where possible to record the past history. Agricultural data include the nature of the crop, pasture, or forest, the methods of tillage, the use of fertilizers, and practices for land protection or improvement such as drainage, terracing, and irrigation. Engineering data include the character of existing roads, buildings, and other artefacts, and of the materials used in their construction.

13.6 Soil profile observations and sampling

Soils are observed by means of pits and auger borings to a depth adequate to penetrate any horizons exploited by plant roots. Pits require more resources to dig, but are needed to obtain a clear view of the horizons in the soil profile. One is, however, enough to characterize each type of terrain being studied. Most observations are made on samples brought up by augers. Broadly, these are either of (a) 'screw' or 'worm', (b) 'bucket' or 'post hole', or (c) 'core' varieties. Screw augers have a typical diameter of 3–5 cm and will raise a few grams of soil. They are quick to use and suited to rapid checking in known areas. Bucket augers usually have a diameter of 5–15 cm and are able to bring up samples large enough for laboratory analysis, although the process of screwing them into the ground churns the sample. A variety of designs are available with different sorts of cutting edge and retaining flange for different types of soil. Core augers are of similar size. They are pressed rather than screwed into the ground and remove an undisturbed sample. They can, however, only be used in relatively soft and stoneless soil. A recent development has been the use of large mechanical augers, mounted on lorries, which can in a few minutes bore a hole 1 m in diameter, large enough to serve as a soil observation pit.

A detailed description of the methods of soil profile recording are given by the USDA Soil Survey Staff (1962, 1976), Hodgson (1976), and FAO (1977). The basis is the systematic recording of profiles by horizons. Each horizon is assessed for texture by the feel of moistened soil and then observed visually to determine the abundance, lithology, and distribution of stones, the Munsell colour, moisture, structure, consistency, cracks, visible salts, or plant roots, and features such as pans (hard layers) or *krotovinas* (former animal burrows now filled with soil).

Soil samples are taken for laboratory analysis. The principle is to sample each horizon of one or two pits in each soil type in detail, but to sample intermediate borings more sparingly. In practice, to save costs, this means that, in a typical project area 30–60 pits are comprehensively sampled by horizons, while the other sites are boreholes, only sampled where particular checks are necessary. Samples may be ordinary or undisturbed cores. Ordinary samples from pits are obtained by using an adze and scoop on a profile face, from boreholes by emptying material brought up by the auger from each horizon into bags. In boreholes where horizons cannot be seen, it is useful to sample by regular steps, i.e. of 50 cm. Ordinary samples are adequate for most laboratory tests, but undisturbed cores are required for physical tests such as specific gravity or permeability. These are obtained by driving metal tubes into the ground surface or steps at predetermined depths in soil pits.

Where the pit or boring penetrates to below the water-table, the latter's depth is recorded. Where the chemical content of the groundwater may be significant to irrigation or drainage, a sample should be taken for chemical analysis.

13.7 Field tests

There are certain tests which only give acceptable accuracy or are most conveniently done in the field. They include testing for soil pH, salinity, moisture, permeability, strength, and radiation.

The pH can be determined with a test kit using colour indicator solutions. This is accurate to somewhat better than 0.5 of a pH unit. Greater accuracy than this can only be obtained by taking samples for laboratory testing with pH meters too delicate for normal field use. Salinity can be measured in terms of the electroconductivity of small soil samples or water bodies. Battery-powered 'salinity meters' generate a small current from one pole in an electrical probe and measure its return through the other pole in terms of milliSiemens per centimetre, which correlates directly with the proportion of dissolved salts in the solution tested.

Soil moisture is normally measured by oven drying and weighing in the laboratory, but for repetitive measurements under field conditions, it is preferable to use a 'neutron probe' lowered down an auger hole to the depth to be tested. This is a probe containing a radioactive source which emits fast neutrons into the surrounding soil. Collisions with the nuclei of the soil atoms, predominantly those of hydrogen and soil water, cause the neutrons to scatter, to slow, and to lose energy. Thus a 'cloud' of slow neutrons is generated within the soil around the source. The density of this cloud, which is largely a function of soil water content, is sampled by a slow neutron detector in the probe. The method is described by Bell (1976). The neutron probe, however, needs to be calibrated separately for different types of soil, is costly, and because of its inclusion of radioactive material, requires protective storage.

Permeability is a general term for the rate of water movement through soils and can vary with the head of water and the degree of saturation of the soil. 'Hydraulic conductivity' is a more reproducible, and hence more useful, measurement which eliminates these two sources of variation. It is most accurately measured below the water-table under conditions in which flow is laminar and the gravity factor can be eliminated. Where the water-table is within about 5 m of the surface, a good method is the pump-out auger hole method described by van Beers (1958) and FAO (1976b).

Where the water-table is below augering depth, i.e. below about 5 m, it is not possible to make a field measurement of hydraulic conductivity, and it is necessary to fall back on empirical measurements of soil surface permeability or 'infiltration rate' which can be calibrated against hydraulic conductivity measurements where these are available. The method is either to measure the rate of infiltration of water into an auger hole (e.g. Sudan, 1963) or into cylinders driven into the ground (Hills, undated; Youngs, 1987; Jarvis *et al.*, 1987). Some of these methods are described in Appendix B.

Soil strength controls the capacity of the land to sustain the passage of vehicles and to support buildings and roads. It is measured in two directions: vertically with penetrometers, and horizontally with shear vanes.

Penetrometers can be pocket-sized or haversack-sized. Both record on a dial the resistance of the ground to the insertion of a rod of known cross-sectional area down to a depth of about 60 cm They are operated manually by pressing down on a handle. The most satisfactory seem to be haversack-sized models of USAEWES (1963b), the Military Engineering Experimental Establishment (MEXE) (1968), and Proctor (Gerrard, 1982) patterns. These are similar except that the last two are housed in stronger casings and are more robust. Readings are taken as graduated marks on their shafts penetrate the ground. Dials are calibrated in terms of weight per unit area of the penetrating rod (e.g. lb in^{-2} for the MEXE pattern) or California bearing ratio (CBR) units, which MEXE (1968) correlate with vehicle performance. The number of readings required at each site and the statistical methods for handling the results have been worked out in some detail by Beckett and Webster (1965d). Pocket-sized penetrometers operate on the same principle but cannot penetrate more than the top few centimetres of soil and give less reproducible results. Shear vanes are essentially vertical plates connected to torque springs recording on a dial when twisted in the ground. Although a number of patterns exist, they have not given sufficiently reproducible results to be widely used.

The capacity of terrain to support the passage of vehicles is variously known as 'going' or 'trafficability'. There is no standard widely used test of this but some empirical tests have been devised. One compared the distance travelled by a vehicle across selected tracts of terrain with that travelled on a smooth road surface, thus showing the combined effects of slippage (lost traction) and sinkage (poor weight support) without distinguishing between the two (MEXE, 1966). The distance travelled over the terrain was expressed as a percentage of that over the road and called the 'going coefficient'.

The vehicle used was a Land Rover which was held in second gear and at constant engine speed, using an impulse tachometer to show the number of engine revolutions and a throttle screw to control them. The distance covered was measured with a revolution counter on a trailing bicycle wheel sprung to hold it down on the ground. In studies in Libya, Bahrein, and the United Arab Emirates, it was found that most terrain gave going coefficients between 70 and 90 per cent, and that immobilizations occurred when these fell below 60–70 per cent (Perrin and Mitchell, 1970). As might be expected, wet saline ground and sand dunes scored lowest and dry, firm, desert plains highest.

A complete empirical assessment of trafficability should also include a measure of 'bumposity', i.e. the frequency and size of bumps encountered, and the distance of diversion necessary round obstacles, on a random traverse. Bumposity can be measured with a 'bump counter' attached to the side of a vehicle. This is a bar, hinged at one end and suspended from a spring at the other so as to record the motions of the sprung end (MEXE, 1966). Diversion distance has been assessed by measuring the percentage of additional traverse imposed on a vehicle by the microrelief when it is driven between two intervisible points of known separation. Tests were performed at Yuma, Arizona (USAEWES, 1962), and in the United Arab Emirates (Perrin and Mitchell, 1970).

Soil erosion has been measured in a number of ways. The comparison of aerial photographs over a lapse of years will reveal major changes. The most accurate methods are based on the use of erosion plots. Changes in their surface can be measured from the bending, tilting, or depth of cover around erosion pins, or from changes in the topographic surface indicated by a profilometer. Some promising results have been obtained from monitoring the movement of traces of isotopes in the soil, notably caesium-137 and iron-59 (Loughran, 1989).

Radiation measurements are sometimes required in order to relate the spectral response of ground features to their appearance on aircraft or satellite imagery. These measurements are made with ground-based radiometers, which have been reviewed by Holmes (1970) and Lee (1975). The laboratory for the Applications of Remote Sensing (LARS), for instance, use a spectroradiometer mounted on a platform whose altitude is adjustable up to about 12 m above ground, known as a 'cherry-picker', connected to a vehicle in which data are continuously logged (LARS, 1968; Miller *et al.*, 1976). Portable field radiometers do not record continuously but are relatively inexpensive and more readily incorporated into survey programmes. They quantify the reflected radiation in band-passes corresponding to particular channels, such as those of the Landsat multispectral scanners (Tucker, 1978; Milton, 1979). The usual practice is to mount the radiometer on a tripod, and to take readings of the underlying ground, continually calibrating them against an inserted standard reference card. Portable instruments are also used for recording thermal IR radiation, such as the Barnes PRT-5 and PRT-10 (Bonn, 1976), and microwave radiation (Chapman *et al.*, 1970; Bryan and Larson, 1973).

13.8 A note on laboratory analyses

There are a number of laboratory tests of the physical, chemical, and mineralogical properties of soils important to agriculture and engineering. These are based on analyses of soil samples taken in the field, and are described in full in modern laboratory textbooks.

The main physical parameters requiring measurement are 'particle size distribution' (mechanical analysis), 'linear shrinkage', 'shear strength', 'compressibility', 'dry density', 'porosity', 'pore size distribution', 'moisture content', the 'moisture characteristic', including 'field capacity' and 'permanent wilting point', and 'Atterberg limits'. These last are the 'plastic limit' and the 'liquid limit' of a remoulded soil. The plastic limit is its moisture content, calculated as a percentage of oven-dry weight, when it changes from crumbling under applied pressure to become plastic. The liquid limit is the moisture content, calculated in the same way, at which it ceases to be plastic and becomes fluid. The plastic index is the numerical difference between the two percentages. Methods of analysis for physical attributes are given by the British Standards Institution (1975).

The chemical properties most usually required are 'soil reaction' (pH), 'salinity' (in terms of electrical conductivity), 'exchangeable cations' (Ca, Mg, Na, and K), 'exchangeable anions' (Cl, SO_4, NO_3, and HCO_3), 'cation exchange capacity', 'available phosphorus', 'total nitrogen', and 'organic carbon'. 'Trace elements' such as boron, copper, molybdenum, and cadmium are sometimes important. In arid soils, total $CaCO_3$, $CaSO_4$, and 'sodium absorption ratio' (SAR) are often required. Established analytical methods are given by Richards (1954) and Avery and Bascomb (1974), and recent technology involving the use of plasma spectrometry to speed routine chemical work by Barnes (1979).

The determination of the mineralogy of soil clay colloids (<0.002 mm diameter) helps to account for some of the physical and chemical characteristics of soils. Determinations are by 'X-ray diffraction' and 'differential thermal analysis' and are described by Brown (1962) and Grim (1968). The arrangement of the colloids around larger particles, indicating their past movements, is known as 'soil micromorphology'. It is best interpreted from thin sections using microscopes with magnifications of 50–500×. The terminology and methods have been reviewed by Chartres (1975).

A technique which has developed of recent years for analysing the genesis of earth materials and waters is 'stable isotope geochemistry', fully discussed by Hoefs (1980). It is based on the fact that many common elements, notably hydrogen, carbon, oxygen, sulphur, and nitrogen, have stable isotopes of slightly different atomic weights from their common forms. The lighter isotopes form compounds which are slightly less strongly bonded and more chemically reactive than the heavier. They are also more volatile, especially at higher temperatures, the difference in volatility increasing with the difference between their relative weights. The relative abundance (expressed as a ratio) of the different isotopes of the same element in a sample may thus give important indications of the genetic processes it has experienced. To take just one example, sulphur occurs as ^{34}S and ^{32}S, the former being about 23 times more abundant than the latter. In general the highest proportions of ^{32}S occur where there has been bacterial reduction and the lowest where there have been chemical exchange reactions. Samples must be submitted to specialized stable isotope laboratories for such tests.

Further reading

Avery and Bascomb (1974), Bell (1976), Bonn (1976), Bryan and Larson (1973), Chapman *et al.* (1970), Chartres (1975), Dent and Young (1981), FAO

(1977), Gerrard (1982), Grim (1968), Hills (undated), Hodgson (1976), Hoefs (1980), Howard and Mitchell (1985), Jarvis *et al.* (1987), Kuchler (1967), MEXE (1968), Milton (1979), Perrin and Mitchell (1970), Richards (1954), USDA Soil Survey Staff (1962, 1976), USAEWES (1963b), van Beers (1958), Youngs (1987).

14

Data processing and geographical information systems: principles

14.1 Introduction

The availability of information about land resources is a vital factor in their understanding and management. Its flow can be measured in terms of 'hubits' per capita per year. Hubits are units of communicated information. In most societies their rate is very large. In advanced societies they can be counted by sample surveys of the frequency with which certain terms are used by the press, radio, or other mass media, multiplied by the number of people who receive them. In poorer societies, their number is smaller and must be estimated from descriptions of public life and social interactions. A comparison of the estimated median per capita incomes of people in the most modern literate societies with those in large cities in the poorest societies (Ethiopia, Indonesia, etc.) showed a ratio of between 25:1 and 75:1. The corresponding ratio for information flow was more than 500:1 (Meier, 1965). The gap between societies in information transmission was therefore 'six or seven times' greater than that between incomes, wide as this was. Furthermore, this gap appears in the earliest stages of the development process. It is therefore reasonable to conclude that it is one of the important preconditions for economic progress. The theory that the rate of economic development is due to it is called the 'information theory of development'.

14.2 Geographical information systems

14.2.1 *Scope and definitions*

Many organizations need to store and process spatial data. These include both public and private bodies, with interests ranging from problems at individual sites to major plans involving international cooperation. The common requirement is for rapid access to statistical data, appreciations, and thematic maps. The amount of data potentially involved is vast, and the variety of location, topic, and format wide. Nevertheless, experience shows that virtually all problems can be reduced to answering one or other of two questions:

1. Finding locations with specified properties;
2. Finding the properties of specified locations.

Large amounts of data cannot readily be managed or understood without specialized equipment. An environmental scientist can normally mentally integrate field data, maps, and aerial imagery. But with four or more data sets, he becomes increasingly confused. Since the 1970s this problem has been overcome by the use of computers, and has led to the development of 'geographical information systems', generally and hereinafter abbreviated to GIS. These can be simply defined as systems for storing, manipulating, and retrieving spatially referenced data.

The data in one GIS are its 'database', which can be composed of a number of 'data planes' (J.A.T. Young's 'data elements' 1986), derived from 'data sources' whose combinations can yield 'data interpretations' (see Fig. 14.1).

Representative data planes   Data interpretation

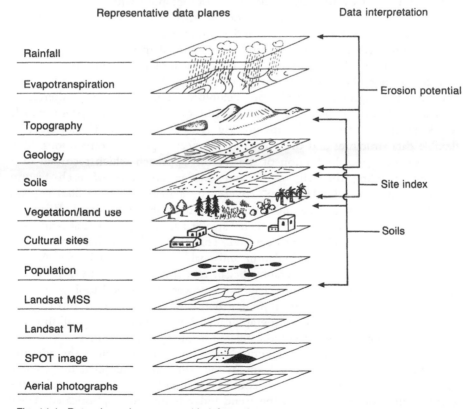

Rainfall

Evapotranspiration   Erosion potential

Topography

Geology

Soils   Site index

Vegetation/land use

Cultural sites   Soils

Population

Landsat MSS

Landsat TM

SPOT image

Aerial photographs

Fig. 14.1 Data planes in a geographic information system

171

Each data plane is of one 'data type'. Examples of data types are 'real valued functions' (e.g. digitized elevation data), 'vector valued functions' (e.g. terrain units), and 'functions taking general symbolic values' (e.g. place-names). The data in one data plane have a 'data structure'. In non-digital data, this can be written text, maps, tables, or photographs. Digital data can either have a 'vector' structure based on lines and polygons, or a 'raster' structure based on a matrix of points. An example of the former is digitized elevation data from which contours can be drawn, and of the latter, multispectral imagery.

14.2.2 *Development of GIS*

The GIS concept has evolved through a number of stages, and is in effect the marriage of three types of spatially referenced data: mapped information, spatially referenced statistics, and remotely sensed imagery, into one integrated system.

The storage of spatial data on maps is ancient, but in the last 20 years has been transformed by the advent of digital cartography, which is discussed in Chapter 17. The first stage was the development of computer hardware to capture digital data sets, edit them, and reproduce them cartographically. The resulting products attempted to match conventional map design in the hope that, with continued development, the digital technology would lead to labour and cost savings. However, the large effort required for this pointed to the necessity of improving speed and flexibility. The next stage, therefore, concentrated on the mapping software and the design of a system which would produce a variety of maps quickly. This required the development of flexible data structures and the avoidance of specific graphic conventions. The result was the general adoption of the 'vector' approach which recorded map data by encoding the left, right, and on-line attributes of each feature perimeter. An area of $60\,000\,\text{km}^2$ covered by a map sheet at 1:50 000 takes about one week to digitize and computer store ready for retrieval. With the increasing power of computers and the greater availability of digital map data, a further stage of development occurred. This moved the emphasis away from map generation to the manipulation and analysis of the spatial data. As this stage developed, digital cartography began to constitute not only a means of graphical representation of spatial data but a powerful analytical tool.

Spatially referenced statistics represent bodies of data which are too large or complex to be represented cartographically. The traditional way of showing them is through the use of volume-related mapping symbols such as 'spheres', 'pies', or histograms, but cartographic limitations have prevented the addition of further tabular detail. The advent of the computer multiplied the volume and complexity of the data which could be manipulated, but did not alter the cartographic limitations. Thus, the trend today is towards sophisticated digital data storage and manipulation linked to a capability for selecting among a large number of relatively simple cartographic options.

Remotely sensed imagery began to be used in systems of spatial data storage when aerial photographs were used for mapping after about 1950. However, the main increase has come with the processing of digital multispectral imagery, especially since the Landsat programme of the 1970s. This has not only provided semi-cartographic coverage of many little-known and poorly mapped parts of the world, but has also provided multithematic and repetitive databases of wide practical importance.

The next stage was the coordination of these three data sources, into what is generally referred to as 'integrated GIS', or IGIS. Data in linear form such as meridia, parallels, route lines, contours, and thematic boundaries are stored in vector format, while data which relate to point features such as bench marks, boreholes, and important sites, as well as remotely sensed data based on pixels, are in raster format. Integration depends on the coordination of vector and raster data sets. In principle, remotely sensed imagery in digital form could be readily incorporated into a vector-based GIS, either by adding it pixel by pixel as a separate data plane or by subdividing it into a number of planes each of which can be combined with the existing data on the relevant thematic plane. In practice, such integration presents difficulties resulting from the need for geometrical correction of the imagery and problems of cloud cover, poor tonal contrast, and poor resolution. However, it is possible to rasterize and overlay any encoded point, line, or polygon-type cartographic data set on satellite remotely sensed data, leading to the integrated processing of the two. Thus, since 1984 the trend has moved away from narrow digital mapping and remote sensing applications towards more generalized spatial environment modelling.

These developments led to the definition of a major programme of research and development of a fully integrated knowledge-based GIS. This involves the investigation of two basic tools: the 'hierarchical model' and the 'intelligent knowledge-based systems' approach.

Research into the hierarchical model is both theoretical and practical. The theoretical approach is to review alternative recursive hierarchical subdivisions of plane spaces known as 'tesselations', and the arithmetics which are needed to manipulate the tesselated data efficiently. The main example of such subdivision is the 'quadtree' and this has proved the most suitable for many applications. This is a conceptual 'tree' which results when a map or remotely sensed image is quartered along its horizontal and vertical axes, each quarter being recurrently requartered until no more subdivision is possible. The subdivisions are known as 'tiles', the ultimate being the 'leaf', which for remotely sensed imagery is one pixel. Quadtree subdivision stops whenever a tile is homogeneous.

The 'hextree' is an extension of the same concept in which every node has six rather than four branches and each subdivision is a nested level of hexagons. The basic approach is to develop demonstration databases to act as 'test-beds' for the new concepts and to provide operational experience. The

173

main test system is designed as a 'quad pyramid' and utilizes tesseral addressing and arithmetic. Intelligent knowledge-based systems (IKBS) aim to increase data volume capacities, levels of computational efficiency, and range of applications (Peuquet, 1984). Their approach is to improve search procedures by heuristic search techniques which limit the portion of the database to be searched, by incorporating a 'learning' capacity to speed second enquiries on a given subject or adapt to new ones, to make the interface more 'user friendly', and to improve the classification of remotely sensed raster imagery prior to its inclusion within the database. There is little doubt that such systems will become the rule rather than the exception in the 1990s (Jackson and Mason, 1986).

14.3 Criteria for assessing GIS

A GIS must be able to present information to users in language and format that is not only accurate but also graphic and comprehensible to the whole range of users, some of whom will be non-specialists. There should be a high ratio of maps and diagrams to written text. When the amount of data is modest and can be presented two-dimensionally, manual or simple graphic methods will often suffice and will be most cost-effective. Otherwise a computer-assisted system will be needed.

The quality of GIS can be measured according to the following three indices (Fig. 14.2):

1. Amount and precision of location identification;
2. Amount of data related to each location;
3. Manipulation capability.

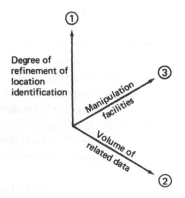

Fig. 14.2 Capability indices of a data storage system

Index 1 represents the number of geographical points or areas about which data are held and the precision of their locations. Points can be derived either from map (vector) data or imagery (pixels) and can be either approximate or defined by exact x–y coordinates. Areas can be tracts identified by standard geographical names such as 'prairie', 'South Downs', etc., grid squares, or polygons defined by limiting coordinates.

Geographical names are the traditional pigeon-holes for manually stored written information. Grid squares have been used on a systematic basis at a variety of scales. Some national census and postal systems are based on a 10 m grid. The US Large Area Crop Inventory Experiment (LACIE) used Landsat cells of 1×1 km (Driggers *et al.*, 1978). Satellite rainfall monitoring in Africa used cells of 50 and 100 km^2, and squares of 7.5×7.5 km and 15×15 km were recommended for a world photographic index (Howard and van Dijk, 1980). The use of a grid allows easier data analysis and manipulation for operations such as overlaying, but it necessitates the inclusion of natural variations within units.

Probably the best areal units are polygons designed to coincide as far as possible with natural units, because they can avoid the internal variations of grid squares. This point deserves emphasis. Natural units, such as land facets and land systems, are especially valuable because they represent an optimal environmental subdivision of the land. When remotely sensed imagery is used, there must be a prior stage of human interpretation. This is because the interpreter can select any part of an image simultaneously and interpret complex patterns (including shadows) on the basis of experience with greater speed and reliability, can separate the types of terrain data under different sun angles and conditions of haze, and can probably do it at lower cost than is possible with purely automated means (Townshend, 1981).

The second index is the amount and precision of data attached to each geographical location. The amount includes both the variety and the temporal frequency. Three levels can be recognized: small 0–100 items per location; medium 100–1 000, and large: over 1 000. For instance, topographic or thematic surveys would generally have a small amount of data for each location, integrated surveys covering a number of resources more, and regularly recording meteorological stations a large amount. The World Meteorological Organization's World Weather Watch, which covers an international network of weather stations, is an example of an organization which would both have large amounts of data for each station and also would have many locations in terms of the first index.

The third index, manipulation, can likewise be considered as having three levels of increasing capability. The basic facility is simple indexing of the data within one data plane. Following from this is the ability to make distance and area measurements, calculations of attributes within defined areas, comparisons between areas, route determinations, summaries, selective searches, scale changes, and mapping. This optimally includes not only labelling but also boundaries around uniform areas, graded diagrammatic symbols (e.g. histograms) at spot locations, or automatic contouring, which may be supplemented by hill or layer shading. The second level is the introduction of one or more further data planes and the application of the same analyses to a combination of them. The third level is that of monitoring. Beyond analysing the data, it is important to keep watch on their temporal changes in order to anticipate the need for external decisions. An advanced capability could even trigger automatic action to exploit environmental opportunities, warn of hazards, or initiate preventive or remedial measures. This is especially important in lands subject to sudden and drastic change through tectonic activity, flooding, or erosion.

Although the GIS benefits by an increased database in either of the first two indices or an improved capability in the third, it is the combined effect of all three which ultimately determines its value, i.e. moving the GIS further from the 'origin' in Fig. 14.2. Manual methods of analysis will in general suffice when there are not more than about 100 sites with less than 100 items of data about each, and only simple calculation and mapping are required. But beyond this, reliance must be placed on computer processing.

14.4 Elements of a GIS for terrain

A GIS consists of six stages which can be viewed as subsystems in the operation of the total system: data acquisition, input, storage, data processing, output, and use, within the context of a seventh, management (Tomlinson *et al.*, 1976; Jackson, 1985). Automation can be used to some extent at all stages, and dominantly in all but the first and last.

14.4.1 *Data acquisition*

Data are collected from all sources including field surveys, maps, remotely sensed imagery, and from both published and unpublished records of all kinds. They may be about terrain as such ('facts') or about its suitability for a particular purpose ('appraisals'). They may be 'raw', as about the relief or site conditions at a particular point, or 'summarized' accounts of water supply or fertility of generalized areas, referring to information which is too voluminous or diffuse to summarize in the store. The data are often only available in remote or inaccessible libraries or files, and field checking is impossible. It is thus important to use analogies between known and unknown areas.

14.4.2 *Data input*

This stage involves preprocessing and formatting the data for storage. In manual systems, it is normally held on standard cards or microfilm; in automated sytems, on computer tape or discs. Simple systems will not justify conversion to automatic and will remain manual, but where this is done, a number of problems must be overcome. The first is the wide variety of spatially distributed data making it difficult to reduce them to a common numerical coding. Cartographic data vary in scale, projection, and in the classification methods and categories used, while remotely sensed imagery varies in pixel size and wavebands. Both vary in date and level of precision. The difficulties of this stage and the amount of potentially valuable data are so great that some must generally remain in analogue or bibliographic form.

It may be necessary to edit and generalize the data from the database before applying spatial processing operations. Difficulties arise at this stage because of a lack of commonly agreed guidelines for such editing and generalization. Thus, unknown forms of data reduction or generalization may have been applied before the data are received into the system. Even where the algorithms used are known, the effect on accuracy depends on local characteristics of the data, for example whether a line is straight or sinuous (Jackson, 1985).

14.4.3 *Data storage*

This is the stage in the system where all data are brought together and made accessible to a user. It includes not only the mechanics of classifying and filing terrain information in all forms, but also its manipulation so that it can be retrieved quickly and easily. This manipulation can be manual or automatic. Manual systems use simple filing. Automatic systems use the data on tape or discs. It is generally necessary to retain separate vector and raster databases because of the practical difficulties of combining them. Raster data, e.g. from satellites, is held as a library of imagery and magnetic tapes and indexed according to geographical area, sensor, cloud coverage, image quality, etc. Vector data are more structured and exist in smaller quantities, and it is generally necessary to split coordinate from attribute data in a pair of linked file structures.

14.4.4 *Data processing*

This stage processes the data into a form which is translatable for the user. Manual systems, discussed in section 15.1, may use punched cards to arrange and organize the data. Computer methods have developed around the two formats for processing spatial data: vector and raster. In vector format, storage is by points, lines, and areas, each defined by pairs of x–y coordinates; in raster format, by pixels. GIS based on mapping have employed vector-based techniques, while those involving remote sensing have used raster systems. Integrated spatial processing has been restricted by this difference.

The options in achieving integration are to use an 'all-vector', an 'all-raster', or an 'integrated' approach (Jackson, 1985). The all–vector approach transforms raster data into vector format, smoothing to eliminate noise. This is usually achieved by classifying the raster image into land classes and transforming their boundaries into vector form. The method has the advantage of compactness and is readily comprehended since it deals with lines, points, and polygons, which is the way the ordinary user conceptualizes spatial data. It has the disadvantage that raster data can be used only after the image processing has been completed.

The all–raster approach incorporates the vector data into the actual image processing. This facilitates interactive processing, notably speeding overlay operations and area calculations. It economizes skilled manpower and requires less powerful vector processing hardware. It does, however, have serious disadvantages. Holding large-scale map data in raster form tends to be very inefficient. Large-scale UK Ordnance Survey maps, for instance, would only use 1 per cent of raster cells. Not only is this very wasteful, but more importantly, it creates serious problems when one is forced to reduce scale to operate at the pixel level. This is because the raster format has lost all the structure held in the vector mode, e.g. size of parcel, containment, line length, neighbours, etc.

The ideal solution is an integrated system. Computer software now exists to allow the necessary registration and integration of encoded vector data from maps with raster satellite imagery (United Kingdom NERC, 1983). After analysing the relative advantages of the two structures, Burrough (1986) concludes that their combined use as components of GISs will increase in the future as demands for high-resolution compact data structures grow and the power of flexible data analyses increases.

14.4.5 *Data output*

Data output is the stage at which information is provided for the user, and its format is governed by his requirements. Normally, it is in the form of a report, maps, annotated remotely sensed imagery, and bibliographical references.

The procedure in manual systems is to use the index to locate the relevant item numbers and then to interpret and collate the information required for issue. There will also sometimes be a need to make predictions in unknown areas from analogous known areas. The index must therefore classify the land systems and land facets in such a way as to identify the degree of analogy between them, so that the user will receive data on terrain units as similar as possible to those in his area of interest.

The trend in computerized systems is increasingly to involve the user at the processing stage, so that he can interact with the system. They must therefore be 'user-friendly'. In computerized systems, the software will contain the computer algorithms and packages for undertaking specific operations such as statistical tests, modelling, classification, change monitoring, and designing the map or image output. The best approach is to maintain a facility for guiding the user around the system and helping him to make appropriate and maximal use of its facilities while enabling him to incorporate his own professional knowledge into the work he is doing. Where the problems are especially complex, they are often soluble if the user is helped to approach them in an incremental fashion. At least, he would be enabled to manage the data so as to avoid major errors or bad cartographic design.

14.4.6 *Data use*

This represents the application of information from the GIS to actual land problems and includes the subsequent monitoring of results. An important component is the feedback from the user about subsequent changes in the project area to refine and improve the GIS. This is a stage that is often neglected but it is a vital input because it is vital in evaluating failures and successes so that land practices can be improved.

Different stages will be carried out by different workers and organizations, although it is also important to have specialists who can provide continuity through all stages. Stage 1 is mainly a survey and research operation and is done by organizations of this type. Stages 2–5 mainly involve data and image processing and require specialists in these fields. Stage 5 requires editing, communications, and public relations skills and brings in the specialist user to a limited extent, while stage 6 depends mainly on this user to make continuous observations of land properties after the completion of development projects and give feedback to the GIS.

14.4.7 *Management*

The system will tend to be operated from a geographic information centre of the type described in Chapter 15. Its management requires an administrative structure. The preceding section has outlined the specialist technical skills involved, but there are several aspects which require further expertise.

179

First, general supervision is required of the whole process to ensure that the data input is relevant to the required output and to coordinate the stages of the process. Second, at the acquisition stage it is important to maintain liaison among the members of the research team, especially between those at headquarters and those in the field, by arranging regular communications and opportunities for discussion. Third, liaison must be maintained with those who provide and those who use the data, to keep pace with the changing needs of the latter and the changing capabilities of the former. Fourth, contact must be kept with governmental, educational, and commercial organizations to publicize the contribution that terrain intelligence can make to their respective fields and to keep them informed of its developments. Finally, management is required of all the financial aspects: payment of staff, purchase of equipment, overheads, income received, etc.

Further reading

Burrough (1986), Howard and Mitchell (1985), Jackson (1985), Jackson and Mason (1986), Meier (1965), Peuquet (1984), Tomlinson *et al.* (1976), Townshend (1981).

15

Data processing and geographical information systems: methods

Processing of geographically referenced information by manual methods has generally been based on grid squares or landscape units and computerized systems have been based on grid coordinates.

15.1 Manual methods

15.1.1 *Grid squares*

Grid squares are normally used only where data are relatively sparse per unit area. The US Army Headquarters Quartermaster Research and Engineering Command use 1° quadrangles of latitude and longitude for storing numerical climatic data (Anstey, 1960).

On a finer scale, Childs (1967) suggested relating data about landscape and buildings in Britain to 'cubic–micro-regions'. These are based on the calculation that the biosphere extends approximately to 4 km in altitude and depth below sea-level and that a surface area of 8 × 8 km forms a useful packet. This packet is subsequently divided into 2 × 2 km 'grid units' and these into 1 × 1 km 'component units'.

The aerial photograph can be regarded as a type of grid square. Its use as a vehicle for data storage has been developed by the French journal *Photo-Interprétation* published by Editions Technip, whose object is to make available a comprehensive library of interpreted aerial imagery representing as many different types of landform, vegetation, human settlements, etc. as possible.

Each bi-monthly issue of the journal consists of about six portfolios of such photographs annotated according to a standard legend. Each portfolio interprets a single aerial photograph on four interrelated sheets. Two of these reproduce the photograph in stereo, one is a transparent interpretive overlay, and one a verbal description, as illustrated in Fig. 15.1. Both of the sheets with aerial

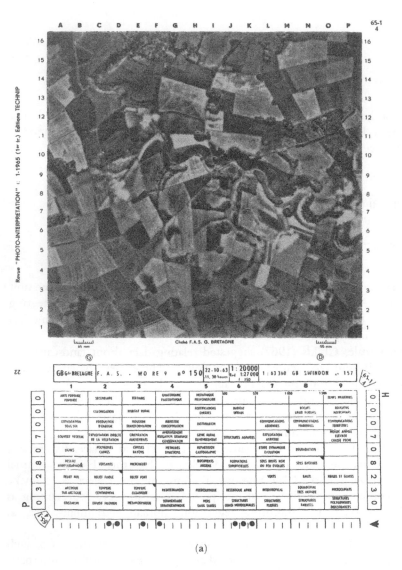

(a)

Fig. 15.1 The Geotechnip method for information storage on terrain: Photo Interpretation portfolio no. 65–1 no. 4 January–February 1965, showing the soils, physiography, and cultural patterns of part of the English Cotswolds

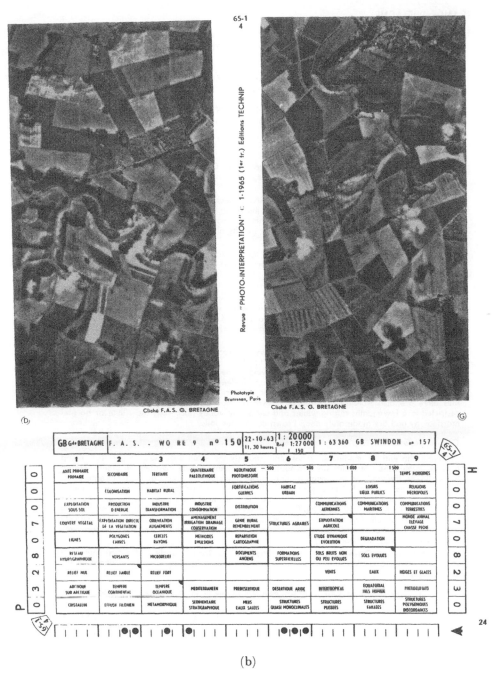

Fig. 15.1 (a) Photography   (c) Verbal description. (R. Webster)
        (b) Stereo pair   (d) Transparent interpretive overlay

GRANDE-BRETAGNE

MISSION FAS WO RE 9   Cliché n° 150 du 22/10/1963 (11 h 30) — Échelle 1:20 000 (réduction 1:27 000 environ)
focale 150 mm   Carte 1:63 360 (1 inch/mile)   SWINDON n° 157

THE SOILS, PHYSIOGRAPHY AND CULTURAL PATTERN OF PART OF THE ENGLISH COTSWOLDS

These photographs show a small portion of the English Cotswolds region around Eastleach, (1° 40'W, 51° 45'N), Gloucestershire. The present climate is typical of South Central England. Mean annual rainfall is approximately 700 mm.

Physiography.

The country rock is Middle Jurassic Limestone, mainly Great Oolite, which is thinly bedded, close jointed and, particularly near the surface, finely fragmented or rubbly. The dip is very gentle, about 1° towards the South-East.

The dominant landscape feature is the extensive plateau about 120 m above sea level, gently sloping and more or less parallel to the dip of the strata. This is dissected to between 30 and 40 m by the sinuous valley of the River Leach, which for most of its course is usually dry. Several smaller valleys, which begin as gentle depressions in the plateau and are now permanently dry, join the main valley.

The landscape took its present form during the Pleistocene period in a periglacial regime. The River Leach must then have been a large river probably fed by melting snow, by extremely heavy rains or a combination of both, incising its meanders into the limestone strata.

Soils.

The valley is floored by alluvial gravel of this period with a covering of finer material more recently deposited. These later became filled to a depth of one to several meters with solifluction material, soil and limestone rubble (locally known as « head ») derived from the surrounding land.

The soils on the plateau are medium to heavy textured, well drained, brown calcareous soils with much limestone rubble below 50 cm. On the steep slopes the soils are shallower, tending towards rendzinas in character, but variable owing to the varying nature of the strata on which they lie. The permanently dry valley bottoms exhibit deep well-drained brown calcareous soils on head. On the main valley floor ground water gleis are associated with the permanent stream and there are also isolated occurrences elsewhere. However, usually where stream flow is restricted to a few months during winter and spring the valley floor soils show little or no glei morphology and are moderately deep brown calcareous soils.

Cultural pattern.

From the end of mediaeval times until recently the area was used largely for sheep grazing. Indeed the Cotswolds were famous for their sheep. Recent advances in plant nutrition have allowed the land use to change to more profitable cereal growing as the main enterprise. But this change has not led to any appreciable changes in field patterns : the large regular fields of the sheep graziers suit the present day management for cereals on the plateau. The dissection slopes are usually too steep for cultivation and these remain either in pasture, or have been planted or allowed to revert to woodland. Similarly the main valley floor is scarcely worth cultivating because of its narrow tortuous configuration, even where the soils are well drained. It too is predominantly pasture.

Incidentally the Cotswold region was settled by the Romans early in the first millenium A.D. A Roman road, Akeman Street traverses the area ; its line is clearly evident on the photograph.

R. WEBSTER
Department of Agriculture
University of Oxford

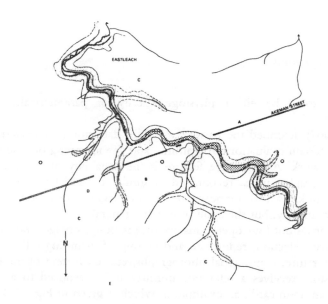

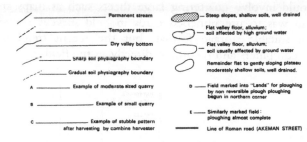

————————— Permanent stream		Steep slopes, shallow soils, well drained
—— —— —— Temporary stream		Flat valley floor, alluvium; soil affected by high ground water
—·—·—·— Dry valley bottom		Flat valley floor, alluvium; soil usually affected by ground water
—··—··— Sharp soil physiography boundary		Remainder flat to gently sloping plateau moderately shallow soils, well drained.
—·—·— Gradual soil physiography boundary		
A ————————— Example of moderate sized quarry	D ——	Field marked into "Lands" for ploughing by non reversible plough ploughing begun in northern corner
B ————————— Example of small quarry	E ——	Similarly marked field : ploughing almost complete
C ————————— Example of stubble pattern after harvesting by combine harvester	————	Line of Roman road (AKEMAN STREET)

d

photographs contain reference information and tabular keys. The latter are designed for edge-punching cards to facilitate filing, which can either be in terms of 'physical' or 'human' features of the landscape, depending on which is the main interest of the user.

Since these portfolios have been published continuously since 1961, several hundred are now available together with cumulative indices called *Tables analytiques* and distribution maps called *Cartes de répartition des photographies publiées*.

15.1.2 *Landscape units*

Data can be stored by either physiographically or parametrically defined landscape units.

McNeil (1967) described the basis of a terrain data store, conceived as part of the MEXE terrain evaluation programme, and the following outline is based mainly on his work. A distinctive feature is the capability of using analogies between physiographic units, recognized as similar on aerial photographs, to predict data from known to unknown areas. The original needs which inspired the work were military, but the applications are general.

The store carries out four operations: input, indexing, storage, and output.

Input involves selection, reduction, and editing of incoming information in the form of literature, maps, aerial photographs, etc. Each item of information entering the store receives a reference number and is reduced to a standard format, called an 'item card', an example of which is given in Fig. 15.2. This is then photographed down on to microfilm for the 'fast access store' (see below). This would involve quartering large sheets such as maps and charts. The original plan was to use punched cards in field surveys convertible by computer into a suitable storage format, but the recent trend is towards carrying portable computers and recording directly in the field.

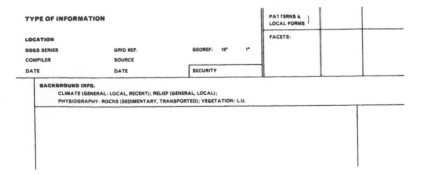

Fig. 15.2 An item card. (Design: P. H. T. Beckett)

Indexing is designed to allow quick random access to stored material. The simplest manual approach is known as the 'uniterm' system using optical coordinate matching. This works as follows: each item of information is analysed for its contents and these are reduced to a number of single terms called 'descriptors'. These are of four types: land units, including abstracts and local forms of both land systems and land facets, grid references, information about the form of items, i.e. map, photograph, table etc., and information about the content of items, i.e. fertility for cotton, trafficability for cars, etc.

To take an example, an item of information might be that wheat grows well on the Andover soil series on the Middle Chalk plateau in northern Berkshire. Now wheat, Andover soil series, the chalk plateau land facet abstract, the specific Berkshire land facet local form of this, the chalkland land system, the Berkshire Downs land system local form, and the grid coordinates, would each be regarded as descriptors. For each of these descriptors, a 'feature card' is, prepared. The simplest form of feature card is an array of numbered punchable spaces. Examples of commercially available types with 1 000 spaces and with 10 000 spaces are shown in Fig. 15.3. When an item relates to a certain descriptor, its number is punched on the relevant feature card. In the example quoted, the same numbered space would be punched out on the feature cards for wheat, Andover series, Middle Chalk plateau land facet abstract and local form. The advantage of using a range of terrain descriptors varying in scale and abstraction for the same item is to maximize the capability of matching different areas. If, for instance, an unknown area cannot be matched exactly with a known land facet local form, it is useful at least to find an analogous land facet abstract, or failing this an analogous land system, etc.

To find an item relating to two or more descriptors, one retrieves their feature cards and superimposes them on a light table. The places where the light shows through identify the numbers which coincide on the feature cards and indicate the relevant item or items.

Feature card retrieval can be operated manually or mechanically. If there are relatively few cards, they can be indexed and placed in alphabetical order. When their number exceeds a few hundred, it is simpler to sort them with needles inserted into sorting holes around the edges. These holes are punched according to a numerical combination whereby a maximum of two holes punched out in a field of four (representing respectively the numbers 1, 2, 4, and 7) can represent any digit (Casey *et al.*, 1958). When needles are inserted in a specified combination of positions, the desired cards drop into a tray. The operation of the needles can be mechanized by the use of a pneumatic actuator and valve system.

As has been indicated previously, the data store itself in which the items, as distinct from the feature cards, are kept, consists of two parts: a fast access store and a slow access store. The fast access store consists essentially of a microfilm data bank with retrieval, viewing, and printing facilities, of which commercial models are available. After processing on to a standard format and indexing, all items are then photographed down on to microfilm. The original material can

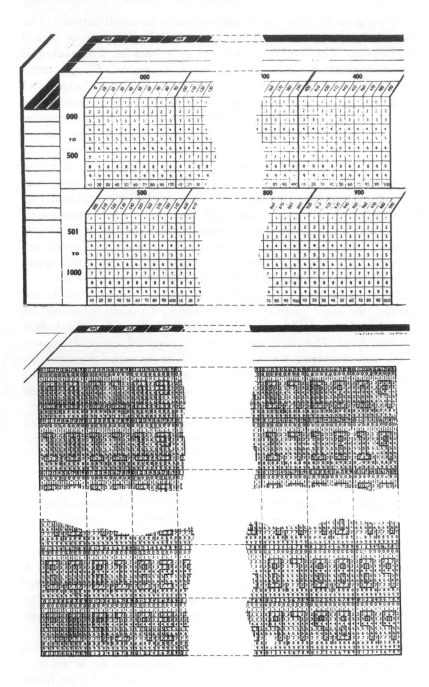

Fig. 15.3 Examples of commercially available feature cards: (a) with 1 000 punchable spaces (British–Vistem pattern); and (b) with 10 000 punchable spaces (Carter–Parratt pattern)

then be destroyed or placed in the slow access store. Updating can then be done either by blanketing the old information and giving the new the same number, or by simply adding the new information as an additional item and punching out the number of the old on a coloured transparent overlay. When search is made, the overlay can be automatically inserted. Superseded items will then appear in the viewer in white light while those not superseded still show the colour of the overlay.

The slow access store is the repository of all original material unsuitable for reduction or for which there is limited demand. It includes books, articles, maps, and remotely sensed imagery. Where the information is less important or inaccessible, or where space must be economized, the slow access store need only contain a bibliographic reference and a note about where the material in question can be found.

The output from the store is in the form of 'briefs', which can be classified as 'national briefs', 'general briefs', or 'specific briefs'.

A national brief provides an overall view of the terrain (relief, materials, soils, and vegetation) of national territory or major region. It consists of a general introduction to the area, a land system map, descriptions of every land system and land facet with aids to the identification of the latter. It may also include air photograph cover diagrams and topographic maps. The national brief thus aims to provide sufficient data for regional planning and, by permitting the identification of the land facets, the key to more detailed information from the store.

The general brief relates to a smaller area, perhaps equivalent to an English county or a single land system local form, and gives more detail than the national brief. It is accompanied by a large-scale topographic map, a land system map showing the area in question at a scale of about 1:1 000 000, and a representative stereo-pair of aerial photographs of each land system annotated to show the land facets, and selected so that the user may be able to identify the land facets anywhere within the land system. The general brief is not aimed at a specific purpose but at providing a generalized appreciation of an area down to land facet level.

The specific brief gives more detail for the specialist user before he has acquired local experience. The normal form of presentation is a table and accompanying transparent overlay for the aerial photographs showing the applicability of each land facet for the particular purpose required, e.g. water supply, trafficability, agricultural suitability. For instance, a specific brief for a water engineer would show the hydrological properties of the land facets, while a specific brief for a fruit grower would evaluate them in terms of orchard potential.

For more detail than this, it is desirable that the user develop his own site data store to guide further reconnaissance and field sampling to provide information from which he can derive his own 'project briefs', giving the fullest detail on each site.

A method of data storage by small composite parametric units which seeks to combine the advantages of the landscape approach with those of parametric definition of land units was explored by the Canadian army (Parry *et al.* 1968). A punched card is made out for each terrain envelope showing the parameters used to define it. The cards are then computer-scanned to find units with any specified attribute or combination of attributes. When a card is found whose attributes meet the required conditions, its number is stored in an array. Once the entire data bank has been scanned and the array formed, the subprogram to print a map of the area with the required attributes is called.

Today, computer spreadsheet systems such as LOTUS 123, dBase IV, and SMART are increasingly replacing punched cards for this type of data storage. Items and features in the fast access store can be numbered in the same way as on the cards, but are stored on discs and displayed on the computer screen rather than being handled manually. Whereas peekaboo indexing overlays punched cards to relate features to items in the fast access store, a computerized system links them by numerical coding. The slow access store remains manually operated, but is accessible from the computerized fast access store.

Local forms and abstracts of both land systems and land facets are categorized as features in the computerized, as in the manually operated, store. Because of the importance of recognizing analogies between different land areas, the classification and coding of all land units must be sufficiently detailed and specific to lead quickly to aerial photographs in the slow access store which can be visually matched. To take an example, it may be important to understand how karstic processes are undermining a road in southern Africa. The data storage system must not only identify usefully analogous terrain units elsewhere in southern Africa and also in say South America and Australia, but must also indicate others whose biophysical characteristics may be less closely analogous but where engineering experience may be greater or more relevant. This matching process requires an indexing of features and items which can give access to hard copy aerial photographs and written records of all the areas involved.

15.2 Computerized methods

Computerized data processing and storage methods are indexed by grid coordinates. These have the advantage that data can be quickly transferred from maps by scanning them with electronic position increasing devices, and feeding the information automatically into storage in digital form. They have been reviewed by Tomlinson (1984).

15.2.1 *Single-purpose systems*

Single-purpose systems can be national or international. National schemes can relate to different land use needs, but most progress has been made for soils. A number of countries have developed such systems, including the Netherlands, Denmark, Canada, and Yugoslavia. The Netherlands Soil Survey Institute, for instance, has a commercially available, minicomputer-based, turnkey graphics system geared to user requirements (Bie, 1975; Sadovsky and Bie, 1978; Burrough and Bie, 1984). The Danish system encompasses not only soils but also a forest inventory (Denmark: Ministry of Agriculture, 1982). An example of a scheme which emphasizes crop yield data under different levels of management is the Comprehensive Resource Inventory and Evaluation System (CRIES) of Michigan University (1979). This subdivides the landscape into units at different orders of magnitude: larger-scale resource planning units (RPUs) and smaller production potential areas (PPAs), on which data are held on environmental factors and crop yields under different management levels. These are cross-referenced with administrative areas. The output gives crop choices and likely costs and yields for each crop and area.

There are smaller-scale studies which include the integration of remotely sensed and mapped data. Examples are those on the distribution of agriculture on soils of different qualities (Cowan *et al.*, 1976) and on temporal changes in land characteristics (Schlesinger *et al.*, 1979).

Some systems are oriented more towards engineering uses. Hicks (1977) and ESRI (1979) deal with residential and recreational land suitability, and the UK Transport and Road Research Laboratory has developed its own microcomputer-based image processing and database management programme for road planning (Thomas and Stewart, 1985). International schemes include those for soils (Bie, 1975; Kloosterman and Dumansky, 1978) and for climate and topographic factors affecting military operations (Anstey, 1960; Pearson, 1979).

15.2.2 *Multi-purpose systems*

The Canada Land Data System, also known as the Canadian Geographical Information System, pioneered a multi-purpose integrated approach which entered routine use in 1968. It is in two parts: a data bank and a set of procedures for manipulating the data. The data bank can accept maps, of which there will be up to 30 000 for the country, and items of data related to grid-referenced points or polygons. A group of programs to convert mapped data into computerized records, known as the Polygon Information Overlay System (PIOS), is used to measure any data in the bank relating to map areas or line lengths or to count point frequencies. Within any given boundary the

191

area can be calculated and different types of data compared. Locations can be found with any specific characteristics. Also included is a capability for adding a reliability factor to all information and for continued modification and updating. Finally, alphabetical and numerical data can be printed out or plotted graphically. The system made important early contributions to general GIS procedures by introducing new techniques for the compact storage of boundary data and the rapid comparison of one map with another (Tomlinson, 1968; Tomlinson *et al.*, 1976; Thie *et al.*, 1979). Its value, however, is necessarily limited by the amount of numerical point data which can be fed into the system, and it does not easily allow for the uses of remote sensing to recognize analogies between areas.

Work in the UK began with the Oxford system of automated cartography (Mott, 1967). It included the development of a prototype instrument for producing maps automatically from data banks of individual 'libraries' by Dobbie McKinnes of Glasgow in collaboration with the Experimental Cartography Unit of the Royal College of Art. Each library contained information on one type of topic, i.e. represented one data plane. The library titles were 'framework' (i.e. the basic map information), hydrology, surface cover, geology and soil, climate, buildings and built-up areas, temperatures, movement or flows, and anthropology. All information was kept on magnetic tape and the libraries were built up into a data bank. When a map is needed, the required information is fed into a computer, which can either pass it to the scanning arm of a 'coordinatograph' to scribe the map, or else transmit it telegraphically any distance to an intelligence subcentre.

The first output was an experimental version of the 1:63 360 Abingdon sheet of the British Geological Survey, using all possible sources of input including satellite photography. After that, developments were rapid and a survey published in 1985 showed that in the UK, 11 government bodies, 7 commercial firms and 9 universities and polytechnics were concerned to an important degree with the development of GIS (Thomas and Stewart, 1985). Among the most important were the National Remote Sensing Centre (NRSC), the Ordnance Survey, the Natural Environmental Research Council (NERC), the Institute of Terrestrial Ecology (ITE), and the Meteorological Office, whose various approaches have been reviewed by J.A.T. Young (1986).

Among these, the work most related to terrain evaluation is that by NERC, which is actively researching the conceptual design of a GIS. It holds a wide variety of spatially referenced data, accumulated over many years, about the natural environment, so that it now has over 50 databases, with a variety of software systems. This has led to the establishment of separate centres for geological, oceanographical, ecological, and biomass data sets. The trend is today towards improving the computing facilities, developing digital data sets with on-line retrieval, and increasingly incorporating remotely sensed data.

Popular interest is served by the BBC 'Domesday' project to mark the 900th anniversary of the original survey. The plan, costing £2.5 million, encapsulates several million pages of information describing the UK, its environment, people, society, and organizations, in the late twentieth century. The basic information includes a 'people's data base', consisting of locally collected information, probably including land use, about small areas such as 1 km grid squares, photographs referenced by geographical position, population statistics, and a detailed gazetteer. All the data are available to users on two computer discs, so the service will be both convenient and cheap (Rhind, 1985). The impetus of these advances has led to a complementary development of GIS software and spatial algorithms for educational purposes (Gardner, 1985).

Developments in the USA have been the most extensive, and have been reviewed by Smith (1985) and Jackson and Mason (1986). The main agencies have been government departments, e.g. the US Geological Survey, the Department of Defense, NASA, and the CIA; universities, e.g. Maryland, California, Rochester, Stanford, Carnegie-Mellon, and Buffalo; and the Environmental Systems Research Institute (ESRI) at Redlands, California (ARC/INFO). Soil survey organizations use microcomputers to store data on the soil pedons of the USA and have developed a system for the automated modular description of soil characters (Reybold and Petersen, 1987). Again, the US Geological Survey, for example, is developing a cartographic database covering all information from 55 000 '7 minute' map sheets covering the entire USA (Peuquet, 1984). A conservative estimate of the digital data volume involved is 10^{15} bits for a single data layer, such as topography or hydrology (Mason and Cross, 1985). The central emphasis for these and other agencies has been towards the development of advanced computational techniques.

The number of agencies involved and their different emphases and areal coverage has stimulated research into methods of coordination. One experiment explored the extent to which it is possible to combine four different local systems: for Minnesota, New York, Tennessee, and the Canadian into the same database. All four included land use data by square grid cells, but they differed widely in the classifications used and in the size of the cells, ranging from 40 acres (Minnesota) to 1 km^2 (New York). The experiment showed that such combinations could be achieved but concluded that the use of grid squares as areal units lost linear features when databases were overlaid, and that polygons were more accurate (Tomlinson *et al.*, 1976).

Considerable research has been devoted to the development of a knowledge-based GIS (KBGIS) using a hierarchical data model (quadtrees). One of the largest and most advanced programmes has been at the University of California at Santa Barbara. It tackles the problems of reducing response times and storage volumes, and increasing the range of applications possible. It combines heuristic search procedures, which speed data retrieval by bypassing irrelevant data with a 'learning' capability for refining the system. Another programme employing KBGIS is being undertaken at Carnegie-Mellon University. This concentrates on the design of a KBGIS for map-guided image interpretation. The system, known as MAPS (Map Assisted Photointerpretation System) is a large integrated database consisting of digitized maps, photographs,

193

and three dimensional descriptions of a Washington, DC test area (Jackson and Mason, 1986).

In other countries, similar research programmes are being undertaken by the Austrian Graz Institute for Image Processing and Computer Graphics (Ranzinger and Ranzinger, 1984), the Swedish National Defence Research Institute (Jungert *et al.*, 1985), the French Université Pierre et Marie Curie (Bouille, 1984), and mainly concerned with soil data, the Australian CSIRO Division of Water Resources (Davis and Nanninga, 1984). In Indonesia, there is a system for managing information on the land resources of parts of the country designated for transmigration from more congested districts (F. Dent, 1980, private communication). In Chile, data from a variety of existing thematic maps are digitized and stored for simultaneous or selected retrieval according to their geographical coordinates. Classified research is being undertaken by many industrial contractors which should ultimately accelerate the commercial availability of such systems.

In the long run GIS hold the key to the solution of many information management problems, and are likely to become the accepted method of operation in the future. They are mainly limited today by their high operating costs, the difficulty of demonstrating their value, and because potential users are often unaware of the services they can provide.

15.3 Geographic information centres.

Output to the user from a GIS will logically come from a geographical information centre, which if worldwide could become a 'world data centre' (e.g. ICSU, 1979). Comparable centres, known by such names as 'information analysis centres' or 'specialized information centres', exist for other topics (AGARD, 1970). They are now proliferating and they meet many needs both nationally and internationally, ranging from relatively narrow fields such as heat transfer and fluid flow in materials (Cousins, 1970) and maritime pollution (Langston, 1970) to wider subjects such as glaciology, vulcanology, and oceanology.

Further reading

Bie (1975), Burrough (1986), Burrough and Bie (1984), Cowan *et al.* (1976), ICSU (1979), Jackson and Mason (1986), Kloosterman and Dumansky (1978), McNeil (1967), Michigan University (1979), Mott (1967), Parry *et al.* (1968), Pearson (1979), Peuquet (1984), Rhind (1985), Schlesinger *et al.* (1979), Thie *et al.* (1979), Tomlinson *et al.* (1976), Tomlinson (1984), J.A.T. Young (1986).

Part IV

Display, reporting and mapping of terrain data

16

Geomorphological and terrain mapping

16.1 Geomorphological and soil mapping

Geomorphological maps have developed from geological maps. These have traditionally been based on the stratigraphy of solid and drift deposits, although specialized maps show such features as tectonics (Unesco, 1968), structure (Demek *et al.*, 1972), metamorphism (Unesco, 1974), Quaternary forms (Unesco, 1967), and engineering geology (Unesco, 1976). Their development was reviewed by Saint Onge (1968). Although first suggested by Passarge in 1919, they only began to be produced in a number of countries after about 1950. At first they emphasized the genetic aspects of landscape, but more recently, emphasis has shifted to serving social and ecological requirements. Among the most notable schemes are the French series under the guidance of Tricart (1965b), and that of the Netherlands by Verstappen and colleagues (Verstappen and van Zuidam, 1968; van Zuidam and van Zuidam-Cancellado, 1978). Westermann (1970–) has produced a variety of colourful and imaginative terrain maps of West Germany at 1:25 000 with examples from each of its seven form regions, such as the Norddeutsches Flachland, Mittelgebirge, and Alpenvorland. The Geological Survey of Japan have since 1980 published graphic slope classification maps, also at this scale.

In order to harmonize the wide differences in practice between the different mapping schemes, the International Geographical Union in 1960 set up a subcommission to standardize legends and ensure that there was an agreed scheme for showing dimensions and slope values (morphometry), age (morphochronology), appearance and surface materials (morphography), and processes (morphogeny) of each form. A schematic illustration showing these separate types of mapping is shown in Fig. 16.1. The size of the problem of combining these themes can be seen from the fact that the Russian scheme, which gave attractive maps and was the most comprehensive, had over 500 items on the legend, but yet lacked any slope data (Demek *et al.*, 1972).

Verstappen and van Zuidam (1968) produced a flexible adaptation of previous schemes which provided codes for general purpose 'standard'

197

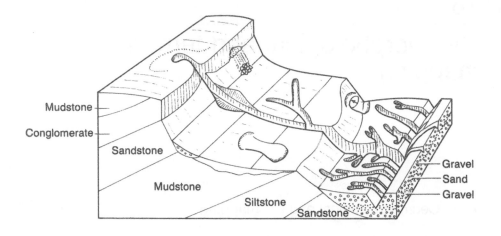

Mudstone
Conglomerate
Sandstone
Mudstone
Siltstone
Sandstone
Gravel
Sand
Gravel

A. MORPHOLOGICAL/MORPHOMETRIC MAP

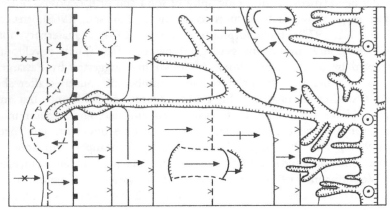

Morphological mapping symbols

⌄⌄⌄ Convex break of slope	4 2	Concave unit
△△△ Concave break of slope	2 3	Convex unit
⌄⌄⌄ Convex change of slope	— 30 —	Contours in metres
△△△ Concave change of slope	120•	Spot height
10 → Slope direction and angle	⊙15	Depth of incision
■■■ Cliff > 45°		
Convex and concave breaks of slope in close association		

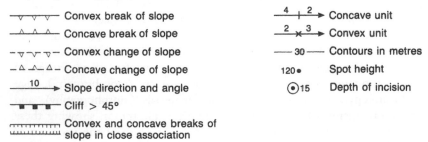

Fig. 16.1 Schematic diagrams illustrating the different principles of geomorphological map representation for the same land surface: Reproduced by permission of the Geological Society Working Party, 1982.

B. MORPHOCHRONOLOGICAL MAP

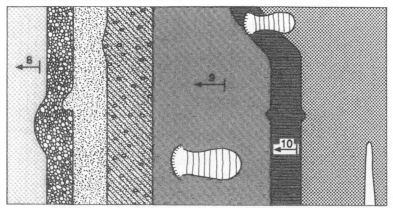

Bedrock succession

Planation surface
– Mid- Tertiary

Conglomerate

Sandstone

Mudstone
(highly weathered)

Siltstone
– Late Pleistocene valley

} Retreating scarps pediment and associated gravels Early-Middle Pleistocene

Unconsolidated sediments

River terrace and infill
– Devension

River sand – recent

Angular boulders
– intermixed
– recent gravel and sand

Superficially disturbed

Landslips – active

10 ⊣ Dip

C. MORPHOGRAPHIC MAP

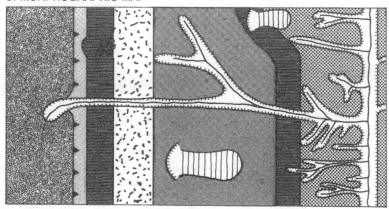

Planation surface

Cuesto scarp face

Rock wall

Scree – debris slope

Pediment

River terrace and valley infill

Incised valley side slope

Landslides

Spring

Permanent stream

Waterfall

Major gully

Minor gully

Fig. 16.1 continued

D. MORPHOGENETIC/DYNAMIC MAP

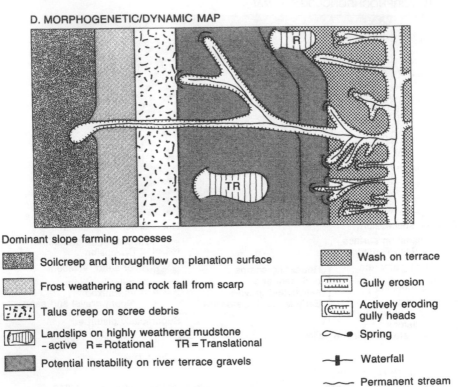

Dominant slope farming processes

Soilcreep and throughflow on planation surface

Frost weathering and rock fall from scarp

Talus creep on scree debris

Landslips on highly weathered mudstone
– active   R = Rotational   TR = Translational

Potential instability on river terrace gravels

Wash on terrace

Gully erosion

Actively eroding
gully heads

Spring

Waterfall

Permanent stream

Fig. 16.1 continued

geomorphological maps which can be considered the small-scale equivalent of the 'Sheffield' method for landforms previously considered. These had a legend of 469 symbols divided among 12 broad lithogenetic types of landscape – volcanic, fluvial, etc. They also provided codes for special purpose 'morpho-conservation' maps for engineering purposes, and 'hydro-morphological' maps for hydrological purposes. Figure 16.2 reproduces a few of the subclasses and their symbols used in the latter. Van Zuidam and van Zuidam-Cancellado (1978) show how aerial photographs can be used to assist such mapping. Generalized small-scale land surface form maps have also been produced of North and South America and of the USA by Hammond (1954, 1964) and of the world by Murphy (1968). These are shown in Figs 16.3–16.6.

Soil mapping is considered in more detail in Chapter 21. It differs from geomorphological mapping in being based primarily on soil profile taxonomy rather than on landscape conditions, although some geomorphological criteria are normally included in legends. It is carried out in almost all countries. For instance, the Soil Survey of Great Britain publishes maps at scales of 1:25 000,

1:63 360, and 1:250 000 and the Netherlands Soil Survey Institute at 1:25 000 and 1:50 000. In the USA, soil maps are on a county basis at scales generally between 1:20 000 and 1:50 000. In most countries, project soil surveys are also executed for specific local needs such as land reclamation, rural development, or urban planning. Because these are concerned only with restricted areas, they may incorporate factors of local importance such as slope angle or drainage conditions which might not appear on the national maps.

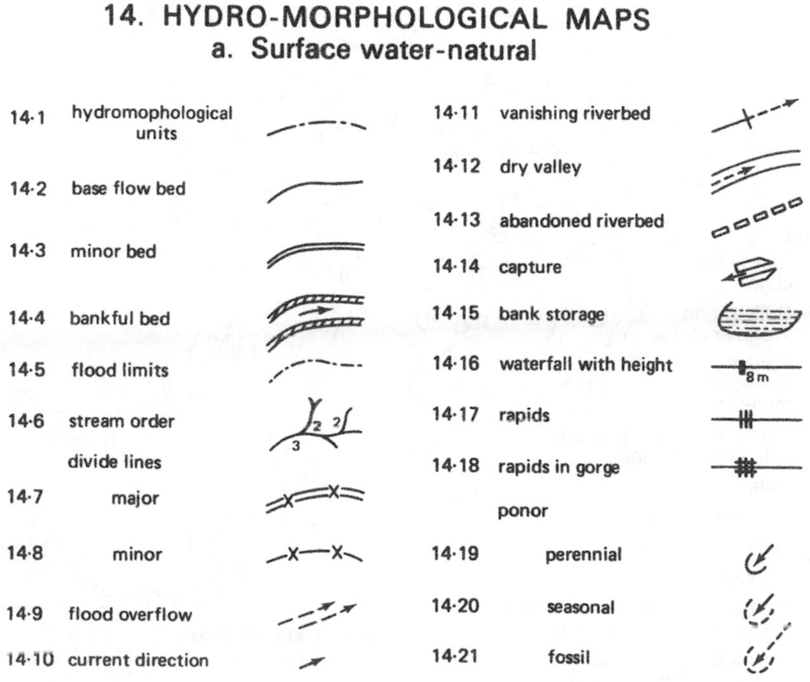

14. HYDRO-MORPHOLOGICAL MAPS
a. Surface water-natural

14·1	hydromophological units		14·11	vanishing riverbed	
14·2	base flow bed		14·12	dry valley	
			14·13	abandoned riverbed	
14·3	minor bed		14·14	capture	
14·4	bankful bed		14·15	bank storage	
14·5	flood limits		14·16	waterfall with height	
14·6	stream order		14·17	rapids	
	divide lines		14·18	rapids in gorge	
14·7	major			ponor	
14·8	minor		14·19	perennial	
14·9	flood overflow		14·20	seasonal	
14·10	current direction		14·21	fossil	

Fig. 16.2 A sample page from the ITC legend for geomorphological mapping.
(Source: Verstappen and van Zuidam, 1968)

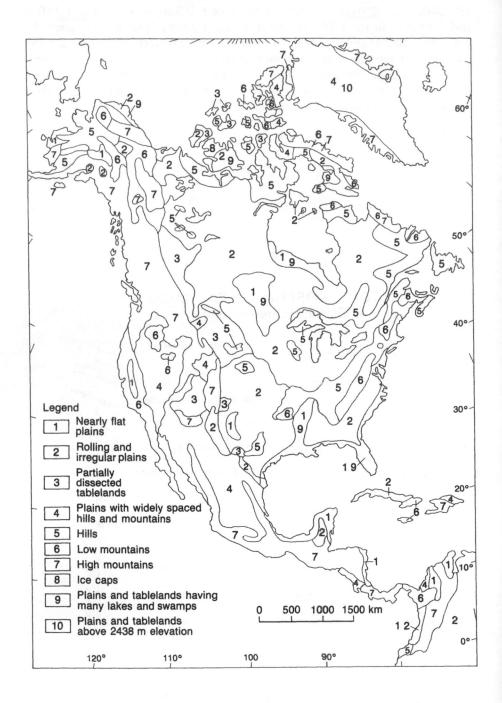

Legend

1	Nearly flat plains
2	Rolling and irregular plains
3	Partially dissected tablelands
4	Plains with widely spaced hills and mountains
5	Hills
6	Low mountains
7	High mountains
8	Ice caps
9	Plains and tablelands having many lakes and swamps
10	Plains and tablelands above 2438 m elevation

0   500   1000   1500 km

Fig. 16.3 Terrain types of North America. (Source: Hammond, 1954)

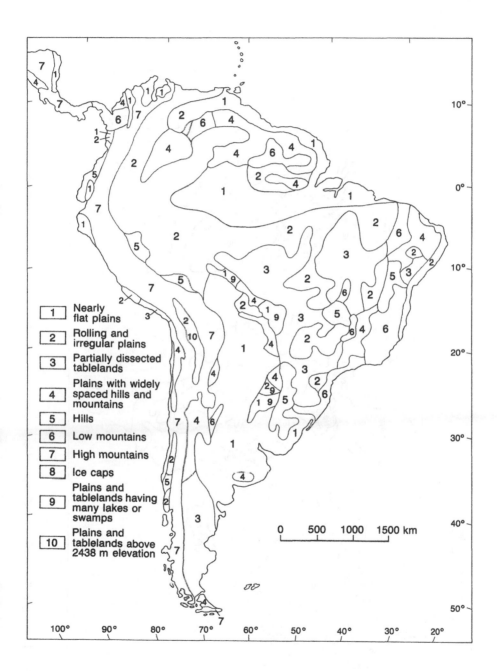

Nearly
1 flat plains

2 Rolling and
irregular plains

3 Partially dissected
tablelands

Plains with widely
4 spaced hills and
mountains

5 Hills

6 Low mountains

7 High mountains

8 Ice caps

Plains and
9 tablelands having
many lakes or
swamps

Plains and
10 tablelands above
2438 m elevation

0   500   1000 1500 km

Fig. 16.4 Terrain types of South America. (Source: Hammond, 1954)

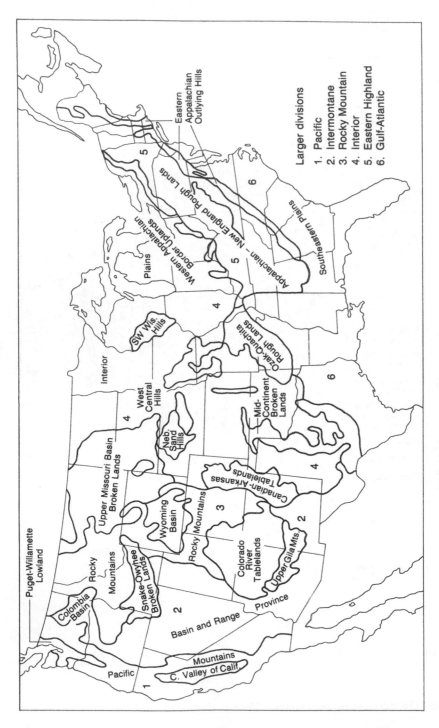

Fig. 16.5 Summary of the principal subdivisions of land-surface form in the 48 states, USA. (Source: Hammond, 1964)

LEGEND:
The 1st letter indicates geological age: A Alpine, C Caledonian and Hercynian,
   G Gondwana, L Laurasia, R Rifted Shield areas, S Sedimentary cover outside
   Shield exposures, V Volcanic.
The 2nd letter indicates landform type: M mountains, W widely spaced mountains,
   T high tablelands, H hill and low tablelands, D depressions or basins, P. plains.
The 3rd letter indicates climate:
   i.   Ice caps at present. ••••••
   w. Wisconsin or Wurm glaciated areas. •—•—•
   g. Pre-Wisconsin, pre-Wurm and undifferentiated Pleistocene
       glaciated areas. — — — —
   h. Humid landform areas.
   d. Dry or arid landform areas.
Division between humid and dry landform areas. ⌒⌒
Major oceanic rift and fault lines. — —⸝— —
Continental shelf. xxxxx

Fig. 16.6 Landform map of the world. (Source: Murphy, 1968)

16.2 Terrain mapping

Terrain maps differ from geomorphological maps in that they emphasize the identification of simply outlined units which can be used for information storage and extrapolation via a comprehensive legend rather than the cartographic representation of a large variety of information. They aim, therefore, to be somewhat more inclusive by incorporating soil and vegetation, as well as geomorphological, data. Terrain maps can be physiographic or parametric. A useful combination is land system–type mapping with diagrammatically illustrated land facets whose attributes are tabulated in quantitative form in the legend. This combines a graphic and readily comprehensible framework relevant to a wide range of land uses with the maximum possible scientific detail (Mitchell, 1987). The mapping must also convey the dynamic and continually changing nature of the landscape. Although a picture at one point in time, it should incorporate a portrayal of morphogenesis and morphochronology by using different tones for different ages and 'active' symbols such as arrows, zigzags, and waves.

The main operational decisions are about scale and legend.

16.2.1 *Scale*

The scale is governed by the purpose of the survey and who will be the eventual users. It should be large enough to show all necessary detail but small enough to avoid an excessive number of map sheets. It also depends on the type of available remotely sensed imagery. Field data are compiled on to airphotos or base maps which are usually at about twice the scale of the final printed map, and the consequent scale reduction must be considered when planning the field survey. Fairly frequently, excessive detail is collected, which is wasteful of time because it cannot all be presented. On the other hand, in exceptional cases, with the newest aerial photography and using improved lenses, filters, and film quality, it has been possible to enlarge to a map scale as much as twice the original photographic scale.

The smallest mapped unit is influenced by the minimum thickness of a drawn line. This is between about 0.2 and 0.5 mm, which at a scale of 1:100 000 represents a band 20–50 m wide. The smallest practicable mapping unit is a square about 3 mm on a side or a strip 2mm wide, which would represent minimum areas of 9 ha and 200 m respectively at this scale, and proportionately more at smaller scales.

The proliferation of thematic maps at large and medium scales has probably increased rather than decreased the need for small-scale reconnaissance mapping such as that which satellite imagery provides at scales of 1:200 000 or smaller. There appear to be three reasons for this. First, large-scale maps will only cover limited areas and it may be economically necessary to generalize outwards from these to much larger areas, even to country-wide or international scales. Second, there is a need to extend terrain mapping to areas for which little or no thematic cover currently exists. An analogous example which shows the importance of such generalization is the FAO/Unesco *Soil Map of the World* at 1:5 000 000 (1974–) which integrates diverse national surveys produced at much larger scales on a worldwide basis. Third, there is an increasing demand for syntheses of complex local studies for broad-scale planning purposes. Satellite imagery, particularly Landsat, is now widely used for summary presentations of earth resource information covered by more detailed thematic mapping.

16.2.2 *Legend*

The legend is the key to the map and shows how its content is organized. It should be as clear and comprehensive as possible and with a minimum of complication. Land systems are the usual units shown, and accord with most other thematic classifications in being hierarchical. They are grouped into land regions, land provinces, etc. while their subdivisions are not mapped but included in the legend.

The basis is geomorphological mapping. This shows the smaller morphometric features of the landscape such as cliffs, slump features, springs, stream channels, etc. with graphic or diagrammatic symbols. These are overprinted on boundaries around areas of homogeneous surface materials or landforms such as limestone tablelands, river terraces, active dunes, etc., which are themselves enclosed within more inclusive boundaries around areas of differing geological ages, i.e. Devonian, Cretaceous, Eocene.

The Sheffield method for describing small landforms was discussed in section 3.5. It is appropriate for mapping at scales of between 1:1 250 and 1:63 360. The legend consists of the features shown in Table 16.1.

Each feature is marked on maps by a special symbol, some of which are shown in Fig. 16.7. Conventions are also supplied for representing boundaries which are themselves made up of complex micro-units.

Parametric legends, either in coordinate or multivariate form, are sometimes required, notably within homogeneous terrain not otherwise easily subdivided. These are 'closed' in the sense that they cover all the variation within each parameter used. For example, the subdivisions of a featureless alluvial plain might be based on overlaying numerical ranges of surface gradient, soil clay percentage, and pH.

Table 16.1 Micro-units of slope form (after Savigear, 1965)

Flat	Slope <2°
Slope	Slope 2°–40°
Cliff	Slope >40°
Facet	A plane surface in the landscape of any gradient
Segment	A curved surface in the landscape of any gradient; can be concave or convex
Irregular facet or segment	Facet or segment with surface irregularities too small to map at field scale
Micro-facet	Facet too small to be separately mapped
Micro-segment	Segment too small to be separately mapped
Morphological unit	Facet, segment, micro-facet, or micro-segment
Break of slope	Discontinuity of ground surface
Inflection	Point, line, or zone of maximum slope between two adjacent segments
True slope	Direction and amount of maximum surface slope of a facet or segment
Apparent slope	Direction and amount of slope measured in any other direction

16.2.3 *Boundaries*

Thematic map drawing is essentially a question of delineating natural units. Where such units merge gradually into one another, boundaries must come at the point of maximum change. Where the units are nodal, the boundaries must separate the outermost ecotones of neighbouring nodes. The exact location is often determined somewhat arbitrarily by using one dominating criterion in one place and a different one in another. The upper limit of a talus slope, for instance, against a steep mountain front may be defined on the basis of the sharp slope increase and the change to solid rock. The lower margin, however, may merge gradually into a detrital plain composed of finer materials or the more stratified alluvium of a river valley. Here the boundary may be a medial line between the two physiographic nodes, based on a maximum rate of change in such attributes as gradient, soil texture, or the degree of profile stratification.

The boundaries around land systems are also those of their constituent land facets. Because these are often vague or convoluted, cartographic simplicity frequently demands the inclusion of small parts of one unit within the boundaries of another. A common practice in the earth sciences is to accept that up to 15 per cent of any mapping unit may not represent the class for which it is named. Where only a single contrasting area is included the standard is higher, the USDA, for instance, requiring less than 10 per cent 'impurity' on soil maps under these circumstances.

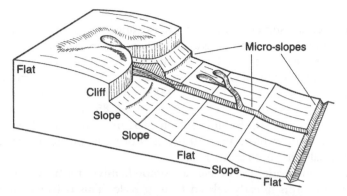

Breaks of slope between concave-convex associations should be emphasised by thickened lines

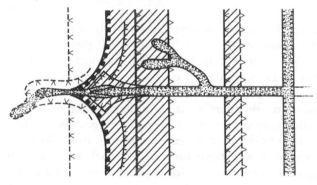

MORPHOLOGICAL MAPPING SYMBOLS

ʌ ʌ ʌ ʌ ʌ Angular convex break of slope

ᴠ ᴠ ᴠ ᴠ ᴠ Angular concave break of slope

ʌ ʌ⁻ʌ⁻ ʌ⁻ʌ Smoothly convex change of slope

ᴠ⁻ ᴠ⁻ ᴠ⁻ ᴠ Smoothly concave change of slope

▪▬▪▬▪▬▪ Cliffs (40° or more)

Slopes more than 5° and less than 40°

Areas of permanent or intermittent water flow or seepage

⊤⊤⊤⊤⊤⊤⊤ Minor angular convex and concave breaks of slope too close together to allow the use of separate symbols

⊤ ⊤ ⊤ ⊤⊤⊤⊤ Minor indefinite smoothly convex and concave changes of slope too close together to allow the use of separate symbols

Micro-slopes

Fig. 16.7 An illustration of the application of the 'Sheffield' symbols to some landscape patterns typical of a part of the southern Pennines, England. (Source: Savigear, 1965)

16.3 Map preparation

The base map for terrain survey should be an existing topographic map or controlled photomosaic because these give the required planimetric accuracy. The accuracy of this map is assessed in a different way from that of the thematic information that is shown on it. 'Horizontal accuracy' on topographic maps is assured when, at scales larger than 1:20 000, not more than 10 per cent of well-defined listed points are in error by more than 0.85 mm (1/30 in) and, at scales smaller than 1:20 000, by more than 0.51 mm (1/50 in). 'Vertical accuracy' is assured when not more than 10 per cent of elevations tested are in error by more than half the contour interval.

Attempts have been made to derive a comprehensive formula for all cartographic errors in terms linearly related to map scale. This is based on an allowable standard error d, defined as:

$$d = e^2/(n - 1) \tag{16.1}$$

where e is the error and n the number of observations (Robinson *et al.*, 1984).

If no suitable base maps exist, or if the scale of those available is considerably smaller than that required for final production, the only alternatives are either to attain the required accuracy by establishing expensive ground control or else to accept some error and portray the terrain features in diagrammatic or sketch form, adding information on their reliability.

Accuracy in thematic mapping differs in being less concerned with the precision of individual map positions than with the general truthfulness of the phenomena being portrayed. It sometimes has to use symbolism at the expense of the strictest cartographic exactness as, for instance, when boundaries to important properties have to be generalized across sites or areas that may have doubtful or aberrant values.

Unless computer-assisted methods are used, the usual procedure is to transfer the results of the field work on to stereoscopic pairs of aerial photographs or photomosaics. This information is then transferred to the base map via a transparent overlay on which the ground control points and the terrain features have been scribed. Alternatively, the photographs may be assembled directly as an 'uncontrolled', 'semi-controlled', or 'controlled' photomosaic and a separate overlay prepared for the terrain data. Place names and similar 'gazetteer' information are marked on to further overlays, and these are combined during printing to provide the final black and white or coloured map.

16.4 Map presentation

The basis of terrain mapping is in geomorphic, hydrological and soils mapping. The general method in all is to give the background human and topographic detail in black or grey and to overprint areal units in relatively unobtrusive area shades or colours. Once the track of boundary lines in terms of $x–y$ coordinates is established, the colouring or area shading can be performed digitally by computer.

In geomorphic mapping, conspicuous landforms are additionally overprinted in black (see Fig. 16.1). Satellite photographs or photomosaics may be used as the base map (Cochrane and Brown, 1981), landforms being emphasized by image enhancement. Although there is no standard code of colours for areal units, it is usual to use shades that suggest the origin or surface appearance of the parent rocks. Warm red or purple colours are often used for igneous or metamorphic rocks, brown for sandstones, cream for limestones, grey for shales, and black for organic materials such as coal or peat. Drift and alluvium are shown in paler colours, e.g. grey for gravel, yellow for sands. Subdivisions of these are shown by 'drop-outs', i.e. bars or dots left blank within the colour wash by graded amounts.

Geomorphic features are portrayed in graphic symbols whose boldness is related to their importance in the landscape. For instance, strong lines are used for mountain crests, toothed lines for escarpments, fan shapes for detrital cones and fans, nested lobes for mud and lava flows, arrangements of dots for dunes, circles with external rays for volcanoes, and with internal rays for depressions such as sink holes and dolines. Where they do not interfere with important thematic information, water features are shown in blue. They range from pecked lines for fingertip streams to washes for rivers and lakes. Examples of these and other symbols are shown in Fig. 16.8. In homogeneous areas where parametric mapping is used, different colours or black and white patterns are used for each attribute and their gradings shown with ranked cartographic drop-outs.

Hydrological maps show groundwater and the surface channel network. There are three main types: 'groundwater depth', 'hydrogeological', and 'geohydrochemical'. Groundwater depth maps usually show the subterranean contours of the water table. Hydrogeological maps have the same basis but add the distribution of the geological formations and their character as aquifers or aquicludes (Unesco, 1970). The extra complication makes it necessary for them to depend on colour for clarity. Because of the relation between soil type and the rate of rain infiltration to supplement the aquifers, the Soil Survey of England and Wales (1977) published a winter rainfall acceptance map of the country. Geohydrochemical maps show the nature and degree of mineralization of the water, generally overprinting contours of water quality on areal hydrogeological units. Their legends for different scales have been standardized internationally by Unesco (1975).

Soil taxonomy groups soils into series or associations on the basis of profile morphology, which are mapped by shading or colouring. Soil complexes are shown by fractioning the symbols of their component units. Subdivisions of mapping units based on factors such as topography and soil texture are represented by letter symbols or drop-outs. In large-scale mapping,

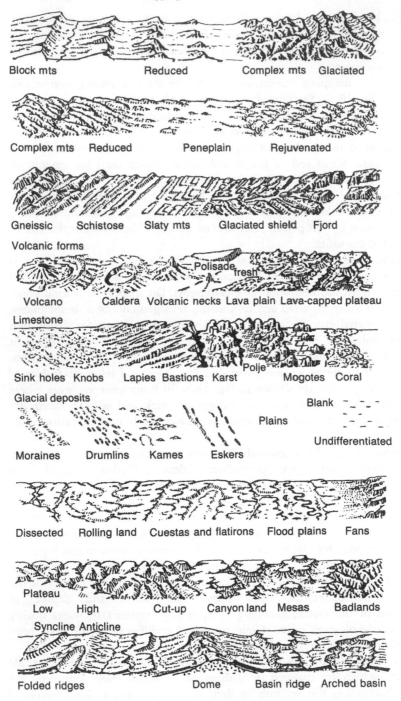

Fig. 16.8 Some graphic mapping symbols for landforms. (Source: Raisz, 1962)

miscellaneous classes such as urban land, beaches, quarries, etc. are outlined and labelled but left uncoloured. The main advances in the past three decades have been in the refinement and definition of soil attributes, the recognition of palaeosols and polycyclic soil evolution, and the greatly increased geographical coverage of detailed soil mapping. This has been synthesized at a scale of 1:5 000 000 in the FAO–Unesco *Soil Map of the World* (1974–), which has introduced a new nomenclature, correlative classification, and standardized mapping code for world soils.

The Unesco Man and Biosphere Programme (Journaux, 1987) has developed a methodology and legend for integrated environmental mapping. It attempts selectively to combine data on simple features or processes with assessments of land capability for defined purposes. The approach is through 13 case studies representing different countries, environments, and population densities, for each of which a map is prepared. The emphasis is on graphic and sophisticated cartographic presentation, using satellite imagery wherever possible.

The classification and mapping of vegetation physiognomy follow the system of Dansereau (1958), widely used and advocated by Colinvaux (1973) and Vink (1983). The essence of the method is the definition of all plants in terms of six different parameters: life form, size, leaf function, leaf shape and size, leaf texture, and ground coverage. A simplified version is shown in Fig. 16.9.

CODE AND LEGEND FOR DEFINING VEGETATION PHYSIOGNOMY

1. Life form
   T. O trees
   S. ♀ shrubs
   H. ▽ herbs and grasses
   M. ∩ mosses and other bryoids

2. Size
   1. more than 25 metres
   2. 10–25 m
   3. 8–10 m.
   4. 2–8 m.
   5. 0·5–2 m.
   6. 0·1–0·5 m.
   7. 0–0·1 m.

3. Function of leaves
   d. ▭ deciduous
   e. ▦ evergreen
   s. ▨ succulent
   l. ▥ leafless

4. Leaf shape and size
   n. ⌒ needle
   g. 0 grass
   m. ◇ medium, small (up to 20 cm²)
   b. △ broad, large
   v. ∿ compound

5. Leaf texture
   f. ▨ filmy
   z. ▭ papery
   x. ▬ hard and tough
   k. ▤ succulent

6. Coverage
   i. Discontinuous or interrupted:
       1. 0–20%
       2. 21–40%
       3. 41–60%
   c. Continuous:
       4. 61–80%
       5. 81–100%
   b. Barren or rare
   p. Tufts or groups

Fig. 16.9 Code and legend for defining vegetation physiognomy (after Phillips, 1965)

213

Sample areas representative of terrain units are chosen according to the size of the plant dominants involved. Squares with sides of 10 m are required for trees, 4 m for shrubs, 1 m for herbs and grasses, and 10 cm for mosses and lichens. The plants within the sample area are counted and represented on graph paper with seven subdivisions of the vertical scale, corresponding to the subdivisions of plant height, and with enough subdivisions of the horizontal scale to accommodate all the plants in the chosen sample area. A diagram is drawn by magnifying the life form, leaf shape and leaf size symbols to accord with the sizes of the plants they represent to give a completed structural chart of the type shown in Fig. 16.10. A more refined and specialized form of this structural chart system, with a larger number of classes for plant height and physiognomy was worked out for the US army by Addor (1963). A study for the Canadian army by Parry *et al.* (1968) related the stem diameter and spacing of the vegetation of Camp Petawawa, Ontario, to its penetrability by military vehicles. Only the spacing could be readily assessed from aerial photographs.

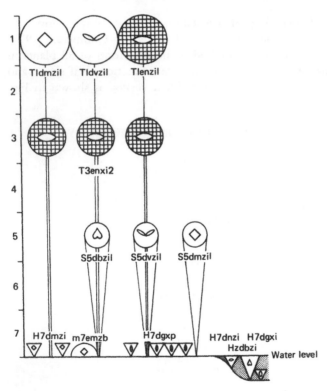

Fig. 16.10 Example of a vegetation structural chart by the method of Dansereau (after Phillips, 1965)

Land system mapping is simpler and cheaper than any of these methods, because its essential purpose is to economize by reducing mapping scale and substituting a locationally unspecific but highly comprehensive legend for one in which detail is represented cartographically. As with other forms of thematic mapping, the background human and topographic data are shown in black or grey and the land systems overprinted on this, usually in colour. However, because they are internally variable, for instance including areas both of a parent rock and its associated detritus, the choice of area shades or colours is more flexible. The land system boundaries generally coincide with significant geomorphic, hydrological, or soils boundaries, but where they do not, for instance where terrain areas do not coincide with drainage catchments, the boundaries of both should be shown. This should be on the main map, but if this is not feasible, the catchment boundaries should be recorded on a supplementary map.

Parametric classifications are shown on the map by giving different colours or black and white patterns for each attribute and showing their gradings with different cartographic 'drop-outs'.

Slope profiles can sometimes be better appreciated when translated from map contours or from parallax measurements on aerial photographs into cross sections by graphical methods. Relief can also be represented diagrammatically by 'area–height curves' and 'hypsometric curves'. The former are histograms showing the proportion of land in each inter-contour envelope, while the latter are cumulative graphs showing the proportion of the total area above or below any given altitude.

16.5 Aids to artistic presentation

Thematic maps should not only be clear and accurate but also make a good visual impact. The general layout, titling, etc. should be well balanced and comprehensive but without such a mass of detail that the artistic impact is lost. The background gazetteer information should be clear but unobtrusive by comparison with the land systems and other thematic data being shown. Colours should as far as possible suggest those which appear in nature, and use made of other forms of graphic presentation.

An important part of the land system method is its use of block diagrams in the legend. Principles to be used in their preparation have been described by Lobeck (1958), Schou (1962), and Lawrance (1966). The two vertical sides to the block portray the underlying geology (see for instance Fig. 24.1). This should where possible use the recognized symbols for strike, dip, folding, faulting, lithology, etc. of rocks. The 'thickness' of the block has the effect of indicating the scale of the area represented: the thinner the block the larger the area. On the upper surface of the block, contrasts between different materials should be emphasized by differences in the drawing, vertical strokes of the lettering should be aligned towards the vanishing point, and shading should follow the convention that the light is coming from the top left.

Finally, it is often desirable to improve clarity or artistic effect by supplementing the mapping and block diagrams with other forms of graphic presentation. Using a vertical viewpoint, Lobeck (1939) and Raisz (1962) developed the use of illustrative physiographic diagrams, and Marchesini and Pistolesi (1964; Geomap, 1968) that of topomorphic maps (Fig. 16.11). Useful methods for representing landscape as seen from ground level are simple sketches (Hutchings, 1960) which may be supplemented by photographs (e.g. Wainwright and Brabbs, 1985) or cross-sectional profiles derived from maps or stereoscopic examination of aerial photographs. The recent growth of tourism has also stimulated the demand for panoramas such as that shown in Fig. 16.12 (Jesty and Wainwright, 1978) and oblique views of landscape often based on satellite or aerial photography (US Geological Survey, 1970).

Fig. 16.11 Example of a topomorphic map at 1:250 000 of Ethiopia. (Source: Geomap, 1968)

Further reading

Demek *et al.* (1972), FAO/Unesco (1974–), Hutchings (1960), Jesty and Wainwright (1978), Lobeck (1939, 1958), Marchesini and Pistolesi (1964), Raisz (1962), Robinson *et al.* (1984), Saint Onge (1968), Schou (1962), Soil Survey of England and Wales (1977), Tricart (1965b), Unesco (1967, 1968, 1970, 1974, 1975, 1976), US Geological Survey (1970), van Zuidam and van Zuidam-Cancellado (1978), Verstappen and van Zuidam (1968), Wainwright and Brabbs (1985), Westermann (1970–).

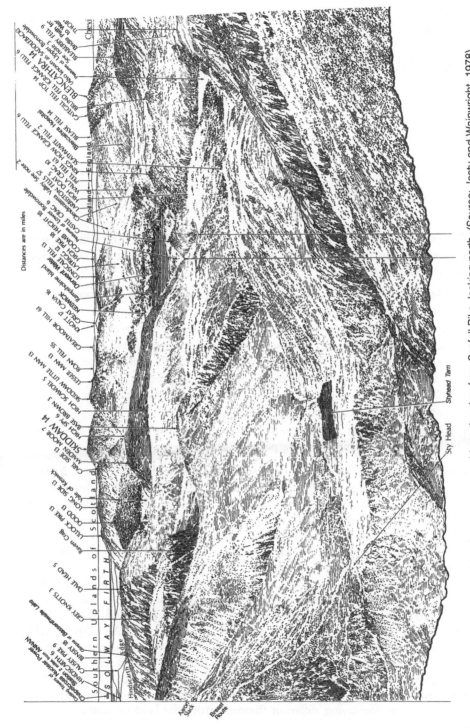

Fig. 16.12 Example of a ground-level panorama: a guide to the view from Scafell Pike, looking north. (Source: Jesty and Wainwright, 1978)

217

17

Computer cartography

17.1 Introduction

The essence of the science of cartography is the combination of different sorts of spatial information on a single map sheet. The traditionally produced paper map has a number of inherent disadvantages:

1. Because of the limitations of space, it is always an abstraction from a much larger amount of spatial data.
2. It is necessary to produce many map sheets because of the difficulty of showing everything required on a single sheet at small scale.
3. Accuracy and clarity must be emphasized, and this often necessitates the elimination or downplaying of some important information.
4. Once a map is produced, it is hard to extract data from it in a form in which it can be combined with other spatial data.
5. A map is a static qualitative document, essentially a snapshot of an area seen by one discipline at one point in time.

These disadvantages were more widely acceptable when maps lasted 20 years or more, but changes are now so rapid that conventional techniques are becoming inadequate for many purposes. Remote sensing is a vital assistance to updating, but is not itself mapping unless related to ground truth, georeferenced, and annotated. Today, the demands for up-to-date inter-disciplinary spatial information have become so pressing that many of the older cartographic techniques are obsolescent.

The name given to the new methods is 'computer cartography'. It can be defined as the process of creating cartographic and graphic images using computers. Its development has been reviewed by Jackson and Mason (1986) and Burrough (1986). An early example was the SYMAP (Synographic mapping system) program of the Harvard Laboratory for Computer Graphics which used various ways of overprinting of lineprinter characters to produce choropleth or isopleth (i.e. contour-type) interpolations (Fisher, 1978). Computer cartography has now become a large-scale activity with wide applications in overlaying different thematic maps for land evaluation purposes

218

(Burrough, 1987). Because of its widespread use by the media, more cartographic and graphic images are now produced by computer than by manual means, so that it should now be regarded as mainstream cartography.

17.2 Principles of computer cartography

17.2.1 *Vector and raster graphics*

The two major systems for creating cartographic images using a computer are graphics based on vector and raster storage systems. Vector graphics are usually employed where it is necessary to integrate manual and computer cartography techniques, where topological structure and annotations are required, and where the efficiency of the storage structure is critical. Raster systems have suffered the disadvantage that necessary coding of every pixel is costly in time and effort, but recent advances have improved their ability in this area (Burrough, 1986). The difference in the amount of storage capacity is not now significant for small images, but may be a major consideration for large. Raster graphics are usually used where it is necessary to integrate topographic and thematic data with remotely sensed data. Image comparison, area filling, and light pen operations (see below) are also best achieved using raster graphics. Microcomputers and many computer screens based on television technology use the raster system. The two systems are, however, complementary rather than competitive because of the availability of vector-to-raster and raster-to-vector conversion routines.

17.2.2 *Resolution*

An important measure of the cartographic capabiliities of a computer system is the resolution. This is normally expressed as the number of discrete units (pixels or coordinate values) along the x and y axes which can be used to store or display an image. Computer cartography systems can be broadly defined as having low, medium, or high resolution according to whether they have less than 100 000, between 100 000 and 250 000, or more than 250 000 units respectively. The present trend is to regard 250 000 units as a minimum, and Tektronix 4010, for instance, which is often regarded as a standard computer cartography terminal, has an array of 1024 × 780 (798 720) pixels.

17.3 Computer cartography hardware

The basic components of a computer system are the keyboard, processor, disc drive, and screen. Computer cartography requires these to be supplemented with a digitizer, joystick, mouse, and light pen for digitizing, and a plotter, camera, and printer for producing hard copy (permanent) output.

17.3.1 *Digitizers*

A digitizer is a device for capturing topographic and thematic maps in a format suitable for storage and manipulation in a computer, and may be used in vector or raster format.

Vector format data may be captured by either electromechanical or semi-automatic line follower digitizers. An 'electromechanical digitizer' is essentially a 'tablet' of electronic graph paper. Attached to it is a pen or tracking cross which can move around the tablet and detect the signal at any intersection of a grid of wires. This signal is coded in the computer into x and y coordinates. Electromechanical digitizers can work in either point or linestream mode. In point mode x–y coordinates are recorded only when the operator gives a signal, for instance by pushing a button. In linestream mode, the digitizer records coordinates at fixed time or distance intervals. The resolution depends on the spacing of the wires in the grid, normally in the range 0.01–0.1 mm. Low-cost variable resistor electromechanical digitizers are available but, although cheaper, they are less accurate, more fragile, and less pleasant to use. A 'semiautomatic line follower digitizer' uses a photosensor or fine laser beam to follow the lines on the map and records coordinates at fixed time or distance intervals.

The two main types of raster format digitizer are 'scanning densitometers' and 'electronic video digitizers'. In a scanning densitometer the map is fixed to a drum which rotates beneath a photosensitive optic. The map is captured in the form of an image which consists of a matrix of pixels. The optic records the intensity of light for each of the pixels along a line as the map rotates, advancing one scan width at the end of each line. Scanning densitometers have a typical resolution of 0.012 mm. An electronic video digitizer essentially uses a video camera to image a whole array of pixels, recording the light intensity of each. Electronic video systems are cheaper than scanning densitometers, but have lower resolution.

Electromechanical digitizers are the most commonly used because of their relatively low capital cost, ease of use, high resolution, and because they produce data in vector format which is still the most widely used in computer cartography. Where large amounts of digitizing must be done, however, as in national mapping programs, semi-automatic vector and raster format digitizers are often used since they are much faster in these circumstances.

17.3.2 *Joystick*

The 'joystick' is a lever which can be moved in two dimensions. Two potentiometers, usually set at right angles in x and y directions, convert its movements to coordinates. Joysticks are often used to position screen cursors, to input commands from a screen menu, or to locate the corners of a screen window.

17.3.3 *Light pen*

A 'light pen' is a small hand-held, pen-like device with a light-sensitive tip. Pressing the tip against the computer screen can be used to interact with images and programs. Light pens may be used to draw or remove lines or shading on the image or to input data to a program. Their main disadvantages are high cost, slow speed, poor accuracy and the fact that their use soon tires an operator. More expensive pens are faster and more accurate.

17.3.4 *Plotters*

Plotters are the main devices for producing high-quality hard copy monochrome and multicolour computer output, and several types exist. 'Drum plotters' are the most widely used in large computer installations. The paper rolls over a drum in the x direction while one or more pens, moving in the y direction, create the image.

'Flat bed plotters' typically have A3 or A4-sized plotting areas and facilities for multicolour printouts. They are distinguished from other plotters in that the paper is fixed to a flat surface and the pen is supported by a gantry that moves in both x and y directions. They have advantages in being able to accept any type of drawing material or pen and some can be used as digitizers. Although relatively cheap, they are comparatively slow to operate.

'Electrostatic' (sometimes called laser) plotters operate in a similar fashion to photocopiers. Electrodes deposit charges on chemically treated paper in the form of an image. The paper then passes through a bath which adheres black toner to the charged areas. They are very fast, taking only a few seconds after the data has been processed to a suitable format, and give medium resolution, but can produce only monochrome images.

17.3.5 *Cameras*

Cameras are used to give hard copy output from computers. At its simplest this involves using a 35 mm camera to take a picture of a screen. The technique can, however, suffer from edge effects, screen reflectance, and lengthy processing times. More acceptable pictures can be obtained by connecting cameras directly to the video or RGB (red–green–blue) socket of a suitable computer. The latter method generates imagery through a program which uses the three colour 'guns' in sequence.

17.4 Computer cartography software

The last 25 years has seen the development of a large number and wide variety of computer cartography programs and packages. Some are in the form of low-level machine-oriented drawing commands and routines, but the most useful are in the form of high-level user-oriented packages.

Computer cartography began in the late 1950s, but the major advance came in the 1960s when an output method was devised for producing easily readable maps on standard text screens and lineprinters. The computer was used to print out an appropriate symbol in each print position in each row, a type of raster graphics. In the 1970s, software was developed which used vector graphics and plotters as the principal hard copy device. This enabled maps to be produced which were constructed of points and lines rather than characters. The software was of two basic types: subroutine libraries and specialist packages.

In the 1980s, the development of microcomputers, and the availability of workstations based on them, reduced the cost of computer cartography by an order of magnitude and heralded its third (and present) phase: the advance to microcomputer systems which are sufficiently advanced for many geographical applications. The changes in technique and availability of equipment have been so dramatic that they can reasonably be called the 'new cartography' (Maguire, 1989). These changes will certainly continue and further increase the cartographer's dependence on the computer, both to supplement and to replace older techniques. It imposes the need to re-examine and advance cartographic theories relating to such questions as projections, mapping themes, and forms of representation, in order fully to exploit the new potential.

The independent development of earlier computer devices and packages has led to such a diversity of computer commands and operations relating to cartography that it is hardly possible to describe them all. There is a need for wider standardization among manufacturers. For instance, it is still difficult to harmonize the origin (the point where drawing begins) in computer systems designed for text, where it is at the top left, with cartographic systems, where it is at the bottom left. To combat the resulting diversity, the graphical kernel system (GKS) has been developed as an international standard that is device independent so that a package written in GKS for use in one system is easily portable to another (Maguire, 1989).

17.5 The advantages and disadvantages of computer cartography

Computer cartography permits existing maps to be produced more quickly and cheaply. Compilation time is much reduced, storage space decreased, and most important, the cost of updating and redrafting maps held in digital form is much less. It facilitates the production of maps which are more closely tailored to user needs, because they can be produced equally easily at any scale or with any sheet boundaries. The advances in computer cartography have led to greater experimentation in the mapping process, especially in relation to the depiction of statistical data. Users can easily try out different class boundaries, shading schemes, and data transformations, and can create new types of cartographic image. For example, three-dimensional perspective plots, stereomap pairs, and new projections all gained in popularity when computers made it possible to produce them at speed.

Computer cartography, however, has some disadvantages. Most systems involve a very high initial outlay, and frequently take longer than expected to reach full production. This may be due to hardware or software faults or else to lack of familiarity on the part of the user, especially because many current systems are user-unfriendly. Also, only the most expensive can yet match the drafting standards achieved by professional cartographers. Computer cartographic systems are rarely cost effective for producing one-off cartographic images, and are best used in situations where images need to be frequently updated, or where many copies of maps are required with only slight differences. Finally, present computer cartography systems can only produce what is programmed and so can easily give results that are geographically incorrect or aesthetically displeasing. The adage 'garbage in – garbage out' applies to this as to all other sorts of computer usage (Rhind, 1977; Maguire, 1989).

17.6 Applications of computer cartography

The principal applications of computer cartography in terrain study are for topographic and thematic mapping and for statistical graphics.

17.6.1 *Topographic and thematic mapping*

Topographic maps show the spatial association of diverse geographical phenomena, such as towns, roads, watercourses, and relief features. Thematic maps, by contrast, concentrate on the spatial variations of single phenomena or the relationship between them, such as soils or vegetation (Robinson *et al.*, 1984). In practice, as far as computer cartography is concerned, the two types of map can be considered together since the same principles apply to both.

The process of converting map data to digital form is called digitizing. In its simplest form, this involves capturing and storing the data in unstructured format, i.e. as a collection of coordinates in vector format or of pixels in raster format. They can then be replotted on a suitable output device. However, with unstructured format, selective reproduction or manipulation is difficult because it is not possible to gain access to individual map entities such as rivers, area boundaries, and point features. Therefore, for most purposes, it is best to store data in structured format.

A number of different schemes have been developed for coding maps in vector and raster format in order to create structured digital map data files (Burrough, 1986). The simplest vector coding scheme in which each point, line, or polygon is coded independently and completely as a series of x–y coordinates is called 'entity-by-entity' coding. Although simple to implement it is inefficient since all common boundaries must be coded twice (once for each entity). In the more sophisticated 'topological' coding scheme, provision is made for coding all common points and boundaries only once. Each polygon boundary is defined by the coordinates of its two end-points and the names of the blocks on each side. The most suitable method for economically storing and manipulating raster data on the irregularly shaped areas characteristic of terrain, is that of quadtrees, although this has the disadvantage of not lending itself to the calculation of shape indices and pattern descriptors.

Digital map data are used in many countries today and there are already a number of important data sets available which provide digital map coverage of the world at various scales (Carter, 1984; DoE, 1987). At the global scale the most well-known data sets are the World Data Bank (WDB) I and II files. WDB II was collected from maps at an approximate scale of 1:3 000 000 and contains 6 000 000 coordinates describing selected topographic features. A world data base for the environmental sciences is being prepared for 1990. This will contain information such as contours, river networks and coastlines digitized from maps at approximately 1:1 000 000. About 30 per cent of the data was available in mid 1986. There are also many government, commercial, and academic bodies involved in the production of digital thematic maps.

In Britain, the Ordnance Survey has completed the digitization of the whole country at 1:625 000 and aims to complete all larger-scale maps by the year 2010. In the USA, the US Geological Survey has digitized the whole country at 1:1 000 000 and aims to complete it at 1:24 000 by the year 2000.

Although the aim of digitization was to improve the efficiency of map production, it soon became apparent that the data could be used in other ways. Matrices of height data, called digital elevation models (DEMs) can be used for many purposes. The use of a DEM, a slope and aspect algorithm, and a model of light reflectance allows the creation of shaded relief maps. Three-dimensional relief models can easily be created, which facilitate such tasks as investigation of soil and water movement on slopes and the evaluation of landscape quality.

17.6.2 *Mapping areally distributed statistical data*

Computers have been used extensively for mapping areally distributed statistical data, such as vegetation indices derived from satellite imagery, river discharges, and population information. Such mapping has been incorporated into many atlases, surveys, and reports all over the world.

One of the most widely used computer packages for this purpose is GIMMS (geographical information mapping and modelling system). It is available for a wide range of computers from mainframes to microcomputers and has been used in many countries. It can create either maps or statistical graphics and is organized as a series of modules each of which covers one aspect of the cartographic image creation process. Thus the GRAPHICS module is used to draw computer graphics, the POLYGON, COMPILE, and PLOTPROG modules are concerned with mapping, and the MANIPULATE and UTILITIES modules deal with data on system organization (Maguire, 1989).

Let us assume that the system is used to map the proportions of stones in the soil profiles of an area. Three files are required, holding respectively locational, non-locational (statistical), and command data. The locational data are x–y coordinates (GIMMS works in vector mode), normally collected using a digitizer. The statistical data are the percentages of stones extracted from a separate file. The command data are GIMMS commands used to combine the locational and statistical data to produce a map with a title, scale bar, north point, and legend.

The map production phase must first define the output device. This can be a plotter or an interactive graphics terminal. The procedure is then to give a series of commands. These first use the locational and statistical files to define the map size. Then, in order, they specify the type of shading, the number of classes and the class boundaries to be shown. For example, a SPACING option defines the distance between the shading lines, an ANGLE option the orientation of the shading, and a USER option the lower and upper class boundaries for each of the soil stoniness classes. A final series of commands add the legend, scale bar, north point, and titles.

17.6.3 *Statistical graphics*

'Statistical graphics' are all pictorial methods of presenting statistical information and include charts, graphs and diagrams. They arouse interest, clarify, simplify, or explain statistics. Although they have existed for a number of years, the recent proliferation of computers, especially of the desk-top variety, has much widened their use.

Statistical graphics appropriate to terrain study may be classified into: (a) bar charts and histograms where the length of each bar is proportional to the value of one data category or class; (b) pie charts where a circle (pie) is divided in proportion to the values of two or more variables; and (c) scatterplots and line graphs where the values of two variables are plotted or shown as a trend line respectively. There are a number of possible variations on these basic themes. For instance, pollen diagrams are a sophisticated form of bar chart and diagrams of three-dimensional objects such as cubes or pyramids can illustrate values in terms of an apparent volume. The latter are especially valuable when large numerical contrasts on the same map are involved. Many commercial microcomputer packages of this sort are available, offering a wide range of graphs and charts in both monochrome and colour (Maguire, 1989).

Means are available for using computers to print out three-dimensional views of grid-numerical data, e.g. of heights, by methods which are known as 'geographics'. These 'digital terrain models' can be either in the form of perspective drawings or vertical or oblique pictures with altitudes shown by hillshading calculated in accordance with a light source of given direction and elevation. They can also show colour-coded information on vegetation and rock types (Macdonald Dettwiler, 1987).

Further reading

Batty (1987), Burrough (1986, 1987), Carter (1984), Dangermond (1988), DoE (1987), Forer (1984), Jackson and Mason (1986), Maguire (1989), Newman and Sproul (1979), Rhind (1977), Robinson *et al.* (1984).

Part V

Applications of terrain evaluation

The classification of terrain evaluation applications in this and the following chapters is based on the main professional types of land user. This subdivision is necessary because, despite the desirability of interdisciplinary integration and some moves towards it, the different professional demands of agriculture, engineering, planning etc. still necessitate a separate consideration of their methods of handling terrain data.

18

Terrain analysis in meteorology and climatology

18.1 Terrain evaluation and climate

Climate and weather near the ground play an important part in determining land values, especially in relation to plant growth and human settlement. They are themselves strongly dependent on terrain conditions. The aim of this chapter is to describe, classify, and evaluate those near-ground climatic conditions which result from the nature of the ground surface and its topography.

The significant effects of terrain on day-to-day human activities are in the boundary layer. These control the local and regional distribution of radiation balance and energy fluxes, temperature and rainfall, the direction and speed of winds, and the incidence of frost, flood, and drought.

The lower part of the boundary layer consists of two zones: the 'laminar boundary layer' and the 'turbulent surface layer'. The laminar boundary layer is essentially a film of dead air, in immediate contact with the ground, within which the transfer of energy is by molecular diffusion only. It varies in thickness with surface roughness, but is at most a few millimetres. Immediately above it is the turbulent surface layer. This is characterized by intense small-scale turbulence generated by surface roughness and convection. By day it extends to a height of about 150 m above the ground, but falls to only a few metres at night. Horizontally, the daily oscillation of the boundary layer within which significant daily thermal differences develop is 50–100 km.

Near-ground effects result both directly from the impact of the daily and annual insolation cycle on specific land areas and indirectly from the air movements caused by differential heating of the surface. These effects are so intimately interconnected that it is most convenient to consider the totality of climate–terrain interactions in terms of scale differences. Broadly, three scales can be distinguished.

Micro-scales refer to terrain–climate interactions over a few metres or tens of metres relevant to a single land surface type or vegetation pattern at the scale of the land element or land facet. The main climatic effects are concerned with the albedo, heat balances, and vertical air movements.

Meso–scales extend up to the characteristic horizontal distance of the boundary layer's daily oscillation because this approximately represents both the span of significant daily thermal differences and the maximum distance at which human beings can normally interact directly with the climate. This is normally 50–100 km, and thus approximates to the scale of the land system. The climatic processes are a combination of vertical with horizontal air movements, resulting mainly from the differential heating and cooling of the surface.

Macro–scales relate to areas over which the terrain modifies the regional and global patterns of wind and precipitation. They thus accord approximately to the scale of the land region or land province. The horizontal component of atmospheric movement is relatively more important because the vertical circulations have at this scale generated the pressure differences which cause significant surface winds.

At very small and very large scales, terrain–climate interactions become less important to land use decisions: over the smallest areas, such as the land element, because their effects are insignificant; over the largest, such as the land division or land zone, because generalizations must be impractically broad.

18.2 Micro-scale terrain–climate interactions

18.2.1 *Solar radiation and the absorption and storage of heat*

Solar radiation reaching the top of the earth's atmosphere comes almost entirely in the 'short wavelength' parts of the spectrum between 0.15 and 4.0 μm. Of this, 51 per cent is either absorbed by clouds and atmospheric gases or reflected back from clouds, and 49 per cent reaches the earth's surface, 21 per cent in the form known as 'diffuse radiation' or 'skylight' and 28 per cent as direct insolation. A proportion of this 28 per cent is reflected back into space. The world average for this is about 4 per cent (i.e. about 14 per cent of the direct insolation). Thus, when this is subtracted from the 49 per cent of the incoming radiation, it leaves 45 per cent which is actually absorbed at the ground surface. The reflected portion depends on the albedo, for which some representative values are given in Table 18.1.

The wide range from extremes of under 5 to over 95 per cent has important consequences. Absorbed radiation is converted into thermal energy which heats the earth's surface, and materials with the lowest albedos have the highest temperatures. If the ground has a low albedo and a high conductivity, the resulting microclimate is mild and stable, since excess heat is quickly absorbed and stored, and as quickly released when temperatures drop. Surface materials of high albedo and low conductivity, on the other hand, make for a microclimate of extremes, since they do not help to balance the swings of the weather. Thus water bodies, wet ground, or grass tend to even out the climate, while the weather over snow, rock, or man-made pavements is more violent: hot in the sun and cold at night (Lynch and Hack, 1984).

Table 18.1 Total surface effect of different types of material with diffuse reflectance (after Geiger, 1965)

Surface	Albedo values	
Snow	20–95	(higher values when fresh and clean)
Light sand dunes	30–60	
Glacier ice	20–46	(higher values when clean)
Sandy soil	15–40	
Meadows and fields	12–30	
Concrete	17–27	
Densely built-up areas	15–25	
Woods	5–20	
Dark cultivated soil	7–10	
Normal water surfaces	3–10	(up to 70 with low sun on a rough sea)
Bitumen	2	

A similar contrast exists between cities and their rural surroundings. Although the albedos of urban materials are not greatly different from those of many rural surfaces, their configuration is vastly different. While the latter tend to be horizontal and to reflect directly, cities are a mixture of horizontal and vertical elements, leading to multiple reflections between streets and buildings with some absorption at each surface before any radiation escapes to the atmosphere. The overall effect is an urban albedo lower than that for a rural area, causing the phenomenon known as the 'urban heat island'.

This heat island influences other climatic elements. It provides a hot spot which enhances convection and leads to increased cloudiness over and immediately downwind of the city. The city's roughness may cause it to act as a barrier to the regional wind flow, forcing the air to rise further, increasing the possibilities of cloud formation. On the other hand, this increased cloud and the urban aerial pollution partially offset their tendency to greater heat absorption by depleting incoming solar radiation.

The heating effect of a low albedo can be obtained by darkening any surface, such as by adding darker-coloured mulches or stones. Absorption can also occur through snow cover, if it is less than about 15 cm thick. Temperatures also vary with the soil's 'heat capacity' (the amount of heat necessary to raise the temperature per unit volume by 1 °C). This is equal to the product of the 'specific heat' (the amount of heat necessary to raise the temperature per unit weight) and the density. Table 18.2 shows these values for some natural materials.

Table 18.2 Thermal properties of natural materials (after Oke, 1978)

	Density (kg m^{-3} × 10^3)	Specific heat (J kg^{-1} × 10^3)	Heat capacity (J m^{-3} × 10^6)
Dry clay soil	1.60	0.89	1.42
Dry sandy soil	1.60	0.80	1.28
Dry peat soil	0.30	1.92	0.58
Water (4 °C)	1.00	4.18	4.18
Ice (0 °C	0.92	2.10	1.93
Fresh snow	0.10	2.09	0.21
Old snow	0.48	2.09	0.84
Air (10 °C)	0.0012	1.01	0.0012

The important conclusion from Table 18.2 is that water requires several times as much energy (and consequently several times as long) to change temperature as does soil. Thus, water bodies act as heat sinks in the daytime and in summer, and as heat stores at night-time and in winter. These effects increase with their volume, being largest for oceans. They are the cause of the alternation of land and sea breezes discussed in section 18.3.6. Wet soils are analogous. They have a greater capacity to store energy and a slower attainment of maximum and minimum temperatures than do dry soils, the size of these effects being directly proportional to the water content. Oases provide a notable example of this when compared with surrounding deserts. Urban areas have the opposite effect. Because their surfaces are designed to ensure the rapid removal of rainwater, they are, except after rainstorms, drier than their rural surroundings and so change temperature more strongly and speedily in response to changes in incoming radiation.

18.2.2 *Surface materials and evaporation*

Terrain influences the level of atmospheric moisture through its control over evaporation from water and land surfaces and transpiration from plants.

Apart from water content, the three factors mainly determining the amount of evaporation from a moist soil are colour, texture, and structure. Dark-coloured materials, because of their higher heat absorption, have higher evaporation than light-coloured. This is difficult to quantify because darker colours are usually associated with higher moisture and organic matter contents whose thermal effects mask those due to colour. Coarse-textured soils have less pore space and hence can store and evaporate less water than fine-textured soils. Sandy soils range from 35 to 50 per cent pore spaces. Medium to fine-textured soils vary from 40 to 60 per cent or even more in cases of high organic matter and marked granulation. Pore space decreases with depth, some compact subsoils dropping to as low as 25–30 per cent. Because their pores form finer capillaries, they can draw water to the surface from greater depth, and thus can cause surface evaporation from a deeper water-table. In hot arid climates, sandy soils permit surface evaporation from a water-table at about 0.5 m, silty soils from one at about 1 m, and clays, in the absence of cracks and fissures, from up to about 2 m.

Soil structure influences the amount of surface exposed to evaporation, through controlling surface roughness. A topsoil with coarse aggregates or plough ridges, for instance, will have especially high evaporation, potentially even higher than open water, because of the bigger surface area. This effect is diminished if the resulting microrelief reduces wind movement by being oriented cross-wind.

18.2.3 *The effects of vegetation*

Vegetation interposes a barrier between the atmosphere and the ground surface which intercepts a proportion of the rainfall and diffuses both inward and outward radiation over the whole zone of plant height. It also brings up large quantities of water from the deeper layers of the soil, so that transpiration from vegetated areas normally exceeds evaporation from bare soil by several times. This effect is proportional to plant volume, trees being more effective than low vegetation. Transpiration may have a drying effect on the underlying soil when water supplies are limited, but at the same time, tree cover reduces the direct evaporation from the ground surface to about one-half the values found on bare sites. Wind strength is also greatly reduced and temperatures are more equable under trees than in the open. Thus vegetation, especially in forests, regulates the whole water regime through the amount of water intercepted, absorbed, and transpired. It reduces wind speeds, but otherwise has a generally 'oceanic' effect on ground temperatures by reducing temperature variations, delaying extremes, and increasing the amount of moisture in air and soil.

18.2.4 *Energy balance and terrain modifications*

The diurnal energy cascade of radiation reaching the earth's surface can be expressed by the formula

$$R_n = H + LE + G \qquad\qquad [18.1]$$

where R_n = net radiation (W m^{-2});

   H = sensible heat flux to and from the atmosphere;

LE = latent heat flux to and from the atmosphere;

G = sensible heat flux to and from the ground.

Advective energy in the form of water vapour and sensible heat transfer through laterally moving air also enters and leaves the system.

This diurnal energy cascade is controlled by the nature of the land surface and the relative abilities of atmosphere and soil to transmit heat. The important variables of the ground surface are its resistance (or otherwise) to evapotranspiration and its thermal conductivity. The effect of these environmental constraints on the energy balance of particular types of site are shown in Table 18.3. The most significant conclusions from Table 18.3 are the importance of water as a reservoir of latent heat and the low net radiation to and from ice.

Table 18.3 Some qualitative generalizations about the energy balance of a variety of environments (© W P Lowry 1989; Thompson *et al.*, 1986)

Type of site	Remarks about energy balance
Mid-latitude vegetated area	$LE > H > G$
Growing crops	LE up to 80–85% of R_n
Meadow	LE dominates during growth, H after grass is cut
Pavement or stonefield	LE = about 0, H approx. = G
Desert	LE = about 0, $H > G$
Ice cap (high latitude)	LE and G about 0; H and R_n about equal, H being positive, R_n negative
Open water	LE and G completely dominate H
Still water	G reduced and both H and LE increased by comparison with open water
Snow (mid latitude)	Large albedo keeps R_n small. While snow and air are <0 °C, G and LE are small and H relatively larger. When snow melts, G increases
Leaf litter	Acts as a thermal insulator keeping G small and, when dry, as a vapour barrier so that $H > LE$

18.3 Meso-scale terrain–climate interactions

18.3.1 *Aspect*

In the absence of cloud, solar energy flux varies regularly through the day, increasing to a maximum when the incident angle is vertical and decreasing thereafter. Relief determines the pattern of solar incident angles for any given point, and in an accidented landscape there is a wide spectrum of conditions.

Maximum insolation occurs on any surface normal to the sun's rays. This means that for any given latitude, and neglecting broad climatic effects such as the seasonal and daily variations in cloud cover, the aspect having the maximum radiation receipt, and exporting the maximum sensible heat back into the atmosphere is a south-facing surface (in the northern hemisphere) inclined so that it will receive the sun's rays most nearly vertically for the maximum time during the year. Since modal conditions may be assumed to occur at the equinoxes, the greatest insolation over the year will be received at sites with an equatorward inclination equal to the latitude, i.e. 41° in New York, 52° in London, etc. Other aspects will have less insolation in proportion to their deviation from these. In middle latitudes, a 10 per cent slope towards the Equator will receive as much direct radiation as flat land 6° closer to it.

The Dumaresq Creek, New South Wales, is an incised east–west valley whose flood plain lies about 100 m below the plateau level. The annual mean of daily minimum air temperatures (at 30 cm height) and the total number of air frosts for 1 year in a transect of five stations from north to south are given in Table 18.4. Thus, over the year the flood plain minimum temperatures averaged 1.7 °C below the hillcrest minima. The differences were greatest when the long stable nights allowed maximum amounts of cold air drainage and strong temperature inversions. In July the mean minimum temperature on the flood plain was nearly 4.5 °C below that on the northern hillcrest, while in the November–March period the shorter and more unsettled nights brought the minimum temperatures of hillcrest and flood plain to within 1 °C of each other. Within the flood plain, the urban areas had a mean minimum temperature about 1.7 °C higher and 6 fewer frost days than the rural parts.

Table 18.4 Annual mean of monthly mean minimum temperatures and number of air frosts in one year recorded at five stations across Dumaresq Creek, New South Wales

	Minimum temperature (°C)	*Number of air frosts*
Hillcrest	8.2	20
Middle ubac*	6.8	65
Flood plain (urban)	6.5	73
Middle adret*	7.2	63
Upper adret*	8.0	38

* Ubac are slopes away from, adret towards, the sun. In the southern hemisphere they are respectively south- and north-facing; in the northern hemisphere the reverse.

Source: Thompson (1973).

Aspect is often reflected in land uses and land values. Contrasts are at their maximum in middle latitudes. In Swiss mountain valleys, adret slopes have higher snow-lines, different land uses, and more favourable conditions for human settlement than ubac. In the Val de Marebbe, Italian Tyrol, the land with the highest insolation is used for wheat, decreasing levels being associated in order with cereals and roots, meadowland, and forest (Garnett, 1935). In the more northerly grape-growing districts of France and Germany, vineyards are located on south-facing slopes to avoid the exposure of the plateau tops and the frost pockets of the valley bottoms.

The adret–ubac contrast is less in low latitudes because of higher elevations of the sun. It is also less marked in high latitudes, especially where cloudy. In the Scottish Highlands, for instance, the contrasts are much less important than in Switzerland. This is because the lower altitudes minimize the contrast between sun and shade temperatures, relief is less, the greater cloudiness diminishes insolation everywhere, and the longer days in summer make for longer solar exposure of north-facing slopes (Garnett, 1939).

18.3.2 *Air movements in valleys and depressions*

Air movements resulting from relief have the most strongly contrasting effects in valleys and depressions. The best developed and most symmetrical wind system might be anticipated in anticyclonic weather in summer in a deep, straight valley with a north–south axis (Fig. 18.1). Even here, there is some asymmetry with time due to the diurnal variation of solar radiation input to west- and east-facing slopes. The valley bottoms will have different diurnal thermal regimes from their surrounding slopes because the reduced time between sunrise and sunset shortens the insolation period and increases the time for outward reradiation.

By day, heating will cause upslope ('anabatic') winds, and these will allow compensating air to descend into the valley centre to complete the circulation. After the warmed air has begun to rise up the valley slopes, a second movement of air takes place up the valley itself. This is known as the 'valley wind'. Where a number of valleys converge upwards in an area of radial drainage, the concentration of upward airflow causes increased cloud formation and rainfall. This can be seen, for example, above the central part of the English Lake District. The circulation within the individual valley 'cell' is completed by a downward counterflow known as the 'anti-valley wind' above the valley sides.

At night these conditions are reversed and cool air, being heavier, moves downslope by 'katabatic' flow into the depressions, usually at a rate of about 1 m s^{-1}. This can be accelerated by ice and snow surfaces, especially on slope tops and valley heads. The downslope convergence in the bottom sites causes a down-valley flow known as the 'mountain wind' which seeps out on to the adjacent lowland. This often comes in intermittent surges rather than a continuous flow, building up to thresholds of pressure behind obstacles before

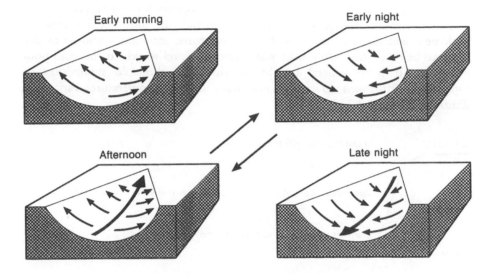

Fig. 18.1 Diurnal variations in air movement in a normal valley (after Geiger, 1965)

it can overpass them. The resulting concentration of cold air in the valley gives a weak lifting motion, causing rising air which completes the circulation by becoming a counterflow up the valley aloft. This is known as the 'anti-mountain wind'. A similar alternation between daytime rises and night-time falls occurs in closed basins, only without the directional winds caused by the valley alignment. Where downflow of air is divergent from an upland centre, there is a tendency to form clearer skies as, for example, can be seen in Fife, Angus, and the Moray Firth when downward-moving air diverges from the Grampians.

The speed of formation of cold air pockets is inversely related to their size and shape. Temperatures in the valley do not fall so low when the depression is steep and narrow as when it is gentle and wide because the latter has a

237

greater volume of downflowing air relative to the size of the depression. If air temperature in the hollow falls below dew-point, radiation fog can form, which may condense *in situ* as dew or sublimate as hoar frost. This concentration of the coldest and densest air causes a temperature increase with height, forming a 'valley inversion', above which the normal adiabatic decrease with height takes over. During nights conducive to excessive outgoing IR radiation, this process can cause significant temperature contrasts even within height differences of <1 m. Cold air pockets may persist over the next day if they are large enough and especially if haze or fog has formed to prevent the sun from warming the ground surface. In general, the most suitable area for settlement and sensitive crops is the contour zone, sometimes known as the 'thermal belt', on the valley sides just above the level to which the cold pool builds up. It is in general possible to discern a threefold subdivision of topographic situations in accordance with diurnal temperature variations, illustrated in Fig. 18.2.

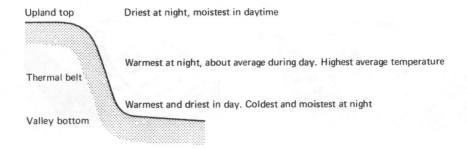

Fig. 18.2 General distribution of heat and moisture according to topographic position

So marked is the contrast between 'warm convex' and 'cold concave' land surfaces that the hollows may at night have temperatures many degrees colder than stations thousands of metres higher. Miller (1946) noted that the lowest temperature then on record in the USA (-65 °F, -54 °C) was reached at Miles City, Montana, lying in a deep hollow in the Great Plains, while Pike's Peak, which is 11 000 feet (3350 m) higher, had never recorded a temperature below -40 °F (-40 °C).

The downward drainage of air at night also occurs on the slopes of isolated hills. In the Sudan, for instance, cool night breezes formed in this way are a welcome relief after the heat of the day, and buildings may be sited at the foot of hills to take advantage of them, for example Sudan Government rest houses, such as that at Jebel Bozi. On the other hand, this phenomenon can cause problems in cold climates where the mountain is high and snow-covered. Here, the rapid descent of cold winds can be unpleasant and when near the coast can cause difficulties to ships. An example of this is the 'williwaw' wind of the coasts of Alaska and southern Chile.

Diurnal alternations of air movement are reflected in similar movements in cloudiness and rainfall. In the valley bottoms, both are at a maximum at night and in the winter, but at a minimum in the daytime and in summer. This is because the greater convectional rise when the sun's heat is strongest carries the moisture to higher altitudes before the dew-point is reached.

When a non-local wind forming part of the regional circulation has passed over a mountain range, it becomes adiabatically warmed and dried as it descends in the lee. This is clearly felt at ground level and has given rise to local names such as 'föhn' in the Alps, and 'chinook' and 'zonda' on the eastern sides of the Rockies and Andes respectively.

18.3.3 *Altitude*

As a general rule, temperature decreases with altitude up to the tropopause, at about 11 km. Wind speeds and potential solar radiation increase in the same direction, but actual radiation is inhibited at lower altitudes by the increase in clouds and rainfall from orographic causes. The fall in temperature with latitude is analogous to that with altitude so that, while lowland sites have attracted most settlement in high and middle latitudes, upland sites have been favoured for resorts in the tropics.

It is of practical importance to quantify some of these effects. In Europe, increasing altitude generally brings climatic disadvantages both for settlement and land use. Although there are considerable variations from year to year, especially in coastal areas, it is possible to obtain some representative values for England and Wales (Smith, 1984). Sunshine decreases by 0.11 h day^{-1} per 100 m in winter (October–March) and 0.18 h day^{-1} per 100 m in summer (April–September) and air temperatures (at 122 cm above ground level) fall by 0.6 °C per 100 m. If it is assumed that plant growth begins at about 6 °C, the effect of this temperature fall is to reduce the annual period during which the temperature exceeds this level. This gives an agricultural growing season which can be calculated from the empirical formula

$$29 T_a - 17 \tag{18.2}$$

where T_a is the mean annual air temperature in degrees Centigrade, and the result is in days.

The grazing season does not coincide with the growing season. In spring, some time has to elapse after the start of appreciable grass growth before animals can find adequate fodder, and this delay increases with height. The end of the season, at least in Britain, is determined more by the external weather and the state of the ground than the continuance of plant growth. The following equation has been found to be more appropriate for this (Smith, 1984):

$$29.3T_a - 0.1R + 19.5 \tag{18.3}$$

where R is the mean annual rainfall in millimetres, and the result again in days per year.

The decrease with altitude varies from place to place according to slope, aspect, and rainfall. The loss is 33–40 days per 100 m altitude in northwest England and Wales and 65–75 days in southwest England with the rest of the country intermediate (Smith, 1984).

The disadvantages of this loss of grazing season length with altitude are somewhat countered by an increase in rainfall and a decrease in potential transpiration in the same direction. The increase in rainfall in England and Wales per 100 m of altitude can be represented by the general formula

$$0.315R - 119 \text{ mm} \tag{18.4}$$

where R is the mean annual rainfall in millimetres and the result in millimetres per year. Potential transpiration from a green crop decreases by about 17.5 mm per 100 m altitude in summer (April–September) and about 10 mm in winter (December–March). The effect of these trends is to give higher lands a shorter grazing season but one which is less subject to a summer soil moisture deficit. This can be seen, for instance, by comparing the agroclimatic areas of the Cumbrian coastal plain (average altitude 109 m) with the nearby Lake District (average altitude 341 m). The former has a mean annual air temperature (30 cm above the ground) of 8.5 °C and a grazing season of 150 days (10 April to 7 September), the latter a mean annual temperature of only 7.1 °C and a grazing season of varying and undefinable length beginning on 12 May (Smith, 1984). On the other hand, the annual soil moisture deficit falls from about 100 mm on the coast to less than 25 mm in the core of the mountain area (Jones and Thomasson, 1985).

18.3.4 *Exposure and shelter*

Exposure can be defined generally as accessibility to climatic effects, notably wind. Shelter from exposure has the effect of reducing temperature variations and delaying their maxima and minima. This can influence both vegetation growth and human comfort.

Shelter belts composed of trees, fences, or other obstacles reduce the effects of exposure. When at right angles to the prevailing wind they change both the energy budget and the moisture balance of nearby ground. They alter wind velocity on both the windward and leeward sides. To windward, they form a small cushion of air, forcing most of the wind either over the barrier or around its sides. A small distance to the lee, the wind velocity is reduced to a minimum. The percentage of reduction at any distance varies with the permeability of the shelter as well as the wind speed and direction. A very dense barrier uplifts a wind current that descends abruptly in the lee. In contrast, a moderately dense barrier acts as a filter rather than an obstruction. Thus with denser belts, the minimum wind velocity is lower but nearer the obstruction; whereas with partially open stands, the minimum wind speed is higher but extends further to leeward. Gloyne (1955), for instance, reported that wind speed was reduced by more than 20 per cent to a distance of 12 times the tree height from a barrier of 30 per cent density, to a distance of 27 times tree height from a barrier of 50 per cent density, but only 15 times tree height from a barrier of 100 per cent density. Other aspects of microclimate are also affected by shelter belts. They increase temperature during the day but decrease it at night, sometimes allowing frosts to occur in their lee. In middle and high latitudes the accumulation of snow-drifts near shelter belts can increase moisture far more than is saved by a lowered evaporation. Agricultural studies have shown increases in crop yields ranging from 7 to 90 per cent in the lee of shelter belts (Chang, 1968). Shelter is also important in river catchments above dams where it is necessary to even and extend the period of spring meltwater flush. The principle is to restrict the number and arrange the distribution of forest trees so as to maximize their sheltering effect while minimizing water loss through transpiration.

No comprehensive index of the exposure factor is available, but some analogy may be seen between the exposure of terrain and that of buildings, whose heat transmission through walls and resulting insulation and heating requirements were the subject of study as long ago as during the Second World War (Dufton, 1940–41). Certain empirical values have been worked out for Britain based on a system of 'penalty points' by which sites are downgraded according to the unfavourability of their orientation and the degree of severity of their exposure to wind.

The values are assessed according to Table 18.5: as the orientation and exposure values are additive, the possible range is from 0, for a sheltered site facing S, to 10, for a severely exposed site facing N, NE, or E. This scheme is not, of course, suitable without modification outside Great Britain, and would be irrelevant outside the northern hemisphere temperate zone.

Table 18.5 Penalty points for the unfavourability of the orientation and exposure of buildings in Great Britain

Orientation:	S	0
	W, SW, or SE	2
	Other	4
Exposure:	Sheltered	0
	Normal	2
	Severe (except N, NE, or E)	4
	Severe (N, NE, or E)	6

18.3.5 *The effect of relief on wind action*

When laminar airflow encounters an obstacle, it accelerates on the upstream side and decelerates on the downstream side. In turbulent flow, which is more normal, a lee eddy or vortex is formed immediately behind the obstacle (Fig. 18.3). Such vortices tend to alternate clockwise and anticlockwise motions when viewed from above, so that the air follows a sinuous path known as the 'Karman vortex street', which gradually dissipates downwind to resume the same path of flow as upstream of the obstacle (see Fig. 18.4) (Henderson-Sellers and Robinson, 1986). The pattern can be observed around an isolated hill or eminence. Light rain or snow tends to be preferentially deposited in the zone of least wind speed behind the hill. Similarly, laminar airflow constricted between two obstacles, as in a narrow place in a valley or mountain pass, speeds up. Deposition of rain or snow is increased where it slows down after the constriction.

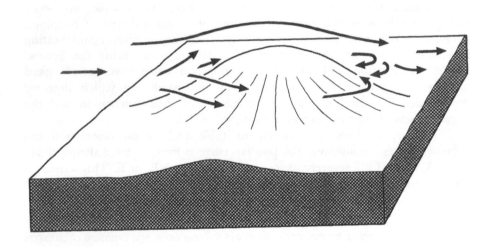

Fig. 18.3 Movement of wind around a topographic obstacle (after Oke, 1978)

The same principle applies to a wind crossing a valley laterally (Fig. 18.5). The enlarged space causes a lee eddy at B, a slowing down of the airflow at C and a speeding again at D to return to normal flow again at E. If a chimney emitting pollution is located at B, the effusion will be circulated there, pressed down to the ground at C, after which the increasing speed of the air will facilitate its removal. These effects will be reduced in a gentler valley or where the wind approaches at an oblique angle.

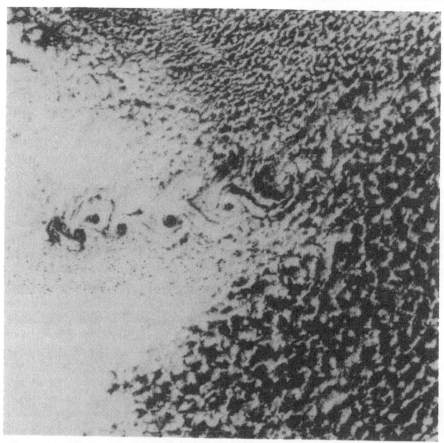

Fig. 18.4 Plan view of a Karman vortex street downwind of Jan Mayen Land on 18 September 1984. (Source: Baylis, 1976, Courtesy of the Weather and the Royal Meteorological Society)

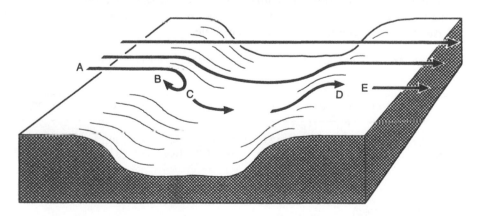

Fig. 18.5 Movement of wind across a valley (after Oke, 1978)

243

18.3.6 The coastal effect

Perhaps the largest-scale effect of land on climate is near coasts and results from the differential thermal response of land and water to heating and cooling. For reasons given earlier (Table 18.2), the sea shows less diurnal temperature variation than the land. The effect of this is increased by the stronger solar radiation and illumination, and the higher potential evaporation and transpiration near coasts than inland. The daytime sea breeze, from the relatively cool sea to the sun-warmed land, usually begins about mid-morning, and may reach speeds of up to 5–7 m s^{-1}. It extends normally up to about 30 km inland, although in southern England it has been recorded as far as 80 km (Simpson, 1964). As the sea breeze front carries cooler air inland, it raises the air it displaces, sometimes causing cloud to form. The opposite nocturnal land breeze, because less energy is available, rarely achieves half this speed, but may be increased by slope effects. Similar effects, but on a smaller scale, occur around lakes.

There are three practical results of these processes: to make temperatures along coasts more equable than inland, to increase winds, and to develop some cloud. The cloud, however, is often carried inland from the immediate coast, so that in England and Wales, for instance, sunshine decreases from coasts towards the interior at rates varying from a minimum of 0.02 h day^{-1} in winter to a maximum of 0.05 h day^{-1} in spring. Also, minimum temperatures on the coast are higher than inland, and maxima generally lower except in midwinter (when the onshore effect is weakest). As a result the winter coastal mean temperatures are appreciably higher, and those in summer slightly below, those inland (Smith, 1984). These effects and the delay in reaching maxima and minima can be seen by comparing a coastal station (Fowey) with an inland station (Cambridge), as shown on Table 18.6. While Cambridge is slightly warmer in summer, Fowey is substantially warmer in winter. Cambridge has clear minima and maxima in January and July respectively, but in Fowey January and February have equal values while its summer maximum is delayed till August, and September is warmer than June.

Table 18.6   Mean monthly temperatures (°C) at Fowey and Cambridge (1911–47)

	J	F	M	A	M	J	J	A	S	O	N	D	Year
Cambridge	3.8	4.2	5.9	8.3	12.0	14.7	16.7	16.5	14.0	10.1	6.3	4.3	9.7
Fowey	6.6	6.6	7.5	9.4	12.2	14.6	16.2	16.4	14.8	11.0	8.7	7.2	11.0

Source: Manley (1962).

The cooling effect of the sea on coastal sites in summer is a disadvantage to tourism in temperate latitudes, but can be welcome in hot weather, as is the onshore cool breeze of Western Australia known as the 'Fremantle doctor'.

18.4 Terrain and the atmosphere: large scale

The earth's climatic patterns are determined by its solar exposure as governed by its orbit and axial rotation. Latitude determines the annual cycle of exposure to the sun's rays. While the insolation near to the Equator is never as great as the maxima achieved at the poles, it is consistently high throughout the year. Polar insolation, on the other hand, is zero during the polar night but peaks during the polar day when the sun is continuously above the horizon. The further a site is from the Equator the lower its total insolation and the wider the variation between solstices. This, combined with the decreasing rotational speeds at higher latitudes give rise to the classical 'three–cell model' of atmospheric circulation (Fig. 18.6), seasonally modified by movements of the inter tropical convergence zone (ITCZ) following the sun's migration between the tropics. This model explains the distribution of the temperate and tropical forest zones which contribute to the earth's climatic balance by their direct effect on surface albedo and the large-scale absorption of carbon dioxide, transpiration of moisture, and emission of oxygen.

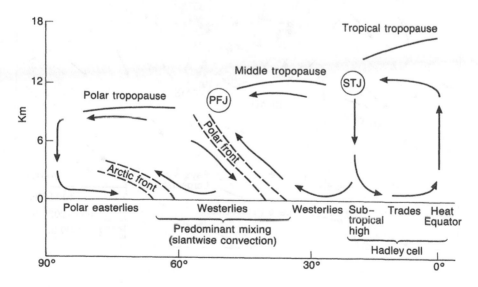

Fig. 18.6 Some features of the global circulation in winter in the northern hemisphere. (Source: Palmen and Newton (1969); Thompson *et al.*, 1986): (PFJ Polar Front Jet Stream; STJ Sub Tropical Jet Stream).

The distribution of continents and oceans imposes the main modifications on this broad picture, because land has a greater and faster response to changes in temperature than sea. This effect is most obvious in mid-latitudes, where continents become centres of low pressure in summer and high pressure in winter, respectively attracting and repelling zonal winds.

Next in importance is the way in which mountain chains divert, concentrate, and block zonal wind circulations. In the Sahara, for instance, mountain massifs like Tibesti and Hoggar divert the main direction of wind movements round their flanks, causing scour where flow is convergent and deposition where it is divergent (Fig. 18.7). High mountain ranges interrupt the wind, generating orographic rainfall zones to windward and rain shadows to leeward. The Andes, for instance, block both the Southeast Trades in central South America and the Westerlies poleward of about 35° S. In the former area they separate the well-watered Amazonian lowlands from the Peruvian coastal desert; in the latter the temperate zone of Chile from Patagonia.

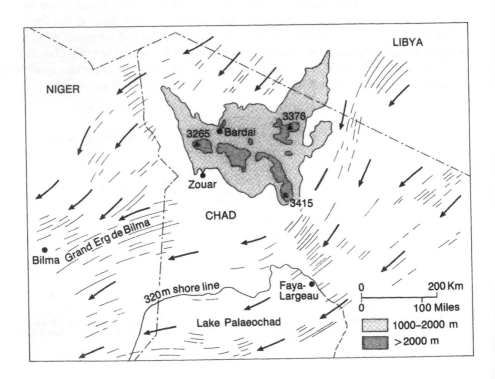

Fig. 18.7 Wind patterns in the Sahara, showing the diversion around the Tibesti Massif and accumulation at Bilma in its lee. (Source: Verstappen and van Zuidam. 1970)

The Indian subcontinent provides an example of the combined effect of differential heating and wind blockage. In winter, the subtropical westerly jet stream divides around the high pressure of the Tibetan Plateau with one branch flowing to the north over central Asia and the other to the south over northern India. The two branches merge east of the plateau to form an immense upper convergence zone over China. In May and June, strong heating of the Indian subcontinent and of the Tibetan Plateau, the latter because of the strong insolation on its high elevation, causes the southern branch to weaken and disintegrate. While this is occurring, an easterly jet stream builds up over the equatorial Indian Ocean and the ITCZ migrates northwards, bringing the monsoonal airflow and rains. In October, the situation is reversed. The equatorial easterly jet stream disintegrates and the subtropical westerly re-forms over northern India. Thus the high relief of Tibet and the Himalayas influences global wind circulations to give India its distinctive monsoonal climate.

18.5 Terrain classification for climate

Climatic differences even within the compass of a single land system can be a significant economic attribute of land, notably in relation to questions such as the locations of buildings and settlements and agricultural potential. There is a climatic difference between 'warm' coarse-textured and 'cold' fine-textured soils, due mainly to the lower moisture content and more rapid evaporation of the former when compared with the latter. Coarse soils, therefore, have advantages for housing and for early crops, while clays have the advantage of higher moisture retention in summer. Vegetation raises the water content of the air and narrows temperature variations.

At a larger scale, the climate of sites is determined by balancing a complex of topographic factors which have different relative importance in different areas. In general, in mid-latitudes, the most favourable conditions are found on slopes oriented towards the sun, at low altitudes, avoiding the opposite hazards of frost hollows and exposed hilltops, and being near to the coast without undesirable exposure to onshore and offshore winds. In practice, an examination of the location of buildings selected by those segments of the population which are influenced by climate in their choice, such as holiday-makers, invalids and the retired, shows a marked bias towards sites where the above considerations are favourable, though an overall quantification of their effects is difficult. Examples of resort locations chosen on the basis of aspect and exposure are the Riviera, the south coasts of England or Long Island, of those based on altitude: hill stations such as Poona and Simla (India), and Azrou (Morocco), and of those based on a combination of these factors: mountain resorts like Beatenberg and Crans-Montana (Switzerland).

The macro-scales of the land region and larger are beyond the range of most land use decisions. They outrun the limits of daily movement of most land users and the control of single political authorities. At these scales, population distributions appear relatively fixed. The climatic control on human settlement patterns is most clearly seen near shifting economic margins, such as those which limit agriculture around mountain, desert, and tundra zones.

Further reading

Chang (1968), Geiger (1965), Gloyne (1955), Henderson-Sellers and Robinson (1986), Howard and Mitchell (1985), Jones and Thomasson (1985), Oke (1978), Smith (1984), Thompson (1973), Thompson *et al.* (1986).

19

Terrain factors in hydrology

19.1 The control of terrain over hydrology

Terrain controls the movement and destination of all water reaching the land from the atmosphere. This is the subject of 'geographical hydrology' (Knapp, 1979) and one of the divisions of Chorley's *Water, Earth and Man* (1969). It includes 'hillslope hydrology' dealing with slope effects on unchannelled water flow and 'hydraulic geometry' covering the relations between channel form and discharge. It overlaps with 'hydrogeology' although the latter is also concerned with water at greater depths in the earth's crust.

The primary unit for scientific and often for administrative purposes is the river catchment or basin. Internal differences in topography, surface materials, and land cover necessitate the separate analysis of particular parts, especially because of the localized character of many rainstorms (Chorley, 1978).

19.2 The relationship of terrain to the distribution of water within a catchment

19.2.1 *Components of flow*

Precipitation which is not transpired or evaporated moves downwards along two paths: 'overland flow' and 'subsurface flow'. The former, also called runoff, is the first water to reach the main stream. If a graph is drawn of the resulting stream flow against time, it is known as a 'hydrograph'. Runoff composes the bulk of the initial 'peak' on this hydrograph, the subsurface water arriving later and forming its falling limb.

The two main models of the way in which water moves from hilltop to valley bottom have been called Hortonian and non-Hortonian (Horton, 1945; Chorley, 1978). The Hortonian model sees water as infiltrating into the soil at the hilltop, emerging at the surface a short way below this and continuing as overland flow, first unconfined but then increasingly confined into rills and gullies until it reaches the trunk stream (path 1 on Fig. 19.1). The non-Hortonian model sees water as infiltrating into the soil and moving downhill to join the stream as subsurface flow with little or no surface runoff (path 2 on Fig. 19.1). Most actual hillslopes represent intergrades between these extremes. Their differences can be quantified in terms of the 'runoff coefficient', defined as the volume of runoff divided by the volume of rainfall, expressed as value between 0 and 1. A pile of gravel, for instance, would have a value near 0, a tarmac road of nearly 1. It tends to increase with storm intensity and to decrease with catchment size.

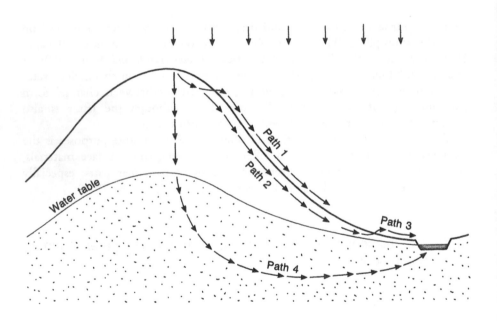

Fig. 19.1 Flow paths on a hillslope

'Infiltration' is the penetration of the surface by water. The ratio between this and surface runoff increases, and hillslopes become less Hortonian, with increasing surface roughness, permeability, density of vegetation cover, degree of topographic concavity and ponding, and convergence of flow lines. The higher this ratio, the greater the delay before the stream reaches its peak flow after a storm. Hortonian flow tends to be characteristic of impermeable thinly vegetated slopes in arid climates, for instance on the much dissected slopes in soft rocks known as 'badlands', non-Hortonian flow of permeable forested slopes in humid climates.

19.2.2 *Runoff*

Surface runoff occurs where rainfall intensity exceeds the infiltration capacity of the surface, and where it occurs at all in a basin, it tends to be widespread. It requires progressively higher rainfall intensities to occur as the covering of soil, humus, litter, and vegetation becomes thicker.

The depth of overland flow depends on whether it is laminar or turbulent. Horton (1945) postulated that areas of one were interspersed with areas of the other. The depth of laminar flow can be estimated from the use of formulae relating channel discharge directly to its catchment area. For turbulent flow, depth may be estimated by combining the continuity equation

$$q = WDv \qquad [19.1]$$

where q = volume of flow $(m^3 s^{-1})$;
W = mean width of channel (m);
D = mean depth of channel (m);
v = mean velocity of flow $(m s^{-1})$,

and the Manning equation

$$v = \frac{0.454}{n} R^{2/3} s^{1/2} \qquad [19.2]$$

where

n = the surface roughness factor (Manning's resistance coefficient). This is an empirical value which varies with type, height, and density of vegetation, roughness of land, and in flow channels, with their size, cross section, alignment, and hydraulic radius, i.e. generally the conditions which increase turbulence and retard flow. It can be assessed as follows (Cowan, 1956): first, a basic n value is selected for the material under consideration – 0.02 is a useful figure for topsoil, 0.025 for rock, 0.024 for fine gravel, and 0.028 for coarse gravel. To these basic values are added numerical assessments in direct proportion to the resistance to flow posed by four further characteristics: surface irregularity 0.02; obstructions (such as debris, stumps, logs, roots, etc.) 0–0.06; vegetation 0–0.05; and degree of meandering 0–0.3. The results of all assessments are then added together to give –n. In general, in smooth materials such as bare rock it lies between 0.012 and 0.024; in bare soil without vegetation it increases from 0.025 to 0.06 with increasing coarseness of texture; and in vegetated channels it increases from 0.04 to 0.2 with increasing size of vegetation (Schwab *et al.*, 1966).

R = the cross–sectional area of flow (m^2) divided by the wetted perimeter (m).
s = the hydraulic gradient (tangent of slope angle or $m m^{-1}$).

The volume of runoff per unit area is the measured as stream flow divided by the area of the contributing catchment.

19.2.3 *Infiltration*

Water which does not run off infiltrates into the soil, and then moves both vertically and laterally. The rate and direction of this movement depends on permeability, which is governed largely by the nature and size of soil pores, gravity, the osmotic suction of plant roots, and in unsaturated soils, also on the capillary attraction resulting from the difference in moisture content between neighbouring parts of the soil.

The added water penetrates to a limit called the wetting front. This is the point at which the hydraulic gradient ceases to be sufficient to overcome the air pressure in the soil pores. Further penetration only occurs when more water is added to the surface, enabling more to percolate downwards by gravity, thus displacing water to lower pores. Thus the wetting front advances spasmodically and the water tends to remain in lenses representing the consecutive periods of surface wetting.

'Field capacity' is reached when the soil has absorbed as much water as it will hold against gravity before drainage loss occurs. After this, percolation attains constant rate. This rate is often less rapid than the infiltration into dry soil because the wetted soil expands to fill cracks and fissures. In the saturated subsoil the velocity of groundwater transmission is defined as the product of the hydraulic conductivity (see section 13.7) and the hydraulic gradient (measurable in terms of a loss of piezometric head), divided by the distance over which this is occurring. Its quantity can be derived by multiplying the velocity by the cross-sectional area across which it passes. Figure 19.2 makes these relations clear.

Infiltrated water either flows laterally downslope within the soil zone or else penetrates to join the groundwater table. The first is known as 'throughflow', the second as 'percolation'.

19.2.4 *Throughflow*

Throughflow is probably the predominant form of hillslope flow in humid regions. It becomes concentrated into horizons permitting downslope seepage known as 'percolines'. It increases relative to percolation with slope steepness, especially where the soil permeability decreases with depth in the profile. The maximum throughflow/percolation ratio occurs therefore at midslope, especially near the point of inflexion between the convex upper and the concave lower parts, and where there are 'argillic horizons' (subsoil horizons enriched in soil clay), soil pans, or permafrost (permanently frozen) layers. On the other hand, the proportion of percolation will tend to a maximum on relatively level sites at the top and the bottom of the toposequence. Throughflow and much of the percolation will ultimately reach the stream channel.

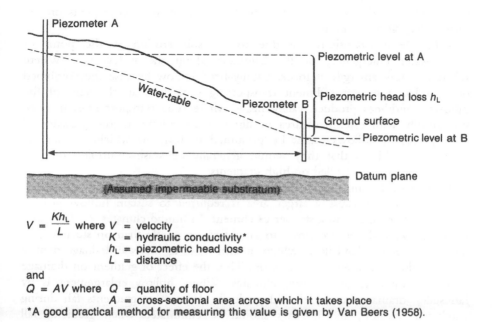

$$V = \frac{Kh_L}{L}$$ where V = velocity
$\quad\quad K$ = hydraulic conductivity*
$\quad\quad h_L$ = piezometric head loss
$\quad\quad L$ = distance

and

$Q = AV$ where $\quad Q$ = quantity of floor
$\quad\quad\quad\quad\quad A$ = cross-sectional area across which it takes place
*A good practical method for measuring this value is given by Van Beers (1958).

Fig. 19.2 Diagram and formulae to illustrate underground water movement

19.2.5 *Saturated overland flow*

In some locations, vertical and horizontal percolation saturates the soil throughout its depth. This causes water to emerge and become overland flow (path 3 on Fig. 19.1). The location and nature of this change are fundamental for the rate and nature of landform evolution. Recognizing where it may occur and cause gullying are of practical value (Douglas, 1977).

Points of emergence are most commonly on concave or hollow parts of slopes, and where soils are thin or impermeable. The resulting stream heads occur in various situations: at springs at geological boundaries, at the edges of swampy zones, at lines of coalescing subsurface water movement, at the foot of masses of rock debris occupying the highest parts of first-order valleys, or in alluvial fans in seasonally wet terrain. They sometimes emerge upwards due to the 'artesian pressure' developed by downward seepage over the higher parts of a slope (Kirkham, 1947).

There is often a variation in their locations following occasional or seasonal changes in the moisture regime. The stream head is extended by back-wall collapse at times of heightened runoff or rapidly raised water-table, while the hollow tends to be filled by debris in drier periods. There is thus a continuing adjustment between the tendency of the channel head extension to form a hollow of increasing concavity and the tendency of mass movements on the surrounding slopes to fill it.

Valley head extensions are due to (a) soil erosion at the point of groundwater emergence, or (b) landslides along the valley axis. Where subsurface flow emerges, it forms a gully-head hollow because the combined outflow has a greater sediment transporting capacity than the sum of the inflows. Landslides produce valley heads where slide transport rates increase substantially with catchment area or distance downslope. In the presence of both processes, hollows might be presumed to form to whichever has the lower threshold, so that they would generally be wash-controlled on low gradients or slide-controlled on high gradients.

In semi-arid climates, decreasing valley gradients lead to a decreasing drainage density because a larger area is required to sustain hollow growth from a gentler than from a steeper catchment. In humid climates, on the other hand, soil water-levels are closer to saturation on gentle than on steep slopes. This increases overland flow, which in turn tends to increase drainage density as the landscape is lowered by erosion. Thus the effect of gradient on drainage density is opposite in the two climates. Where hollow enlargement is by landslides, drainage density should gradually decrease as gradients fall during catchment erosion, and hollows should be eliminated when gradients fall below the ultimate geotechnical threshold (Kirkby, 1987).

Precipitation falling on the zone of saturation will be unable to infiltrate and will be added to the overland flow. It is difficult to separate water from these two sources, and so they are usually considered together as 'saturation overland flow' (Dunne, 1978). The zone of saturation can also produce runoff early in a storm period while infiltration is still occurring higher on the slope, and individual soil layers may produce separate seepage hydrographs following storms (Whipkey, 1965).

19.2.6 *Percolation and storage*

Some infiltrated water percolates to the groundwater table and can either reach the stream or penetrate to deeper subsoil aquifers (path 4 in Fig. 19.1). The process can be slow, often taking years. Such aquifers are composed of rocks or unconsolidated materials which hold and transmit water through pores and other voids. Their underground extent and configuration can sometimes be deduced from surface outcrops. Their effective storage can be calculated from the volume of fissures and pores over about 30 μm in diameter through which water can move relatively rapidly. Rocks differ widely both in porosity and permeability, some representative figures being given in Table 19.1. Among those that can be aquifers, igneous rocks hold most of their water in fissures, and granular materials such as conglomerates, sandstones, alluvial deposits, and sand dunes hold most in pore spaces. However, most flow in sandstones is along cracks and fissures (Mainguet, 1972). Calcareous rocks are porous but usually non-granular and have a network of internal voids resulting from solution.

Table 19.1 Representative porosities and permeabilities for some common materials (after Waltz, 1969)

Material	Representative porosities (% void space)	Approximate range in permeability at hydraulic gradient $= 1$ $(1 \text{ day}^{-1} \text{ m}^{-2})$
Unconsolidated		
Clay	50–60	0.0005–0.05
Silt and glacial till	20–40	0.05–500
Alluvial sands	30–40	500–500,000
Alluvial gravels	25–35	500,000–50,000,000
Indurated		
Sedimentary:		
Shale	5–15	0.000005–0.005
Siltstone	5–20	0.0005–5
Sandstone	5–25	0.05–5,000
Conglomerate	5–25	0.05–5,000
Limestone	0.1–10	0.005–500
Igneous and metamorphic:		
Volcanic (basalt)	0.001–50	0.005–50
Granite (weathered)	0.001–10	0.0005–0.5
Granite (fresh)	0.0001–1	0.000005–0.0005
Slate	0.001–1	0.000005–0.005
Schist	0.001–1	0.00005–0.05
Gneiss	0.0001–1	0.000005–0.005
Tuff	10–80	0.0005–50

19.2.7 Shallow aquifers and springs

Groundwater is most often exploited where accessible at springs or from shallow wells, whose distribution is related to terrain features. Springs, which normally form the 'fingertips' of the hydrographic network, can be categorized as hydrostatic, artesian, deep-seated fracture, or tubular (Bryan, 1919). Hydrostatic springs, which form the majority, occur where the land surface intersects the water-table in porous rocks either because of a break in slope or at the contact with an underlying impermeable bed. Common locations are above natural constrictions or steps in the stream profile due to the outcrop of harder rocks. These 'back up' the water behind them, increasing upstream infiltration. Water may also seep laterally into neighbouring permeable rocks along the stream course. Artesian springs are the special case where the water emerges under pressure through an overlying aquiclude. Deep-seated springs are associated with volcanic landforms and emerge under pressure from vents or deep fissures. They may be hot and include gaseous emissions, i.e. geysers. In arid areas, evaporation of mineralized water from artesian springs forms 'spring mounds' of evaporite materials to the height of the piezometric head. They often have a central crater containing water. They occur in North America and the Middle East, but the largest recorded examples, over 30 m high, are in Tunisia and Australia. They indicate present and past hydrological conditions (Roberts and Mitchell, 1987; Watts, 1975). Fracture springs emerge from fractures, fissures, and joints in impervious, usually igneous, rocks but differ from the deep-seated type in that they result from gravitational, rather than volcanic forces. Their discharge is often related to the greatest concentration of fractures, whose pattern of occurrence is often interpretable from surface topography and vegetation. Tubular springs are similar except that they follow more or less rounded channels due to features such as solution pipes in limestones, caverns in lava, or pipes along decayed tree roots. Some landforms, such as sand dunes, which are both permeable and water retentive, can absorb and store rainwater in higher positions than the surrounding water-table.

19.2.8 *Streams*

From the hydrological point of view, the most important geomorphic characteristics of streams are those which control discharge. Their characterizing elements, however, are so interwoven that it is difficult if not impossible to describe them independently. The mobility of the stream bed and the amount of discharge both vary over very wide ranges and result in exceedingly complex phenomena of progressively changing nature. Rivers seldom reach a state of equilibrium even over short distances. A state of change is the rule rather than the exception. For each varying condition of flow there is a new regimen which the river approaches but never reaches because the influence of time-lag does not permit the establishment of equilibrium before a new and changing condition of flow is encountered.

Certain generalizations are, however, possible. In river stretches not influenced by intermediate lakes or tidal fluctuations, there is, at least in humid regions, an increase in flow from source to mouth, mainly due to the entry of tributaries. Discharges are approximately proportional to the square of the width of the channel and, given uniform climate, with catchment area. In Virginia and Maryland, for instance, discharge in cubic feet per second approximately equals the catchment area in square miles (Hack, 1957). There is also a positive correlation between discharge and the square of the length of individual meanders.

Velocity of flow remains approximately constant throughout the course of a river irrespective of its size or gradient. This constancy is maintained in the face of changes in volume and sediment load mainly by changes in stream geometry. The narrowing or diminution of channel size, for instance, tends to be accompanied by an increase in gradient, while widenings or subdivisions are associated with a decrease.

19.3 Feedback in terms of drainage density and erosion

Where overland flow is produced at uniform volume per unit area over a hillside, the erosive force of the flowing water increases with both distance from the divide and with the gradient. Therefore, on a hillside with uniform gradient, the erosive force is greatest at the base of the slope; but on a convexo–concave slope it is greatest a short distance downslope from the region of steepest gradient. Rills begin here, the deepest and widest forming where the length of slope below the divide is greatest. These grow into gullies by micro–piracy of smaller rills, and once established, become the foci of secondary drainage channels developing in a direction normal to them by the process known as cross–grading (Horton, 1945). Figure 19.3 illustrates the process. Gullies and streams have concave–upwards longitudinal profiles with diminishing gradients. This gives them a logarithmic form which is ascribed to the progressive downstream increase in the efficiency of their channels in adjusting discharge, gradient, and sediment load.

When viewed in plan, landscapes vary in the ratio between overall length of streams and catchment area. This ratio is positively correlated both with that between surface runoff and infiltration and with the rapidity of the achievement of peak flows after storms. The stream length/catchment area ratio has been quantified by indices such as 'drainage density' and the 'texture ratio'. The former, also known as 'drainage texture', is the ratio between the total length of the stream channels in a basin and its area, expressed in kilometres per square kilometre. The 'texture ratio', suggested by Smith (1950), is defined as the number of crenulations in the basin contour having the maximum number of crenulations, divided by the perimeter of the basin. Strahler (1957) has shown that these two indices are related. A wide range of values for drainage density has been recorded: 3–4 for the sandstones of Exmoor and the Appalachian Plateau, 20–30 for the scrub-covered coast ranges of California, 200–400 for the shales of the Dakota badlands and up to 1300 for unvegetated clays (Chorley, 1969). The lowest values generally occur in areas of hard, permeable rocks under deciduous forest and the highest in badland-type areas of soft impermeable clays and shales under semi-arid conditions.

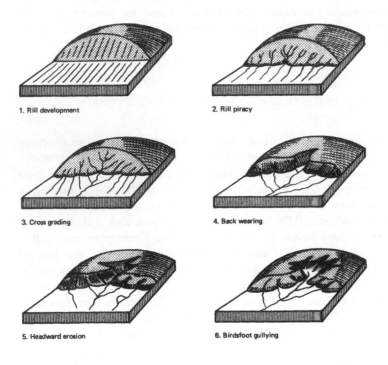

1. Rill development   2. Rill piracy

3. Cross grading   4. Back wearing

5. Headward erosion   6. Birdsfoot gullying

Fig. 19.3 Sketch to illustrate the development of a drainage channel on an original cut slope in uniform materials (adapted from Horton, 1945)

The rapidity in attaining the peak hydrograph also increases with the circularity of the basin. Various indices have been devised to quantify this, among the most widely used being those which relate the length of the master stream either to the average width of the basin or to the diameter of a circle equal to its area (Rodda, 1969).

19.4 Catchment classification

19.4.1 *Types of classification*

A terrain classification for hydrology must use drainage catchments as the basic units, within the framework of the hierarchy of 'stream orders' proposed by Horton (1945) and simplified by Strahler (Chow, 1964). This framework has stimulated much research, summarized by Haggett and Chorley (1969).

Stream order 1 represents fingertip streams receiving no tributary. Stream order 2 includes those formed by the junction of two first-order streams and receiving only first-order tributaries. Stream order 3 includes those which are formed by the junction of two second-order streams and only receive first- and second-order tributaries, and so on (Fig. 19.4). Both the number and the overall length of the streams within a drainage basin decrease geometrically with stream order.

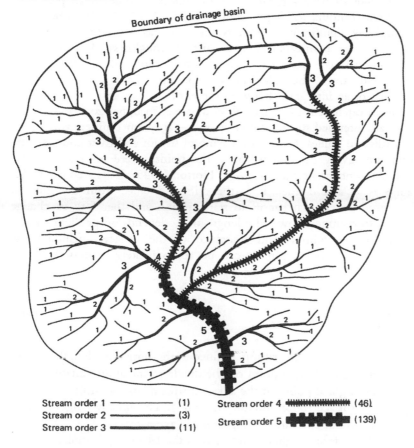

Stream order 1 ——————— (1)   Stream order 4 ⫻⫻⫻⫻⫻⫻⫻⫻ (46)
Stream order 2 ——————— (3)
Stream order 3 ══════════ (11)   Stream order 5 ▓▓▓▓▓▓▓ (139)

Fig. 19.4 Imaginary drainage basin showing Strahler's stream ordering system

259

The 'basin order' is defined numerically by the highest order stream it contains. This gives a hierarchy that parallels that of stream orders and forms the basic subdivision of the terrain. In any given catchment, there is a positive quantitative relationship between basin order both with the discharge of its stream, and the square root of its total drainage area. The hydrological analysis of a major catchment is a synthesis of those of its constituent lower order catchments.

19.4.2 *The subdivision of order 1 catchments*

The ultimate hydrological subdivision of the catchment is the order 1 basin. In general, it has a greater proportion of convex slopes than a larger catchment and its basal concavities only show convergence near to the valley axis. Therefore, it tends to have a greater area of concave profiles (>25 per cent) than of concave contours (*c.* 20 per cent) (Kirkby, 1978a, b). Its subdivision into hydrologically significant land units must be based on the interrelated factors of gradient, permeability, and surface roughness, because these determine the balance between evaporation, runoff, throughflow, and percolation, and thus the classification of the basin as Hortonian, non–Hortonian, or an intergrade between the two.

In Hortonian catchments, overland flow traverses an upper zone where it is unconfined, an intermediate zone where it is partly unconfined and partly confined in rills and gullies, and a lower zone of stream and river courses. The upper zone rims every catchment above the point where both runoff and infiltrated water concentrate into channels. Horton (1945) designated the width of this zone as x_c and defined it as the critical distance from the hillcrest required to produce enough runoff to start erosion and therefore to change from overland to confined flow. The length of x_c depends on gradient, runoff intensity, and the roughness and resistance of the surface, including the vegetation mat, to sheet erosion. It can be approximately expressed by the formula

$$x_c = \frac{0.8}{q_s n} \left(\frac{R_i}{f(s)} \right)^{5/3} \qquad [19.3]$$

where x_c = the width of no erosion, defined as the horizontal length of the flow path from a point on the divide to the approximate orthogonal point on the adjacent stream channel (m);

q_s = the surface runoff intensity (mm h^{-1}). This usually ranges from 12 to 50, 25 being an approximate average;

n = the surface roughness factor, as in the Manning formula;

R_i = the initial surface resistance to sheet erosion (N m^{-2}). This varies from about 5 for high resistance, i.e. where there is high infiltration capacity and a dense vegetation cover, to 24 for low resistance;

$f(s)$= the effect of slope on x_c. This is given by the formula

$$f(s) = \frac{\sin a}{\tan^{0.3} a}$$ [19.4]

where a is the angle of slope.

In first-order drainage basins, the x_c distance, sometimes designated as L_g (Chorley, 1969), seems generally to be equal to about one-half the reciprocal of the drainage density.

The second zone in the basin begins with the commencement of more or less equally spaced rills starting at springs or points of runoff concentration on the hillside below x_c. It includes the zone between the rills where the still unconfined flow is partially deflected laterally into them by the new 'micro-base level' that they impose. The third zone is the stream and river channels themselves.

In non-Hortonian catchments, most of the precipitation reaches the stream via throughflow. In the normal convexo-concave hillside profile, the simplest subdivision is into hilltop, mainly a zone of soil moisture recharge; midslope, a dynamic zone of throughflow; and footslope, characterized by the emergence of groundwater to join the stream and the main initial contributing area of overland flow to the stream hydrograph after storms (Fig. 19.5) (Dunne, 1978).

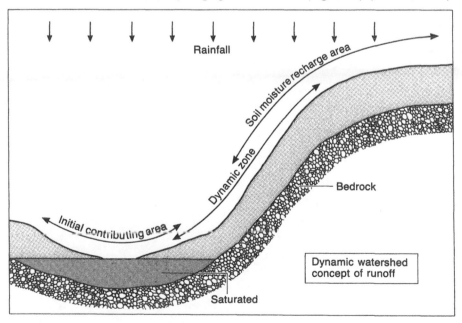

Fig. 19.5 Water flows on a non-Hortonian hillslope. (Source: Dunne, 1978)

Further reading

Chorley (1969), Farquharson *et al.* (1978), Kirkby (1978a), Knapp (1979), Narasimham (1979).

20

Terrain evaluation in geological and mineral survey

20.1 Terrain and geology

Terrain is the surface expression of deeper geology. Its interpretation, especially on remotely sensed imagery, is basic to the understanding of underground structures, stratigraphy, and the disposition of commercially important minerals. Such interpretation is most straightforward in areas of competent rocks undergoing degradation, but is usually also possible where they are buried under drift or covered by forest. The last decades have seen a number of studies of the surface forms of specific rocks in different climates, such as those by Mainguet (1972) on sandstones, Twidale (1982) on granites, and Sweeting (1972), Fenelon (1975), Nicod (1982), and Ford and Williams (1989) on limestone karst.

Certain relations between rock and soil are almost invariable. Higher ground is formed by more competent materials because of greater cohesion or permeability while lower ground indicates softer or more impermeable types. The former dominantly experiences erosion and degradation; the latter, deposition and aggradation. Promontories and bays along a coastline reflect harder and softer rocks respectively. However, the same rock can vary in competence from place to place, assuming different surface forms. It can even form a hill in one place and a valley in another depending on its hardness in relation to neighbouring rocks.

20.2 General relations of landform to geology: linear and circular features

One feature which aids in the recognition of geological information from the terrain is the interpretation of shapes as viewed from above. These can be broadly classified as linear or curved. Table 20.1 uses these categories in summarizing some of the ways in which terrain can be used to make geological interpretations.

263

Table 20.1 Some relations of landform and geology to minerals

Landforms	Geology	Minerals
1. Linear features		
1.1 *Igneous and metamorphic*		
Hills with constant slope angles and rhombic or rectilinear distribution	Plutonic rocks	Minerals and water along fractures; deep-seated rock bodies can contain kimberlites, metallic ores (Cr, Ni, Pt, Au); surface constructional materials
Dike ridges	Hypabyssal rocks; structurally aligned vulcanism	Mineralization, e.g. of gold
Strongly aligned areas	Metamorphic rocks; boundaries of tectonic blocks or rock types; plate junctions, aulacogens*	Mineralization along shear zones, marble, slate
1.2 *Sedimentary*		
Banded elongated rock zones, anticlines, synclines, periclines†	Alternations of hard (often limestone or sandstone) and soft (often shale or marl) rocks; fold directions and intensities indicate stresses	Water in porous aquifers; building materials; fossil fuels; ironstone; Pb and Zn ores in carbonate sediments
1.3 *Unconsolidated and secondary*		
Valley or coastal alignments	Linear fluvial tracts and beach ridges	Gravel and sand for construction

Landforms	Geology	Minerals
2. Curved or circular features		
2.1 Extrusive:		
Colours dark if ferromagnesian, lighter if siliceous, whitish if kaolinitic	Volcanic craters, calderas, ring dikes, necks, plugs, diatremes[‡] hot springs, geysers (often in clusters); lava flows with lobate edges	Mineral deposits in surrounding alteration zones, especially toward their outer edges. Fe shown by ochreous colours, other ores light from hydration and carbonation
2.2 Intrusive and structural	Plutons[§], domes, ambient zones of contact metamorphism, sometimes surrounded by annuli of lighter colour	Kaolin for ceramics; Cu, Pb, Zn, Mo, Sb, Hg
2.3 Impact	Meteor craters, astroblemes[¶]	
2.4 Denudational	Inselbergs and surrounding pediments (in tropics)	
2.5 Solutional	Sink holes and karst	
2.6 Depositional		
Conic alluvial spreads Fan-shaped alluvial spreads Scroll-patterned alluvial spreads	Talus, scree Alluvial fans River alluvium	Sorting by material size away from the source causes separation of building materials: sand and gravel; separation of minerals into residual lag deposits (Al and Ni), placer deposits (Au, Pt, Sn) and precipitates from solution (Fe, Ca, Mn, U)

Table 20.1 continued

Landforms	Geology	Minerals
Curved geometrical forms light in tone	Dunes	Constructional sand
Wind-oriented undulations with near-vertical valley sides	Loess	Brick-making materials
Hummocky surface, wavy ridges, leminiscate hillocks	Glacial drift, boulder clay, moraines, eskers, kames, drumlins	Brick clays
Wavy homogeneous level tracts	Lake beds; claypans in arid areas	Clay and salts

* An incompletely developed rift valley between major geological plates.
† A crustal fold structure in the form of a dome or basin.
‡ A volcanic vent through crustal rocks.
§ A cylindrical mass of granitic rock emplaced at high level.
¶ An ancient crater-like feature on the earth's surface, thought to be due to collision with a meteorite.

Linear alignments can be broadly categorized as elevated or depressed. Elevated alignments include hillcrests, plateau edges, dike ridges, and the boundaries of tectonic blocks or rock types. Depressed alignments include faults, coasts, and drainage lines, the latter often acting as pathways for mineral movement and as loci for their concentration. Natural lines are usually discontinuous, and a series of uplands, depressions, or gaps can betray the underlying structural control.

Most areas have a directional 'grain' in their topography, reflecting the alignment of faults and structural features, especially when viewed on aerial or satellite photography. A rectilinear or rhombic distribution of uplands, where slope angles are relatively constant, indicates igneous and metamorphic rocks. The fault lines sometimes are associated with deep-seated rock bodies of economic importance such as kimberlites.

On a continental scale, mountain chains usually include aligned fault systems and associated vulcanism which may be related to plate junctions. While most of the latter are planar, the earth's spherical shell requires that there be joints where three plates meet, forming a three-armed rift. Sometimes two of the arms become active rifts and open to form an ocean while the third remains as an aborted rift valley, known as an 'aulacogen', which as other valleys, may become filled with sediment and provide a favourable site for water, mineral, or petroleum accumulation, sometimes indicated by springs or seeps. A number of examples of aulacogens are found in Gondwanaland between the American and African plates, e.g. the Ouachita Mountains, Oklahoma, and in the East African Rift System, e.g. the Afar Triangle (Siegal and Gillespie, 1980).

Geological phenomena associated with curved or circular geomorphic features can be due to igneous, structural, fluvial, or impact forces. Both volcanic and plutonic rock masses are often subcircular in form when viewed from above. Circular features in level topography may result from internal drainage. These include ponds and sink holes in humid climates or claypans in arid. Meteoric impact craters are circular, and may be relatively fresh such as the Great Meteor Crater in Arizona, or degraded in form, to which the name 'astrobleme' has been given, of which examples occur in Canada (Siegal and Gillespie, 1980).

20.3 Terrain, rock types, and economic minerals

20.3.1 *Volcanic rocks*

Volcanic landforms are generally recognizable from the distinctive shape of volcanoes, calderas, and cinder cones, often occurring in clusters, and of lava flows. These last are distinguished by their dark colours due to the prevalence of ferromagnesian minerals, and by a lobate surface when fresh, which is relatively quickly weathered. Hot springs and the whitening of surface materials by kaolinization indicate geothermal activity. They may be surrounded by alteration zones in whose outer fringes are oxidized and carbonized minerals containing metallic ores of economic significance.

20.3.2 *Plutonic rocks*

Landforms on plutonic and hypabyssal rocks include domes, diatremes (plug-like intrusive bodies), and ring dike complexes. Plutonic rocks, because they harden in the earth's crust, mainly outcrop where they are exposed by circumdenudation, reflecting climatic changes in evidences of multistage evolution. As they are usually more acidic than volcanic rocks, they are more resistant and show fewer evidences of surface weathering. They have relatively constant and repetitive slope angles without directional orientation, and summits which are often accordant in height. Their distribution and limits reflect the underlying fracture patterns which are usually rectilinear or rhombic. Plutonic landforms differ somewhat between climates. In the temperate zone, the combination of thick vegetation cover and soil mantle gives them relatively smooth convex–concave slope profiles. In old, tectonically stable parts of the tropics, they have steep erosion slopes of uniform gradient surrounded by relatively level and unaccidented 'etch plains' or 'pediplains' from which they are often separated by sharp nick points (Fig. 20.1). In arid and semi–arid areas, they are indicated by four major types of landform: boulders and tors, inselbergs, intricately dissected terrain known as 'all–slopes' or 'polyconvex' topography (in contrast to dissected sedimentary tablelands which are 'polyconcave'), and plains (Twidale, 1982).

Metal ores, constructional materials, and water supplies are obtained from igneous rocks, and volcanic contact zones are often recognizable from a 'whitening' of the surface materials through geothermally induced kaolinization. Ores in igneous rocks are known as 'primary' to distinguish them from those deposited as sediments which are known as 'secondary'. Ores of such metals as chromium, nickel, and platinum and to a lesser extent the base metals are frequently deposited as sulphides from cooling hydrothermal solutions in magmatic chambers which are often at the edges of an igneous outcrop. Lithium is restricted to pegmatites. Hydrothermal quartz veins can indicate the presence of metallic gold. Where such zones are exposed to the air above the water-table they tend to weather into oxides or carbonates, detectable from the red and yellow colours of the iron compounds which accompany them. Surface occurrences of basalt and granite suitable as building stones, road metal, and ornamental masonry, can be recognized from their distinctive outcrop patterns. In igneous areas groundwater supplies are associated with the trenches formed by fractures, especially where emphasized by the presence of unusually lush vegetation. This method was used, for instance, to locate possible water supplies on satellite photographs of South Yemen (Travaglia and Mitchell, 1982).

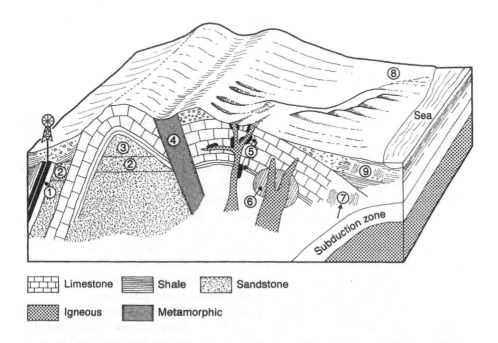

Fig. 20.1 The relation of terrain to the locations of some mineral ores (schematic): (1) coal; (2) oil; (3) gas; (4) slate and marble; (5) lodes along dikes and sills. e.g. precious metals; (6) contact metamorphism: non-ferrous metals; (7) ascent of calc./alkaline magmas including metals; (8) placer deposits; (9) secondary groundwater-borne accumulations, e.g. iron

20.3.3 *Metamorphic rocks*

The terrain of metamorphic rocks resembles that of igneous except for the regional lineations imposed by shear stress, generally expressed in a topography of subparallel ridges and valleys. The intersection of different lineations indicates crustal block tectonics and guides in the location of mineralized zones. In mountainous or shield areas these are usually related to igneous outcrops caused by the same diastrophic forces. Where the two meet there is a zone of 'contact metamorphism' causing physical and chemical changes to the rock structure and forming new compounds including copper, lead, and zinc ores. Where such compounds are schistoid or kaolinitic, the junction zone will be softer and more vulnerable to weathering than where they are siliceous, and

269

this may initiate a topographic hollow. Kaolinitic clays may also, as in Cornwall, be usable for ceramics. Metamorphic rocks of economic value, such as marble and slate, occur in lineated areas subject to strong lateral pressures. Degraded mountain or shield areas may guide to the recognition of Pre-Cambrian rocks which are associated with certain minerals: nickel and sulphur in stratiform chromite, and titanium in anorthosites. Mountains formed by the orogenesis of Mesozoic and Tertiary rocks are a source of copper, molybdenum, antimony, and mercury.

20.3.4 *Sedimentary rocks*

Sedimentary rocks are usually more readily recognizable than igneous or metamorphic and yield more lithologic information because of the interpretability of alternating strata, even where these are elevated, folded, or faulted. Their alignments can be distinguished from those of metamorphic rocks by being longer, less numerous and more evenly spaced. Characteristic landforms are anticlines, synclines, and uniclinal structures. Erosion differentiates them so that the harder or more permeable rocks, such as limestone and sandstone, form the hills and mountains, usually in the form of cuestas and anticlinal ridges, and the softer shales and clays form the valleys. Along coasts, the competent rocks form the headlands, the soft ones the inlets. Polyphase deformations are indicated by complex outcrops or irregular patterns of rock types, especially in the form of periclines and geoflexures. The former are folds in which the dips are away from, or towards, a centre – i.e. domes or basins – and the latter are curved fold belts. They can be interpreted from unconformities and the bedding surfaces of marker horizons visible at the surface. It is also notable that, following Pumpelly's rule, the axes and planes of macroscopic (i.e. regional) folds are generally parallel to, and can be deduced from, the mesoscopic folds (i.e. those visible in sections and hand specimens) formed by the same tectonic event (Turner and Weiss, 1963).

Sedimentary rocks yield fossil fuels, ores, and building materials. Fossil fuels occur generally at some depth. The underground extent and distribution of coal beds can be inferred from surface outcrops and verified from borings. Oil and gas, being lighter than the water with which underground rocks are saturated, tend to migrate upward into aquiclude–covered domes and other structural traps (see Fig. 20.1). Their location can often be inferred from the nature and size of fold structures visible at the surface. Certain ores are associated with particular rock types, such as those of lead and zinc with carbonate sediments. The presence and distribution of building materials derived from sedimentary rocks may be inferred from the terrain. The scarp and vale landscapes of southeastern England and northern France reflect parallel outcrops of clay, sandstone, limestone, and chalk with flint, each of which has had constructional uses (e.g. Clifton-Taylor, 1982).

20.3.5 *Unconsolidated and secondary deposits*

Secondary mineral accumulations are recognizable from the form of the terrain resulting from the depositional processes which formed them. The downwind and downstream sorting of materials by grain size concentrates them into useful deposits of gravel, sand, silt, and clay in locations which are predictable from a study of the connections between source areas and depositional processes. Ice deposits are recognizable from indications of past movements of glaciers and ice sheets. The mixed sizes of their constituents reduce their value as a source of constructional materials, although included boulders and stones are sometimes useful in areas otherwise impoverished. Colluvial, alluvial, aeolian, and marine deposits are sometimes recognizable from evidence of their modes of origin. Gravel and sands for use in construction are normally obtained from river terraces or beaches, and sand also from dunes where these are accessible.

Fluvial processes concentrate some economic minerals. Some, such as aluminium, iron, and nickel ores are left behind in residual landforms when more easily weathered materials are removed. The concentration of residual iron oxides, in particular, accounts for the ferricrete shields and associated terrain which dominate the landscape in parts of the humid tropics. Some ores such as gold, platinum, copper and tin can be carried by rivers, sorted by specific gravity, and left as 'placer' deposits in slack water areas. Where rivers run past mine workings or oil wells, they may redistribute the spoil or slick into recognizable 'tailings' which may have economic value. Diamonds often occur in alluvial deposits, as in Sierra Leone (Thomas and Thorp, 1985), and have been distributed by longshore drift around a Quaternary lake in Botswana (Cooke and Shaw, 1986).

Many economic compounds are taken into solution, transported, and precipitated where the solution evaporates or undergoes chemical change. The commonest examples are the soluble salts of the alkali metals, but it is also true of iron, manganese, uranium, and phosphorous compounds. Such chemical movements can often be interpreted from terrain conditions and the nature and extent of commercial deposits inferred. Iron, for instance, can be carried in solution in groundwater and precipitated in the interstices of its aquifer, sometimes forming beds of 'ironstone' which may cement the surrounding matrix, making a building stone, or where strongly concentrated, an ore.

Further reading:

Allum (1966), Clifton-Taylor (1982), Cooke and Shaw (1986), Ford and Williams (1989), Howard and Mitchell (1985), Jensen and Bateman (1979), Mainguet (1972), Siegal and Gillespie (1980), Sweeting (1972), Thomas and Thorp (1985), Twidale (1982).

21

Terrain evaluation in soil survey

21.1 Soil and terrain

Soil is a narrower concept than terrain, because it does not include the deeper underlying materials or the surface form. Soil survey is the preliminary stage to the evaluation of the economic qualities and limitations of land which determine its uses. It is based on determination of the spatial distribution of soil profiles.

The soil profile extends downwards to the lower limit of the common rooting of native perennial plants to a diffuse boundary which is shallow in deserts and tundra and deep in the humid tropics. The 'weathering profile' is a vertical section through both soil profile and regolith. The two last coincide except where the regolith is subject to clearly non-biological weathering, as for instance in barren parts of the Arctic and the lower parts of deep weathering profiles in the humid tropics. Since weathering leaves varied traces through time, both sorts of profile are palimpsests of the changes which have occurred within them since their formation.

21.2 The role of terrain in soil formation

21.2.1 *Relief*

The influence of relief on soil formation comes through two aspects: slope and altitude. The effects of the former are mainly geomorphological, those of the latter, climatic.

Slope
The importance of slope angle on soil formation and the differences between soil profiles on different parts of a slope have long been recognized, for example by Milne's concept of the 'soil catena' (1935), which has been used by many workers, e.g. Pallister (1956) in Uganda and Ruhe (1960) in Iowa.

272

The 'soil association' has been used as a mapping unit, notably in Scotland, for slope sequences which dominantly vary in their moisture conditions. Although soils on plateaux usually have more moisture than those on slopes, the amount of soil water generally increases downslope to reach a maximum in valley and basin sites. In humid climates these are often waterlogged or marshy, while in arid areas they may become salt-pans.

When parent materials on a slope are complex, it is necessary to analyse the geological and geomorphic origins in interpreting the soil profiles. Figure 21.1 illustrates this. The slope has evolved as a result of several periods of alternating erosion and deposition. Thus, although the slopes on the two sides of the incised valley appear similar, they in fact reflect quite a different sequence of strata, and will give rise to different soils.

In addition to the effects of water, wind, and human uses, gullying and sheet erosion depend on slope steepness and vegetation cover. Gullying destroys the soil and tends to be at a maximum where its surface is soft and impermeable, as in 'badlands'. Sheet erosion, known as 'congelifluxion' in permafrost environments and 'solifluxion' elsewhere, moves soils downhill, generally causing them to be thinnest on steep slopes and thickest in valley bottoms. Jenny (1941) quotes the example of a Canadian chernozem on a slope from which the black topsoil was completely removed to the lower part of the slope, leaving the brown B horizon exposed on the upper. These processes are at a maximum where the soil readily becomes saturated and where it overlies a smooth surface, especially when this is lubricated with groundwater.

Soils on slopes are less deeply leached than analogues on more level surfaces. This is because they experience relatively more runoff and less infiltration. Daniels *et al.* (1971) quote examples from a loess area in Ohio from which they calculate a quantitative ratio between gradient on the one hand and both the thickness of the solum and the depth to a carbonate layer on the other. This latter is significant in giving a numerical relationship between the depths of normal water penetration and of the carbonate layer, and in showing how much both the total depth of the soil and the depth of the carbonate layer increase with a decrease in the surface gradient. Beckett and Furley (1968) showed that on 10 slope transects on different materials in the Oxford area on all soils the contents of carbon and nitrogen, on acidic soils the pH, and on calcareous soils the percentages of silt plus clay, decreased with gradient. The only attribute showing an increase with gradient was the pH of calcareous soils.

Altitude

Although there are some variations due to anabatic and katabatic winds, mean temperatures generally decrease consistently with altitude. Rainfall, due to orographic effects, increases up to a certain altitude above which it again decreases. This causes a vertical zonation of soil types which is somewhat

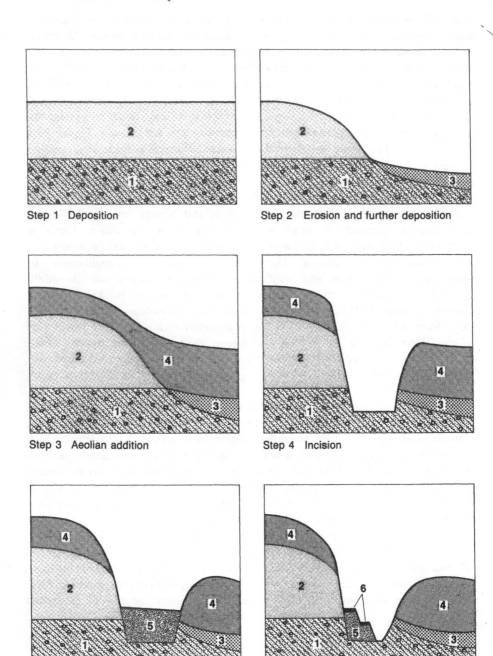

Step 1 Deposition

Step 2 Erosion and further deposition

Step 3 Aeolian addition

Step 4 Incision

Step 5 Alluvuiation

Step 6 Terrace formation

Fig. 21.1 Model of a complex slope evolution. Note the strong differences which can develop between the two sides of a valley

analogous to a latitudinal transect from a relatively dry warm climate through a cooler wetter one to a cold dry one. This zonation is reflected in soil profiles.

Examples of soil transects on hillslopes are given in Figs. 21.2 and 21.3. Fig. 21.2 records an area near Kelso in the Southern Uplands of Scotland, which can be compared with the comprehensive model of hillslope units by Dalrymple *et al.* (1968) in Fig. 3.2. The units in both are at approximately land facet scale. The Kelso soils are developed on arkosic Ordovician and Silurian rocks and drifts derived from them. A subalpine podzol, the Merrick Series (not shown), is common on hills above 600 metres. The hill peat is on an ill-drained 400–600 metre plateau surface. The podzolic Dod and Minchmoor Series, generally above about 300 metres, have deeper water tables. Below these are brown forest soils which become increasingly gleyed downslope as the water table shallows. The hillfoot Ettrick Series is a surface water gley and the alluvial soils in the valleys, at altitudes falling from about 250 to about 60 metres, are humic gleys induced by marshy conditions (Ragg 1960).

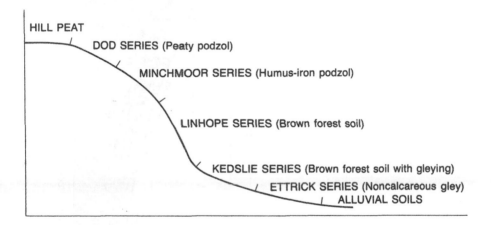

Fig. 21.2 Soil catena to illustrate the effect of altitude: the Ettrick Association in the Kelso area, Scotland (Source: Ragg, 1960)

At a broader scale, soil transects in mountain areas may sometimes be long enough to straddle major climatic zones. Jenny (1941) quotes Thorp's example of the west slope of Big Horns, Wyoming, which ranges from podzolic soils at over 8000 feet (2438 m) down to desert soils at about 5000 feet (1524 m) (Fig. 21.3). Here the soil units are at the approximate scale of land systems.

The position of a soil in relation to the depth of the water-table can lead to the prediction of pedogenesis. In a transect through seven closely related soil series in the Kinston, North Carolina, area of the southern Atlantic coastal plain, the C horizons of the Ruston, Bladen, Norfolk, and Portsmouth series consist of similar unconsolidated sedimentary material (Fig. 21.4) (Jenny 1941).

275

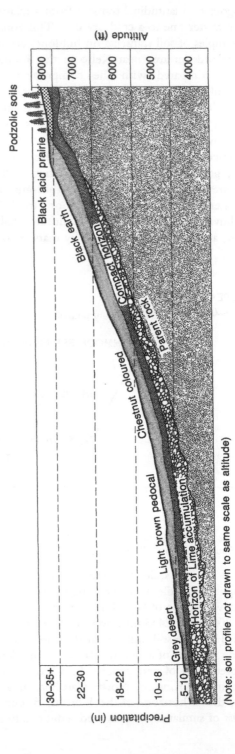

Fig. 21.3 Gradation of soil profile from desert slope to mountain top – west slope of Big Horns, Wyoming. (Source: Thorp, 1931)

(Note: soil profile *not* drawn to same scale as altitude)

The topography is relatively flat and the climate uniform. The Ruston and Norfolk series, which lie adjacent to the main drainage channels, have a relatively deep groundwater table, while the Bladen and Portsmouth series, which are more remote, have a shallower one. The well-drained soils have lower organic matter and silica:alumina ratios, but higher pH levels and clay lessivage (downward translocation of clay particles in the profile) from A to B and C horizons than the poorly drained soils. Variations in drainage consequent solely upon topographic location thus clearly differentiate the profiles.

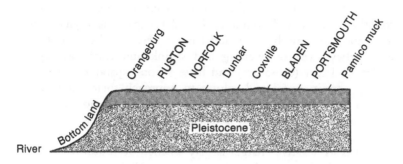

Fig. 21.4 A part of the Appalachian Piedmont illustrating the relationship of water-table and soils to relief. The water-table is the pecked line. (Source: Jenny, 1941)

21.2.2 *Lithology*

The other terrain factor which controls soil formation is lithology. This results from the effects both of the rocks as a whole and of their constituent mineral particles individually.

The resistance of rocks to weathering depends on their fracture pattern, the size and arrangement of particles, and their chemical composition. In igneous rocks, internal stresses caused by temperature and moisture changes are more destructive when the crystals are large. The rocks thus crumble relatively fast, but the constituent crystals are more resistant once detached because of their smaller total surface area. Sedimentary rocks generally consist of grains in a matrix arranged in strata whose alignments may be perpetuated in the soil. Calcareous rocks yield soils through the solution and carbonation of the lime. Sandstones and shales consist of relatively insoluble particles in a cementing matrix, whose nature determines the rate of weathering and the soil product. Although rocks are changed, their relatively resistant constituent grains, notably quartz in temperate and cold climates, survive in the soil. Some minerals, such

277

as zircon and tourmaline, are so resistant that they will survive through several weathering cycles and can be used as an index in calculating the rate of loss of accompanying minerals.

Physical weathering yields the coarser materials: stones, gravel, sand, and silt, whereas chemical weathering destroys minerals and leads to the formation of colloids. The former are relatively inert but provide the skeleton of the soil. The latter are largely responsible for determining its fertility.

Soil colloids consist mainly of clay minerals and iron oxides. They are softer than the parent rocks although they occupy a greater volume. There are three main types of clay minerals, depending on chemical composition: 'montmorillonite', 'kaolinite', and 'hydrous micas'.

Montmorillonite particles are very small ($c.0.01-1$ μm), with a large 'cation exchange capacity'. The small size of these particles gives a large total surface area and makes them plastic and cohesive. The 2:1 expanding crystal lattice structure permits them to absorb water and solutes between their tetrahedral layers. Therefore, they swell on wetting and shrink and crack on drying. Kaolinite particles are larger (0.1–5 μm) and have a non-expanding 1:1 crystal lattice which only permits water and solutes to approach their outer edges. Thus they have a much lower effective surface area and cation exchange capacity than montmorillonite, swelling and contracting much less on wetting and drying. Illite, the commonest hydrous mica, although it has a 2:1 expanding lattice, has its negative charges largely satisfied by K atoms, and so is intermediate in properties between the other two types of clay.

Montmorillonite tends to be formed under conditions of high pH, a good base supply, and poor drainage, whereas kaolinite occurs in strongly leached, acid, low-base environments. Montmorillonite is thus relatively more frequent in arid areas and kaolinite in humid, with illite occurring more widely, especially where potassium is abundant. The practical significance of these distinctions is that montmorillonites are more fertile, but kaolinites are more stable for supporting roads and buildings and better suited to uses such as brick making and ceramics. Illites are intermediate in these respects.

21.3 Rates of soil formation

The rates at which different types of terrain are weathered to soils depend on lithology, temperature, and the amount and chemistry of the percolating water. Attempts to measure these have been reviewed by Ollier (1975), Trudgill (1976), and Coates (1987). Methods have included assessing the rate of destruction of tombstones, Cleopatra's Needle and buildings of known age, the laboratory testing of rock samples in simulated field conditions, and the comparison of the physical and chemical characteristics of soils of unknown age with those of known age. Examples of the last have included soils developed on old stone fortifications, volcanic ash, moraines, sand dunes, coastal polders, and podzols (Jenny, 1941; Cruickshank and Heidenreich, 1969). Results are very variable. In calcareous rocks the rates of soil formation in terms of profile depth range from extremes of 0.003–11.5 mm yr^{-1}, with some tendency for humid areas to be more rapid than arid. There are too few results from other rocks to make valid generalizations. Radiocarbon dating of organic matter in humid temperate soils gives ages ranging from 200 to 2800 years (I. Douglas, 1989, private communication). The evidence generally indicates that parent materials can be weathered into mature soils in the humid temperate zone in a few centuries, and in the tropics in a few decades.

Another way of assessing the rate of soil formation is to link it to the weathering rates of different rocks by analysing the mineral content of their drainage water (Perrin, 1964). A few studies have been done in different environments. As with measurements of surface removal, the results are very variable. In the semi-arid climate of New Mexico, values of 14.6 m^3 km^{-2} yr^{-1} were obtained for sandstones, 3.3 m^3 km^{-2} yr^{-1} for granites, and 0.9 m^3 km^{-2} yr^{-1} for metaquartzites (Miller, 1961). In the cooler climate of Wales, the solution losses of all dissolved solids from impermeable greywackes were 1.55–2.05 m^3 km^{-2} yr^{-1} (Oxley, 1974). More research has been done on calcium loss in drainage waters from limestone areas and this has been supported by studies which have traced the water to its surface origin, measured its flow across weirs and statistically calculated its changes through time. Calculations by 19 authors from various karst regions give values for solutional erosion between 9.0 and 102.0 m^3 km^{-2} yr^{-1} (Atkinson, 1971), but Smith and Newson (1974) suggest a minimum of 50 m^3 km^{-2} yr^{-1} as more realistic. This is an order of magnitude faster than for other types of rock and represents a degradation of the surface by an average depth of about 0.05 mm yr^{-1}.

21.4 The relation of soil to terrain classification and mapping

The classification of soil for mapping purposes has always been quite distinct from that of terrain, although recently there has been a convergence of the two, partly as a result of the increased use of remotely sensed imagery. The historical development of soil classifications has followed three lines: (1) *ad hoc* schemes to meet immediate practical needs; (2) genetic schemes attempting to explain the worldwide distribution of major classes; and (3) attempts to fit both into comprehensive general schemes with a framework of quantitative definition. Although the three lines have to some extent developed together, they nevertheless broadly represent three stages in the overall development of soil classification.

Early soil classifications created classes using either rule-of-thumb local terms such as 'limon' or 'loam' or were based on single attributes such as clay or humus content. Such artificial systems are still widely used today at local level. The first recognition that soils were independent natural bodies with distinctive morphologies based on the effects of local and zonal soil-forming agents was by Dokuchaiev and his school (Afanasiev, 1927) in the late nineteenth century. This made possible a 'natural' classification of soils which has formed the basis of the development of the science to this day. The basic unit of classification is the soil profile and the major classes broadly represent zonal types created by the effects of climate. The accumulating data on soils since about 1950 have led to a third phase in which there has been progressive definition and systematization of soil classes based on natural morphological criteria.

The classification systems which aim at a comprehensive worldwide applicability are those produced by the USSR, France, the USA, and the Food and Agriculture Organization of the United Nations (FAO) in collaboration with Unesco. The Soviet and French schemes use traditional nomenclature and emphasize soil genesis and processes (Rozov and Ivanova, 1967; Commission de Pédologie et de Cartographie des Sols, 1967). The USDA Soil Survey staff's *Soil Taxonomy* (1976) presents a hierarchy of exactly defined mutually exclusive soil classes and also indicates the relations of these to the other major international systems. The definitive criteria of the classes generally become increasingly morphological as one ascends, and increasingly practical as one descends, the categorical sequence: order, suborder, great group, subgroup, family, series, and type. Their conceptual, though not their geographical, parallelism to terrain classification is given in Table 5.1. The FAO coordinated many different national soil survey classifications to produce the 17-sheet *Soil Map of the World* at a scale of 1:5 000 000 (FAO/Unesco, 1974–). This provides an international correlative framework with a new nomenclature based on all the previous systems. It has two higher categoric levels, both called 'soil units', and approximating to the broad-scale USDA suborders and great groups.

These classifications differ from those of terrain because they are based on the soil profile and its horizons rather than on geomorphic form and spatial relations. They are thus somewhat less readily mapped. Geographical relationships appear most clearly at the highest and lowest categoric levels.

The 11 soil orders are the largest units of the USDA scheme, and although they can occur anywhere in the world under the same environmental conditions, they nevertheless broadly represent the soils of the major climatic zones. Spodosols, alfisols, ultisols, mollisols, oxisols, for instance, approximately correlate with cold, intermediate, and warm temperate forest, mid-latitude grassland, and tropical rain forest climates respectively. These orders are thus both conceptually and geographically analogous to the land zones described in Chapter 6, because of the key importance of climate in defining both.

The accordance of soil with terrain classes at intermediate scales of mapping is less clear and it is, for instance, not uncommon for several great soil groups to be found within a relatively small area if the region is geologically or geomorphologically complex and of variable climate. The reason appears to be associated with a problem common to all comprehensive hierarchical classification schemes: the need to harmonize categories derived by deductive subdivision of the global population with aggregations of local units obtained by inductive classification. This problem has often led to the geomorphological grouping of soil units for mapping at intermediate scales, as in England and Wales at 1:1 000 000 (Avery *et al.*, 1974), the Netherlands at 1:50 000 (Soil Survey Institute, 1960), and Belgium at 1:20 000 (IRSIA, various dates). In Australia, the whole framework of soils mapping is now based on terrain units (Australia CSIRO, 1983).

The smallest soil unit is the 'pedon'. This is defined as the smallest area that can be called 'a soil' and is represented by a block 1–10 m square extending downward to the bottom of any pedological horizons present. Its definition is therefore largely geographical and large-scale mapping is based on the grouping of pedons, which can often be related to microrelief features. Similarly, soil types and soil series can be related to land elements and land subfacets respectively. Land facets usually contain at least two and often three or more soil series, whose boundaries seldom transgress theirs. Terrain is thus a good framework for large-scale soil mapping.

Further reading

Brady (1984), Coates (1987), Dalrymple *et al.* (1968), Daniels *et al.* (1971), Dent and Young (1981), FAO/Unesco (1974–), Gregory and Walling (1974), Jenny (1941), Ruhe (1975), Trudgill (1976), USDA (1976), Vink (1983).

22

Terrain evaluation in archaeology

22.1 The role of terrain in archaeological investigation

Terrain provides a key to many aspects of archaeology, since ancient land uses were dependent on particular site conditions recognizable both on the ground and from the air. For example, roadways follow bare hill crests, forts are on isolated hills, and towns and villages are almost always beside surface-water sources. Terraces along important rivers such as the Nile, Rhine, and Thames, occur in flights showing progressively more recent artefacts and remains down the sequence.

Archaeological sites can be considered at two scales: cultural units generally covering areas the size of land regions or land systems, and local settings for families or small groups, associated with land catenas and land facets.

At regional scale, populations have 'activity loci', such as river valleys or plains, within which social and economic interchanges are relatively rapid. In the ancient world, for instance, Mesopotamia, the Nile Valley, and the mountain-rimmed coastal plains of Greece, gave rise to distinct societies with a high degree of internal communication. Within these, there was often a close relationship between particular activities and certain land units, such as agriculture on alluvial areas, pasture on hills, markets at crossroads or bridging points, harbours in sheltered coves with relatively deep water inshore, and irrigation systems in arid areas starting at the debouchment of upland watercourses on to plains. The use of aerial photographs helps unravel the resulting palimpsest of tectonism, erosion, deposition, and changing human settlement patterns, which make up local history.

Many regions have such a potentially large amount of archeological data that it is necessary to use a sampling programme to guide investigations. The approach is either to select areas representing a microcosm of the region as a whole, or else to seek all the sites of a particular activity, such as agriculture, communications, or specialized crafts. In either case the choice will be of appropriate terrain units. A microcosm will be obtained, for instance, from transects across the 'grain' of the country, and particular activity sites from associated landforms such as riverside meadowland for pasture, waterfalls for mills, and rock outcrops for quarries especially in flint-bearing chalk.

Individual sites may be visible or invisible at the surface or from the air. In recent years there has been a rapid increase in the rate of discovery of hitherto invisible sites, largely through the recognition of settlement and vegetation patterns on aerial photographs. It is reasonable to assume that the surface indications of most ancient land uses resemble those of similar land uses today. Thus, to find new sites, it is advisable to seek traces on those similarly used today. These are sometimes so clear that, for example, Wheeler and Woolley were respectively able to predict the locations of the Roman fort at Dover and the royal palace at Atchana, Syria, on environmental grounds alone (quoted by Vita-Finzi, 1978).

On a wider scale, fertile soils, water sources, and ore bodies have always attracted human settlement and prima facie provide sites worthy of archaeological investigation. Field patterns normally conform to the landscape, with strips running up hills or along contours. Where they do not, instructive anomalies can appear. The rigid Roman grid of *centuria quadrata* and their *actus* subdivisions cover considerable areas in Tunisia, Yugoslavia, and France (Bradford, 1957), and settlement in parts of North America is in square plots. The imposition of a grid can cause revealing anomalies. It can force roads and bridges to cross rivers at awkward angles, divert the drainage system into artificial directions, and distort settlement patterns away from alignments most suited to the terrain. On the other hand, strong terrain features sometimes force a distortion of the grid, as when an ancient road is diverted around a large obstacle. Sometimes the obstacle disappears, leaving only the curve in the road to show where it once was. Anomalies of this sort therefore both reveal terrain factors and illuminate human responses to their constraints.

22.2 Terrain types associated with ancient land uses and their recognition

Certain types of terrain can sometimes be used to infer the occurrence of specific archaeological sites. Polar ice and permafrost act as a giant 'deep freeze' in preserving remains of all sorts. These can range from artefacts associated with hunting to the almost unaltered bodies of creatures such as mammoths. Remains can be buried under deposits of volcanic ash, as at Pompeii, or hidden by moving dunes, as at Skara Brae in Orkney, where the Stone Age village was exposed by a storm. Gravel surfaces were popular sites for early man because of their good drainage and the availability of workable stones, riverain alluvium because of its fertility, and limestone areas because of the presence of caves. In Britain, the treeless openness and supplies of flint in chalk landscapes made Salisbury Plain the focus of Stone Age settlement. Some Bronze Age sites are associated with tin ores in areas of contact metamorphism in Cornwall, and Iron Age forges are found in ironstone outcrops in the Weald and the Forest of Dean.

Some artefacts, such as the pyramids of Egypt, the ziggurats of Mesopotamia and Silbury Hill in England, and the Great Wall of China are large enough to be minor landforms. The *tels* of parts of the Near East and southeastern Europe are town and village sites formed into mounds by the accretion of the debris of human existence over many centuries. The *terps* in northern Holland and *wurts* in northwest Germany have somewhat similar form but differ in their mode of origin. Like *tels*, they are village sites, but unlike them, they have been heaped up deliberately by throwing clay, turves, or dung over successively abandoned levels, possibly to cope with the onset of coastal subsidence (Clarke, 1964).

Some large archaeological sites, notably those associated with irrigation works, are visible on satellite photographs. An example is the ancient rainwater concentration structures in the Egyptian desert west of Alexandria (Allan and Richards, 1983; Richards, 1989).

Many surface modifications resulting from human activities show up on aerial photographs. Water-control works such as irrigation canals in Iraq (Hunting Technical Services (1954–), and moats in Cambodia (Moore, 1989) are notable. Smaller features can be broadly classified into 'shadow sites' where there is sufficient topographic variation to cast shadows, and 'crop marks' or 'soil marks' which are due to differences in tone on a homogeneous surface (Atkinson, 1970). Shadow sites show up particularly well under oblique lighting. They include strip lynchets, headland ridges, and terraces from ancient field patterns; necropoli including barrows; and military breastworks. Broken ground can also indicate where a search for such raw materials as lime, clay, marl, and building stone has scarred the surface of the land. Aboriginal shell middens indicating gathering points are recognizable on large-scale photographs of Tasmania (Howard and Mitchell, 1985).

Crop marks, resulting from differences in plant growth, often indicate former ground disturbances. Abandoned settlements include kitchen middens which show a luxuriant growth of weeds, but this effect diminishes rapidly within an archaeological time-scale. Ancient excavations such as pits, trenches, and post holes deepen the soil and are picked out by better plant growth or darker soil tones due to the darker colour of moist soils. Buried foundations, floors, and roads have the opposite effect. The resulting contrasts are most marked when soil moisture tension or surface temperature contrasts are at a maximum. In temperate latitudes, maximum soil water tension is during the latter half of the growing season when crops have well-extended roots and are growing vigorously. As a result, gradations in moisture conditions are often clearly distinguishable, especially during droughts. Significant archaeological discoveries, e.g. of Roman and Norman sites in Britain, have been made at such times. In dry countries, the extra moisture retained in disturbed soils can be especially revealing. In the Fayum (Egypt), for example, Ptolemaic irrigation channels were discovered from the richer growth of desert plants they supported (Clarke, 1964). Soil depth patterns can also be revealed when a thin snow cover is melting, because this tends to be more rapid over shallow droughty soils than over deeper, more moisture retentive, ones.

Burrowing animals can also play their part. When they avoid an area of turf, it can be an indication of a hard substratum at shallow depth. For instance, Stone noticed that a patch of ground at Easton Hill, a few miles east of Salisbury, had been untouched by moles and rabbits. This led him to the discovery of a shallow layer of tightly packed flint nodules covering an urn field (Clarke, 1964).

Further reading

Allan and Richards (1983), Atkinson (1970), Bradford (1957), Clarke (1964), Howard and Mitchell (1985), Vita-Finzi (1978).

23

Agriculture, range and forestry: land suitability and land use evaluation

23.1 Terrain requirements for plant production

Economic plant production includes all forms of cropping, grazing, and forestry in the widest sense. The variety of such activities and of the terrain they occupy is almost infinite. Terrain evaluation aims to optimize a match of the two. It must take account both of the natural controls on the growth of plants and animals, and of those imposing technological limitations on land management. All must be within an economic context. The terrain forms the natural framework within which cropping and marketing decisions are made.

Slope is an important controlling factor, especially in mechanized farming, as shown in Table 23.1. Furthermore, all crops require such basics as adequate soil depth, water, aeration, and plant nutrients, a pH near neutrality, an absence of toxic salts, and textures which are neither too coarse nor too fine, i.e. avoiding the extremes of sand and clay on Fig. 23.1.

Table 23.1 The effects of gradient on farming operations

Gradient (°)	Effects
2	Limit to irrigation by surface flow
3	Limit to unrestricted farming operations; difficulties begin for precision seeding machines
7	Limit for precision seeding machines
8	Limit to unrestricted use of combine harvesters
10	Limit for centre pivot irrigation systems
11	Limit of two-way ploughing; use of combine harvesters becomes difficult because of spillage loss from trays
15	Limit to normal rotations, cultivation costs are high, therefore mainly suitable to permanent grass; trailer loading limited to downhill movement only; combine harvesters become unusable; limit for centre pivot irrigation systems
20	Practical limit for wheeled vehicles for ploughing and fertilizer spreading; occasional tillage possible, but mainly by hand
25	Limit to mechanical operations without specialized machinery

Source: J.B. Whittow (1989, personal communication)

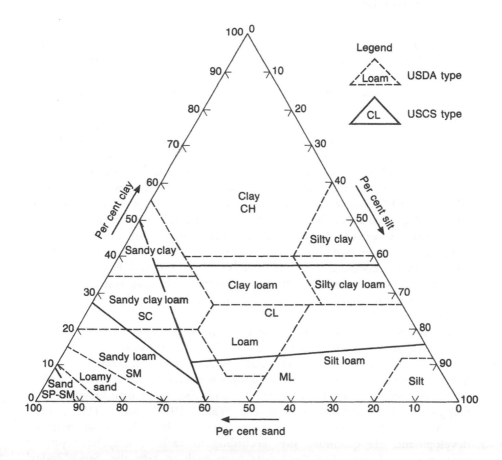

Fig. 23.1 The US Department of Agriculture and Unified Soil Classification System standards for classifying soil textures: (CH) inorganic clays of high plasticity; (CL) inorganic clays of low to medium plasticity; (ML) inorganic silts and very fine sands; (SC) clayey sands; (SM) silty sands; (SP–SM) poorly graded sands or gravelly sands. (For further information see Table 24.1)

Plants, however, differ markedly in their environmental demands. Trees will grow best on deep soils but can tolerate quite steep slopes, while field crops are more suited to level land but do not require the same soil depth. Most cereals need well–drained land, but swamp rice can only thrive under waterlogged conditions.

Technological requirements do not always accord with these natural requirements. Apart from steep slopes, the main restrictions on the operation of agricultural machinery are uneven microrelief and soft ground, neither of which is necessarily associated with poor fertility. On the contrary, softness is usually associated with wet fine–textured soils which can be highly fertile.

23.2   Approaches to agricultural land evaluation

The traditional term for assessing the agricultural productivity of land was 'land capability' which gained general acceptance from its use by the US Bureau of Reclamation. The term 'land use potential' was used, for instance by the FAO in a study of Namibia (Mitchell and King, 1985). The term 'land suitability' to express fitness for a defined land use was introduced by the FAO (1976a).

Land capability classifications marry physical subdivisions of the landscape with assessments of agricultural return. Most use a combination of physical parameters, crop yields, and financial returns, but workers vary in their relative emphasis on physical and socio-economic factors, from 'hard line physical parameter men' at one extreme to 'far-out liberal holists' at the other (Murdoch, 1972). Between these extremes, classification systems can be most simply categorized as 'single crop', agronomic, 'natural', and 'economic'.

In areas dominated by one crop, 'single crop' land classification and mapping systems were developed, such as those for cocoa soils in West Africa (Smyth, 1966), coconuts on Christmas Island (Jenkin and Foale 1968), and oil-palms in Gambia (Hill, 1969). More general multi-crop systems were also adapted for single crops in specific areas, such as that for early potatoes in parts of England and Wales (Tarran, 1984).

Where the range of crops is wider, 'agronomic' systems were developed which sought to classify lands in terms of their general value for a variety of crops. For instance, England and Wales was subdivided into 80 'agroclimatological areas' of differing crop productivity (Great Britain MAFF, 1976; Bendelow and Hartnup, 1980), and the bioclimatic regions were mapped (Soil Survey of England and Wales, 1978). Storie (1954, 1964) proposed a wider index of land productivity in relation to irrigation development. He quantified sites according to each of nine basic physical parameters, and multiplied the values from each to give the 'Storie index', which related the agricultural value of each site to a standard good soil with a notional index of 100. The method has been applied in Nigeria by Steele (1967) and in several other African countries by Sys and Frankart (1971).

Most land capability systems have, however, been based on 'natural' classifications of terrain and soils into units subsequently related to the whole range of agricultural uses for a given area. They can be conveniently grouped under the general term 'landscape schemes' and are based on the assignment of 'penalty points' to the various physical limitations affecting land use. Among these, the overwhelming importance of water has from the first differentiated schemes for rainfed and irrigated plant production.

The first system for rainfed agriculture was that of the USDA Soil Conservation Service (Klingebiel and Montgomery, 1961), which is still widely used today. Quoting Bibby and Mackney (1969): 'it assesses land capability from known relationships between the growth and management of crops and the physical factors of soil, site, and climate'. There are eight classes ranging from 1, the best arable with a wide range of other uses to 8, permanently unusable. Capability subclasses are defined on the physical factor or factors limiting production, each of which is indicated by the use of a letter subscript attached to the relevant class number, such as 'e' for erosion, 'w' for excess water, 's' for soil limitations, etc. These are the lowest category of the classification, and group the soils capable of growing the same kind of crops and requiring the same management. Long-term estimates of crop yields for individual soils are qualified by the statement that they should not vary from the true values by more than about 25 per cent. Table 23.2 summarizes the criteria for the main classes.

The same scheme was adopted both in the Canada Land Inventory (1965) and by the Soil Survey of England and Wales (Bibby and Mackney, 1969). Both countries eliminated the fifth class, which allows mainly for wet soils in level sites poorly adapted for arable crops. In England and Wales, a further subclass was added for gradient and soil pattern limitations on land use. The classification was used to categorize the units on the soil survey maps, and in the national land use capability maps at 1:1 000 000 (Soil Survey of England and Wales, 1979).

The demand to release agricultural land for building forced an acceleration. The Ministry of Agriculture, Fisheries and Food, beginning in 1965, published the *Agricultural Land Classification* maps of England and Wales, whose 184 sheets at a scale of 1:63 360, supported by 7 synthesizing sheets at a scale of 1:250 000 were completed in 1977 (Great Britain MAFF, 1965–77). The classification scheme was simple, using only five grades from 1 for top quality agricultural land with very minor or no limitations to 5 for rough grazing land with very severe soil, relief, or climatic limitations. The principle is, on behalf of agricultural interests, to resist all development on grades 1 and 2, but not to defend grades 4 and 5. The main local debates centre around grade 3 land, necessitating its subdivision into 3a, 3b, and 3c to reflect a decreasing order of quality. A complementary set of maps of 'overall agricultural significance' at the smaller scale emphasize the type of land use (Great Britain MAFF, 1964), and the Soil Survey of England and Wales produced a summary map of 'land use capability' at 1:1 000 000 (1979). In Scotland, land capability mapping consists in attaching evaluations, according to the system of Bibby and Mackney, to the units of the 1: 63 360 soil maps (e.g. Great Britain, Soil Survey of Scotland, 1972).

Other western European countries, including Ireland, France, Italy, Denmark, West Germany, and the Netherlands, have somewhat similar schemes (Haans *et al.*, 1984). There is as yet no effective coordination of these, although the Dutch and German workers separately surveyed the same two areas in both countries and compared results. Their differences were not irreconcilable and mainly concerned the amount of detail mapped and the extent to which each had followed the FAO (1976a) *Framework for Land Evaluation* (Steur *et al.*, 1984). In the USSR, land evaluation for regional planning uses two types of classification: basic and applied. The former is used

289

Table 23.2 Summary of criteria for USDA land classes (after Klingebiel and Montgomery, 1961)

	Land classes							
	I	II	III	IV	V	VI	VII	VIII
	(suited to cultivation and other uses)				(generally not suited to cultivation)			
Type of land use								
1 Crops	+	+	+	+				
2 Pasture	+	+	+	+	+	+	+	
3 Range	+	+	+	+	+	+	+	
4 Woodland	+	+	+	+	+	+	+	+
5 Wild Life	+	+	+	+	+	+	+	
Land attributes								
6 Slopes	Level	Gentle	Moderately steep	Steep	Gentle	Steep	Very Steep	Very Steep
7 Erosion hazard	None	Moderate	High	Severe	Severe	Severe	Severe	Severe
8 Overflow danger	None	Occasional	Frequent	Frequent	Frequent	—	—	—
9 Soil Depth	Ideal	Less than ideal	Shallow	Shallow	Shallow	Shallow	Shallow	Shallow
10 Soil structure and workability	Good	Somewhat unfavourable	—	—	—	—	—	—
11 Drainage	Good	Correctable by drainage	Very slow	Waterlogging	—	Waterlogging	Waterlogging	Waterlogging
12 Water-holding capacity	Good	Moderate	Low	Low	Low	Low	Low	Low
13 Salinity	None	Slight to moderate	Moderate	Severe	—	Severe	Severe	Severe
14 Nutrient status	Good	Moderate	Low					
15 Climate	Favourable	Slight limitation	Moderate	Moderately adverse	Unfavourable	Unfavourable	Unfavourable	Severe
16 Management practices required	Ordinary	Careful	Special	Occasional cultivation only possible		Cultivation not possible		
17 Stoniness	—	—	—	—	Some	Severe	Severe	Severe

for surveying the main land properties, while the latter arranges these into functional land use units (Ignatyev, 1968).

The earliest land classification for irrigated agriculture was produced by the former Bureau of Reclamation of the US Department of the Interior (1951), now called the US Power and Water Resources Service. Its system of surveys, described by Maletic and Hutchings (1967) and summarized by Vink (1975), strongly emphasized comprehensive planning of irrigated land use. A Unesco symposium followed with a study of the environmental problems and effects of irrigation schemes (Worthington, 1977). The Bureau of Reclamation approach has since been used in modified form in a number of countries. It recognizes six classes defined in terms of quantified physical attributes: 1-3 represent irrigable land of decreasing suitability; (4) lands usable for special purposes only; (5) those requiring further investigation; and (6) those which are permanently unirrigable. Most countries modify the class definitions to meet local conditions, e.g. Iran (Mahler, 1970) and Saudi Arabia (Ministry of Agriculture and Water, 1984, (private communication). A comparable, though somewhat simpler scheme was developed for Morocco by Geoffroy (1978).

'Economic' land capability schemes have been based on calculations of financial return. The 'Cornell' system, used over a period of more than 25 years in New York State, is a simpler and more empirical method. It ignores the environmental factors and concentrates entirely on the financial yield of farms (Conklin, 1959). The basis is the 'farm business' and the smallest mapping unit is three farms. There are four classes depending on the incomes yielded: (1) too low to attract operators; (2) just sufficient to attract operators; (3) high; and (4) the best. The maps have been popular with farmers because they tackled the question of income potential.

The two parts of Germany have somewhat different approaches. The East German agro-ecological system is based on a three-phase study aimed at optimizing land uses in local habitats. First, they are mapped into ecological units. Then they assess each habitat and the factors controlling its agricultural yields. Finally, there is an analysis based on agro-economics which makes specific proposals (Vink, 1983). The West German scheme first subdivides the country into arable and pastureland, and then calculates a numerical value for the yield potential of every hectare, based on the main land and soil attributes. This is a level of detail which makes it outstanding among national systems. The values per hectare are aggregated first for fields and then for farms. After making adjustments for such factors as proximity to markets, the nature of the farm buildings, and the risks of flooding or pests, a valuation for sale or tax purposes is made (Weiers and Reid, 1974; Weiers, 1975).

An effective compromise between physical and economic schemes was introduced in the FAO *Framework for Land Evaluation* (1976a). This involved two conceptual changes, which have profoundly influenced subsequent work. First, they view land evaluation only in relation to specified uses, whose definitions must embrace both the objectives and the means for achieving them. This puts emphasis on precision in the terms of reference of interpretation and on the recognition of the possibilities of change in land for better or worse. This change is underlined by substituting the term 'land suitability' for 'land capability' for general use. Second, they squarely accept economic rather than physical criteria on the ground that land can be made suitable if the cost is justified. Evaluation thus requires a comparison of inputs and outputs on different types of land. The authors also reiterate the need for an interdisciplinary approach which considers alternative types of land use, the importance of the local context, and the requirement for sustained production.

Therefore, instead of using comprehensive land capability classes, the FAO use a two-pronged approach. First, they recognize 'land mapping units' based on the physical attributes of the surface. They distinguish 'land characteristics', which are measurable physical attributes such as slope and soil texture, from 'land qualities', which are descriptors of land in relation to land use, such as moisture availability or erosion resistance. The former generally form the basis for the mapping units, but the latter must be included in assessments of land suitability. Second, the suitability of each unit for each 'land utilization type' (LUT) is overprinted on this base map of land characteristics. The LUTs describe the land uses relevant to a given area. The suitability of each to each mapping unit is separately assessed according to a standard code. 'Orders' reflect kinds of suitability: 'S' for suitable and 'N' for unsuitable. These are each subdivided into 'classes', numbered in order of decreasing suitability. 'Subclasses', indicated by a lower case letter, reflect the kinds of limitations or improvement measures required within classes, and 'units' reflect minor differences in required management between subclasses. Thus, S2t(4) represents management area no. 4 which has some topographic limitation and lies in the second grade of suitability for arable cultivation. The detailed application of the *Framework* to the separate requirements of rainfed agriculture, irrigated agriculture, and forestry is described in three Guidelines (FAO, 1983, 1984, 1985). A fourth on rangeland is projected. The method is well suited to categorizing units of land system and land facet type.

23.3 Agro-ecological zones and their population-supporting capacity

The FAO also assessed climatic factors in relation to crop yield, first in Africa (1978a) and then in the Third World as a whole, leading to the recognition and mapping of 'agro-ecological zones', defined in terms of the annual period with adequate soil temperature and moisture for crop growth (Higgins *et al.*, 1984). When these maps are overlaid on the *Soil Map of the World* (FAO/Unesco, 1974–), they allow local, national, regional, and global estimates of the yield potential for each of the world's main food crops at specified levels of management.

The agro–ecological zones were related to potential population-supporting capacity and compared with actual populations. Each area was analysed to ascertain which major crop (or grassland for livestock) yielded the best calorie–protein production. The results were then totalled to give food-producing potentials for each zone based on length of growing period within each country. Inclusion of country-specific dietary requirements allowed computation of the potential population-supporting capacities in each zone in each country. These data were converted to persons per hectare, and compared with present and projected population densities. Critical zones were thus identified where, according to the level of management input envisaged, potential production is insufficient to meet the needs of the populations either at the time of calculation (1975) or as projected in the year 2000 (Higgins *et al.*, 1984).

Although not critical for the present discussion, the results of this analysis are striking. Although the global picture shows relatively adequate land resources, the forecast is that in AD 2000, in the absence of food from elsewhere, of the 117 countries identified by the FAO as being in the Third World, 36 will be unable to feed themselves under intermediate levels of management input or 65 at low levels, compared with 1975 figures of 24 and 54 respectively. Moreover, if soil erosion remains unchecked, the area of potential rainfed cropland will fall by 18 per cent and its overall crop productivity by 29 per cent in the same period. Thus, even if they achieve substantially improved levels of management, including the wide introduction of irrigation, the inadequate land resources of many Third World countries will condemn them to a declining agricultural self-sufficiency until the end of this century and with no prospect of improvement thereafter (Higgins *et al.*, 1984).

23.4 Terrain evaluation systems

The pioneering work in the use of terrain classifications for agricultural purposes was the Australian CSIRO Division of Land Research and Regional Survey recently reorganized under the Division of Water Resources (hereinafter abbreviated to WR).

Following a Unesco conference in Toulouse in 1964, the method became more widely used, especially by the ITC–Unesco Centre for Integrated Surveys at Enschede (Netherlands), the British Overseas Development Natural Resources Institute (ODNRI – formerly Directorate of Overseas Surveys Land Resources Development Centre, and before that Land Resources Division), and a number of consulting firms. The ITC–Unesco Centre has done research on the problems of execution and presentation of the results of integrated surveys (Nossin, 1977). The WR and ODNRI can be taken as representative.

(14) BALBI LAND SYSTEM (30 SQ MILES)

Active or recently active volcanoes.

Geomorphology.—Very recent volcanic land forms including lava flows, debris slopes, and scarps forming the margins of explosion craters or of spines. Mt. Bagana is an active craterless lava cone; Mt. Balbi is a group of cratered active or inactive ash or scoria cones and spines; Lake Billy Mitchell is a large explosion crater; an unnamed peak east of Mt. Bagana is a dissected lava-flow cone. Lakes and ponds occur both in craters and in valleys blocked by lava flows.

Terrain Parameters.—Altitude: H.I, V; min., 1000 ft; max., 8500 ft. Relief: very high (1500 ft). Characteristic slope: very steep. Grain: very coarse (5000 ft). Plan-profile: 4.

Geology.—Andesite, hornblende–andesite, and hornblende-bearing basalt, as lava and ash; Recent.

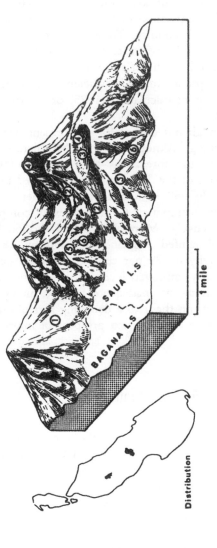

SAUA L.S

BAGANA L.S

1 mile

Distribution

Land Unit	Area (sq miles)	Land Form	Soil	Land Class; AASHO Soil	Vegetation
1	8	Lava flow: sinuous stream of rock up to 2 miles long and 1000 ft wide expanding distally to 3000 ft; steep axial slope (17–30°), increasing at the terminus to very steep (42°); cross-section convex except in upper parts which are irregularly concave; commonly bounded by narrow, very steep, marginal ridges up to 50 ft high	Blocks	$VIIIst_3$ B	Bare except for terminal slope with mixed herbaceous vegetation (*Lycopodium–Gleichenia*)
2	3	Debris slope: long (2000 ft); concave; mainly gentle or moderate (2–20°); traversed by stream beds up to 50 ft wide, locally incised 100 ft	Mainly stones and boulders Locally ash soils: grey fine sands (stony phase) or alluvial soils: shallow grey mottled sands	$VIIIst_3$ B+C VIa_3st_3,n_3 A3+C IVd_3,f_3,n_3 A3	Bare; mixed herbaceous vegetation (*Lycopodium – Gleichenia*, mountain herbaceous vegetation); savannah; and scrub (*Cyathea-Bambusa*, mountain scrub)
3	½	Crater floor: nearly level; up to 500 ft diameter; 1 ft channelled microrelief; ponds	Ash soils: grey fine sands	$VIId_7,n_3$ A3	Bare or with mountain herbaceous vegetation; probably some bogs
4	2½	Scarp: medium length to long (500–2000 ft); irregular; precipitous; commonly with waterfalls	Rock outcrop or ash soils: grey fine sands (stony phase)	$VIIIr_2$ R-N $VIIIst_3,n_3$ A3+C	Bare or with mountain herbaceous vegetation
5	1	Ridge crest: knife-edged or very narrow; very uneven; steep crestal slope	Ash soils: grey fine sands (stony phase)	$VIIIst_3,n_3$ A3+C	As unit 2, locally palm and pandan vegetation (*Gulubia-Pandanus*)
6	15	Erosional hill slope: short or medium length; straight; very steep to precipitous	Boulders or ash soils: grey fine sands (stony phase)	$VIIIst_3,n_3$ B or A3+C	As unit 2, locally mountain low forest

Population and Land Use.—Nil.
Forest Potential.—No forest. Access category III.
Observations.—2, plus 4 aerial observations.

Fig. 23.2 An example of the CSIRO Division of Land Resources and Regional Survey method of defining and describing terrain: the Balbi Land System covering 75 km² (30 miles²) on Bougainville Island. (Source: Australia, CSIRO, 1967)

23.4.1 *Australian CSIRO (WR)*

(division concerned with land resources was called the Division of Land Research and Regional Survey until 1973, the Division of Land Use Research until 1982, the Division of Water and Land Resources until 1988, and the Division of Water Resources since then. It is herein abbreviated to WR.)

Although the work of the CSIRO is here considered under agriculture, pasture, and forestry because of its major thrust in these directions, it really pioneered the concept of 'integrated survey' by viewing terrain in the context of all land uses, including engineering, water supply, minerals, wildlife, fisheries, harbours, scenery, tourism, and recreational attractions.

The background and methodology of integrated surveys carried out by the WR (and other bodies) were summarized by Christian and Stewart (1968). They began extensive resource surveys in undeveloped parts of Australia, Papua, and New Guinea in 1946, using the aerial photography and techniques developed during World War II. These surveys were based on 'land systems' and 'land units' which aimed at being both basic and functional divisions of landscape. Land systems were recurrent patterns of landforms, soils, and vegetation recognizable on aerial photographs, and land units were their constituent subdivisions. Land system surveys were presented in the form of maps at scales of 1:250 000 and 1:1 000 000 accompanied by block diagrams showing the interrelations of the land units and tabulated summaries of their form, soil, and vegetation properties. Figure 23.2 gives an example of one land system from the report on Bougainville (CSIRO, 1967). Shortage of staff, funds, and information made it impracticable to map land units at their relevant scale of about 1:50 000–1:100 000, and so although they are the main vehicle for data storage, land systems are the basic mapping unit. More than 40 such reports have been published up to mid 1990, covering most of northern Australia, and all of Papua New Guinea and associated island territories such as Bougainville (Fig. 23.3). The advantages of the method for Australian conditions are manifest. The nature of the landscape, in which there is a clear and close relationship of landforms to soil, water, and vegetation, and the relatively low economic value of many parts, favours a physiographic reconnaissance approach, based on aerial photographic interpretation.

The use of the land system method integrates a team of specialists to give an overall rather than a single-use appraisal of an area by coordinating interests and phasing, interpreting, and compiling data on the same aerial imagery, exchanging on-the-spot information in the field and collaborating in the editing of maps and reports. In general, the team contains one geomorphologist, one soil surveyor, and one botanist, with other experts such as hydrologists, foresters and agricultural economists being called in where necessary.

Reviewing the progress of the method, Christian (1983) has shown the way it has come to be modified. It has been refined by making the units more uniform and more precisely defined, adapted to relatively densely populated

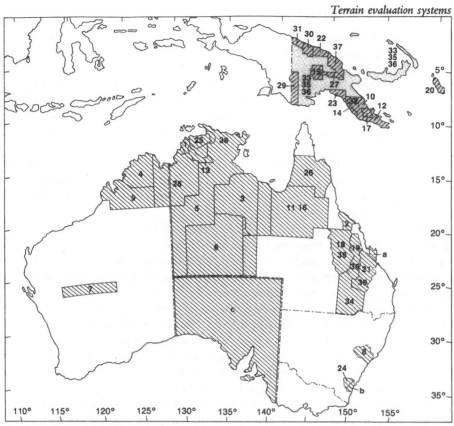

PUBLISHED IN LAND RESEARCH SERIES

No.	TITLE		
1	Katherine–Darwin	15	Wabag–Tari (PNG)
2	Townsville–Bowen	16	Geomorphology of
3	Barkly		Leichhardt–Gilbert
4	North Kimberly	17	Safia–Pongani (PNG)
5	Pasture Lands of the NT	18	Nogoa–Belyando
6	Alice Springs	19	Issac–Comet
7	Wiluna–Meekatharra	20	Bougainville–Buka (PNG)
8	Hunter Valley	21	Dawson–Fitzroy
9	West Kimberley	22	Wewak–Lower Sepik
10	Buna–Kokoda (PNG)		(PNG)
11	Leichhardt–Gilbert	23	Kerema–Vailala (PNG)
12	Wanigela–Cape Vogel	24	Queanbeyan–Shoalhaven
	(PNG)	25	Adelaide–Alligator
13	Tipperary	26	Mitchell–Normanby
14	Port Moresby–Kairuku	27	Goroka–Mt Hagen (PNG)
	(PNG)	28	Ord–Victoria

29	Morehead–Kiunga (PNG)
30	Aitape–Ambunti (PNG)
31	Vanimo (PNG)
32	Eastern Papua (PNG)
33	Geomorphology of PNG
34	Balonne–Maranoa
35	Vegetation of PNG
36	Land Limitations and
	Agric. Land Use Potential
	of PNG
37	Ramu–Madang
38	Alligator Rivers
39	Fitzroy Region

OTHER AREAS SURVEYED NOT PUBLISHED IN LAND RESEARCH SERIES

a Shoalwater Bay
b South Coast of NSW
c Ecological Survey of SA

OTHER AREAS STUDIED BUT NOT MAPPED

1 Coastal lands of Australia
2 Land Systems of the Simpson Desert Region
3 Land Units of Chimbu Province, PNG

Fig. 23.3 Index to areas covered by CSIRO surveys. (Further information on CSIRO work in this and other regions is available from CSIRO, GPO Box 1666, Canberra ACT 2601, Australia)

297

and well-known regions, such as the Hunter Valley, New South Wales (Story *et al.* 1963; Renwick 1968), and developed by remapping previously surveyed areas at larger scales (Aldrick and Robinson, 1972; Fogarty and Wood 1978), and by reclassifying units into groups within which there is a degree of recurrence (Gunn and Nix 1977). A pilot study of South Australia mapped a hierarchy of land units on Landsat imagery to develop a methodology for an ecological survey of the whole continent (Laut *et al.* 1977) It has been used with success in central Australia as the framework for assessing rentals and avoiding abuses in the management of grazing areas (Condon, 1968) and in northern Australia for evaluating the degree of difficulty for mustering cattle across long distances for the control and eradication of bovine tuberculosis (Laut and Nanninga, 1985).

The South Coast Survey of New South Wales was an exceptionally intensive and detailed study of a small area in order to determine the various land use options for each subdivision and demonstrate how to select a preferred one. It recognized 'unique mapping areas' (UMAs), comparable to land systems, and 'functional units', somewhat larger than land facets. It used computer storage and data processing to answer complex questions about 150 attributes of almost 4000 functional units, either in tabular or cartographic form. This relatively high-technology approach appears relevant to other areas of intensive development, but specialist surveys would still probably be necessary at the level of the functional unit for local land use management purposes (Australia CSIRO, 1978; Christian, 1983).

Land unit boundaries have sometimes conformed to those imposed by socio-economic decisions, because these are the planning units about which statistics are collected. In areas where large-scale information is required for defined objectives, surveyors have sometimes used special classifications of smaller components or themes which are not always mutually compatible. Examples are the mapping of 'rangeland types' at land system scale in Western Australia and pastoral areas of land unit scale in Queensland.

The land system method in Australia has stood the test of time. It has, at the time of writing, completed almost half a century of use without substantial change, and developed into a hierarchical physiographic framework for all Australian soils which has replaced other systems (Australia, CSIRO Division of Soils, 1983). It has been used by state authorities as well as CSIRO. The Victoria Land Protection Service has mapped the whole state at least to land system level, and both Queensland (Department of Primary Industries 1974, 1976) and Tasmania (Richley, 1979) have conducted surveys.

23.4.2 *United Kingdom Overseas Development Natural Resources Institute (ODNRI) and other systems*

The work of the ODNRI resembles that of the WR in using interdisciplinary teams and the land system method to plan agricultural development in extensive little-known regions. It covers, however, an even wider range of geographical environments and survey objectives. Since it can be involved in the development process, it sometimes includes specialists such as agronomists in the research team. Reconnaissance land system surveys are carried out at scales of 1:250 000 or 1:500 000, intensive land resource assessments at 1:50 000 or occasionally 1:25 000, and development studies at 1:10 000 or larger. These can be done on an 'integrated' basis, i.e. including a survey of all resource characteristics, or on a 'single aspect' basis where only one is studied. Integrated surveys have been made in recent years of more than 30 areas, largely in African countries but also of the Falkland Islands and parts of Nepal, Malaysia, the south Pacific, and the Caribbean. Single-aspect studies tend to concentrate on specific problems in which the land resource assessment is a subsidiary component. Examples are forest inventories in Botswana (Langdale-Brown and Spooner, 1963) and an irrigation survey in Ethiopia (Makin *et al.*, 1976). Intensive resource assessments can likewise either be conducted on an integrated or single-aspect basis. Studies in Malawi and Fiji are examples of the first and others in New Hebrides, Fiji, and Kenya, of the second, type. Detailed development studies generally relate either to the cultivation of specific crops on relatively good soils such as coconuts on Christmas Island, overhead irrigation of sugar-beet in Egypt, fertilizer control in India or else to a 'working plan' for forest exploitation, as in Bangladesh.

Other individuals and organizations have used the land system method to guide rural development. Taylor (1959) applied it in Nicaragua, retaining the term 'land unit' but using 'land type' instead of 'land system'. In Pakistan, Thirlaway (1959) used it in the Isplingi Valley and Wright (1964) on a Unesco project in the Nagarparkar area. The Photographic Survey Corporation (1956) classified all the Indus plains on a landform basis. Parts of India have been covered by the Basic Resources Division of the Central Arid Zone Research Institute at Jodhpur, India, and Hunting Technical Services (1954–) have employed the landscape approach for a regional development survey in Jebel Marra, Sudan and for a range classification in Jordan (see Fig. 8.1). The latter was developed into a land system classification of the whole country, based on Landsat imagery, by the FAO Remote Sensing Centre (Mitchell and Howard, 1978), who also did similar studies of south Yemen (Travaglia and Mitchell, 1982), northern Iraq (Mitchell, 1985), and Namibia (Mitchell and King, 1985).

23.5 Terrain in land use survey

Since most land use on a world scale is agricultural, it is most appropriate to consider its relationship to terrain in this chapter.

Land use is the pattern of society's exploitation of its environment. Land use surveys are carried out in almost all countries. The first such mapping was the Land Utilization Survey of Great Britain. This was produced in the years following 1930 under the inspiration and direction of L. D. Stamp who, with E. C. Willatts, summarized the methods and first results (1934). The whole of Great Britain was covered at a scale of 1: 63 360, with a synthesizing map at 1:625 000 and a detailed report for each county. The six mapping units were: forest and woodland; meadowland and permanent pasture; arable or tilled land; heathland, moorland, commons, and rough hill pasture; gardens, allotments, orchards and nurseries; and land agriculturally unproductive. An example of part of the 1: 63 360 map is given in Fig. 23.4. The Second Land Utilization Survey of Great Britain was commenced in 1960 under the direction of Alice Coleman (Coleman and Maggs 1965). It was produced at 1:25 000 and so required 843 sheets to cover England and Wales alone. Only 115 were

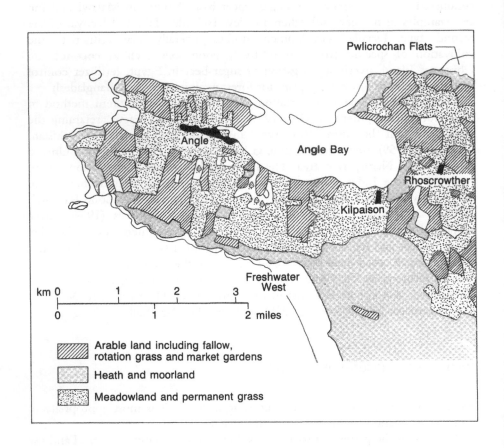

Fig. 23.4 First Land Utilization Survey mapping. Part of Pembroke and Tenby sheet

300

completed when the project was suspended after the Plymouth sheet (No. 15) in 1977, although data for the rest of the country are held (at King's College, London). The basic classification adopted is essentially the same as in the first survey but each class has been subdivided into a number of subclasses so that the map can be interpreted at two levels of generalization. For instance, arable land is subdivided into cereals, roots, and legumes, and woodland into deciduous and coniferous types. A small part of the maps is reproduced in Fig. 23.5. Similar surveys have, for example, covered parts of Africa, e.g. the Gambia at 1:25 000 (Great Britain LRDC, 1958), and Ghana (Hunting Technical Services, 1954–) and parts of Canada (Canada, Department of Mines and Technical Surveys, 1969).

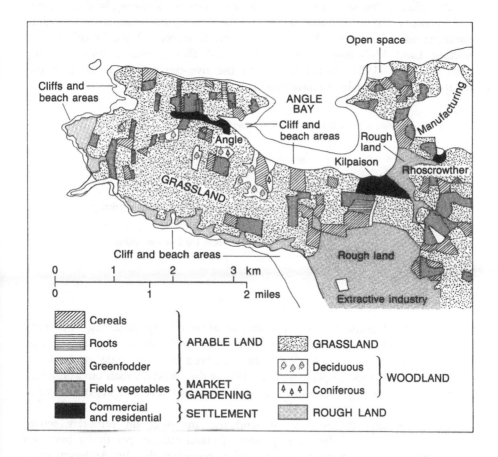

Fig. 23.5 Second Land Utilization Survey mapping. Part of Milford Haven sheet

Land uses, however, vary in intensity as well as in 'nature', and so it is important to classify them as far as possible from an 'activity-based' rather than

301

an 'area-based' viewpoint. For instance, although areas of subsistence and commercial types of wheat cropping might appear the same on a land use (crop) map, the differences between management practices and yields make them distinct forms of land use. This is especially in small-scale mapping where it is impossible to distinguish individual fields or forest plots. This necessitates the development of maps delimiting areas of different agricultural activity. Examples are those which show generalized units such as 'farming types' (e.g. Great Britain: Ordnance Survey 1939; MAFF, 1967–73), or at very small scales identify both dominant and secondary farming types within the same mapping unit (ICI, 1965). To the extent that agriculture is environmentally dependent, such mapping units accord with agroclimatic and terrain regions.

At smaller scales, schemes have been developed for showing land use 'type' and 'intensity' 'layers' simultaneously, for instance by elaborating a hierarchical classification of land uses which includes both concepts in logical relation. The Second Land Utilization Survey, mentioned above, was a move in this direction. Fox (1956) increased this to the five-tier approach for land use mapping in New Zealand reproduced in Table 23.3.

Table 23.3 Five-tier activity-based land use classification

A	Cardinal	Agricultural, forest, pasture, etc.
B	Ordinal	General vegetational response in terms of crop type: cereal roots, legumes, etc.
C	Generic	Intensiveness: subsistence, commercial, market garden, etc.
D	Specific	Crop type: wheat, barley, rye, etc.
E	Varietal	Type of particular crop: hard or soft wheat

Source: Fox (1956).

Long (1974) made a more quantitative approach by including both rural and urban land uses in a common concept of 'artificialization'. Areas can be mapped according to numerical values derived from a scale of land use intensities from 0 for empty desert to 100 for the densest urban environments. This is shown in Fig. 23.6, and applied to different levels of vegetal cover in Table 23.4.

The increasing dependence of land use mapping on remotely sensed imagery has prompted the development of classifications specifically based on its interpretation. The most widely used is probably that by Anderson *et al.* (1976) designed hierarchically so as to be appropriate for use with satellite or aerial imagery at a variety of scales. Its two highest categoric levels are reproduced in Table 23.5.

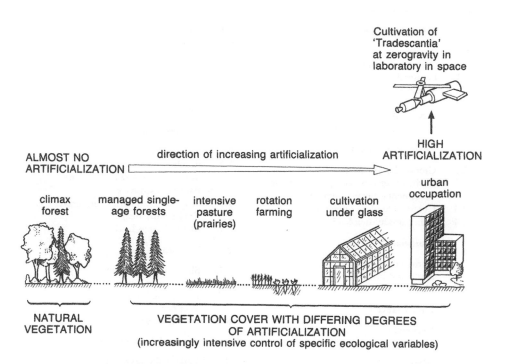

Fig. 23.6 The artificialization transect. (Source: Long, 1974)

Table 23.4 Degrees of artificialization and associated land uses and vegetation types (after Long, 1974)

9	Very extreme: dense urban without vegetation
8	Extreme: medium urban with planned trees, shrubs, and small gardens
7	Very strong: suburban, with parks and sports grounds
6	Strong: intensive orchards, small fruits and market gardens
5	Slightly strong: intensive tree and shrub plantations, fodder crops
4	Moderate: semi–intensive forest, managed shrubland and grassland
3	Slightly feeble: extensive to semi–intensive managed natural forest, improved shrubland, controlled wild grazing
2	Feeble: very extensive old forest, shrubland, or rough grazing
1	Nil: unexploited climax forest, shrubland, alpine meadow, or swamp.
0	Bare ground

Table 23.5 A land use cover classification for use with remote sensor data

1	**Urban built up**	5.2	Lakes	
		5.3	Reservoirs	
1.1	Residential	5.4	Bays and estuaries	
1.2	Commercial and services			
1.3	Industrial	**6**	**Wetland**	
1.4	Transport, communications, and utility			
1.5	Industrial and commercial complexes	6.1	Forested	
1.6	Mixed urban or built up	6.2	Non–forested	
1.7	Other urban or built up	**7**	**Barren land**	
2	**Agricultural land**	7.1	Dry salt flats	
		7.2	Beaches	
2.1	Cropland and pasture	7.3	Sandy areas other than beaches	
2.2	Orchards, groves, vineyards, nurseries, ornamental horticultural areas	7.4	Bare exposed rock	
		7.5	Strip mines, quarries, and gravel pits	
2.3	Confined feeding operations	7.6	Transitional areas	
2.4	Other agricultural land	7.7	Mixed barren land	
3	**Rangeland**	**8**	**Tundra**	
3.1	Herbaceous	8.1	Shrub and brush tundra	
3.2	Shrub and brush	8.2	Herbaceous tundra	
3.3	Mixed	8.3	Bareground tundra	
		8.4	Wet tundra	
4	**Forest land**	8.5	Mixed tundra	
4.1	Deciduous			
4.2	Coniferous (evergreen)	**9**	**Perennial snow and ice**	
5	**Water**	9.1	Perennial snowfields	
		9.2	Glaciers	
5.1	Streams and canals			

Source: Anderson *et al.*, (1976)

Further reading

Anderson *et al.* (1976), Christian (1983), Australia CSIRO Division of Soils (1983), Dent and Young (1981), FAO (1976a, 1983, 1984, 1985), FAO/Unesco (1974–), Higgins *et al.* (1984), Long (1974), Morgan (1986), Trudgill and Briggs (1981), USDA Soil Survey Staff (1976), Vink (1983), Weiers and Reid (1974), Young (1976).

24

Terrain evaluation in civil engineering

24.1 Engineering and terrain

The importance of terrain conditions to engineering practice has been recognized for centuries. The discipline of engineering geology evolved in the nineteenth century but it was not until the 1920s, when Terzaghi appreciated the role of pore-water pressure that analyses to formulate the behaviour of soils became possible. Soil mechanics did not become widely taught until the 1950s. Engineering geology became a separate subject in the late 1950s and early 1960s. Engineering geomorphology is younger still. Its contributions began with the engineering implications of superficial sediments and with slope stability, and have more recently included a wider range of geomorphic processes (Fookes and Gray, 1987).

The need to maximize the advantages of the ground and avoid natural hazards influences and often determines both the location and the design of engineering structures. Also, the character of the land surface governs the choice of vehicles to be used in traversing and carrying out constructional tasks, and may impose modifications to their design.

Engineering works are concerned with routes, areas, or sites. Routes include roads, railways, pipelines, canals, and tunnels, and require narrow, elongated strips of land which, in national or international projects, may extend over thousands of square kilometres. The primary requirement is for data on the distribution of upland and lowland. Areas and sites need relatively compact areas but differ in scale. The former include such features as reclamation and irrigation works, airports, harbours, and coastal protection, requiring the broad assessment of substantial tracts. The latter relate mainly to features such as individual buildings, bridges, and dams, requiring detailed surveys of relatively small sites.

The construction of engineering works also varies in duration, from a day to many months. In general, it can be divided into four stages.

The first stage is that before construction begins, and includes feasibility studies and site investigations. It is necessary, for instance, to decide on the route corridor for a road or a pipeline and whether a river crossing should have a bridge or a tunnel. The second stage is engineering design. This involves the preparation of contract documents and specifications and the procedures for tendering, and requires more detailed data about a more limited number of sites. The results obtained lead to the design of foundations and substructures. The third stage is construction. The basic decisions have been taken, but a continuing watch is needed to deal with unforeseen problems as they arise. The fourth and final stage is post-construction maintenance. This involves a continued monitoring of the impact of the works on the environment. In general, most terrain data are needed at the first stage. The later stages require essentially the same kinds of information but about specific locations, in greater detail, and with more emphasis on soils.

Although some of the terrain data required for engineering works are 'tangible', some are 'intangible'. The former concern the form of the ground, such as the dimensions of natural features, and the locations, characteristics, and reserves of specific materials. Intangible information relates to estimates of landscape processes and engineers' experiences of related areas, especially when these include impressions of success or failure. A terrain evaluation system for engineering has two basic objectives:

1. To make reasonably accurate predictions about terrain conditions in all parts of a large area of interest, even though ground access is limited; and
2. To provide a framework for the collection and handling of precise on-site data.

The FAO (1973), and countries such as the USA (USDA, 1971) and the Netherlands (Westerveld and van den Hurk, 1973), have proposed guidelines for the interpretation of soil maps for engineering purposes. These are sometimes given as annotations on soil maps, as in studies of parts of Mississippi and Virginia (Covell and Shaffer, 1966; Pettry and Coleman, 1973), and Unesco have proposed guidelines for their preparation (1976). There are also specially produced thematic maps of such topics as flood risk and water quality (Working Party, 1982).

Although maps are useful, however, they suffer from the disadvantage that they have had to be prepared for a range of purposes and for a whole region (requiring medium to small scales) and for average or a few specified weather conditions. Against this, the need for information about terrain is usually specific both in purpose and season – for detailed information relevant to a particular problem in a limited area at a definite period. There is, therefore, a need not only for a capability of providing generalized information, but also a mechanism for dealing with specialized requests.

The only way of determining the potentialities of a site with absolute certainty is to go there and try. This is often expensive and sometimes impossible. The cost of predicting terrain conditions with certainty over all parts of a large area would be astronomical. The only alternative is to accept predictions with less than absolute certainty. Any practicable system of terrain evaluation represents the best attainable compromise between cost and truth.

24.2 Terrain characteristics of engineering importance: the case of highways

Highway construction is a common engineering activity which is crucially dependent on terrain information. It can be taken as representative of 'linear' engineering in general, which also includes such works as railways, canals, pipelines, and tunnels, some of which may indeed be included in the same project. In locating a route, the engineer aims at the most economic solution within the major physical constraints: gradient, natural hazards, and the need to balance cut and fill. The first requirement is to minimize the distances between the fixed end-points of the road. Within this constraint it is necessary to minimize gradients: these cause construction and maintenance costs to multiply in direct relation to their steepness because they necessitate greater lengths, larger amounts of cut and fill, protection against landslips and other forms of erosion, and additional constructions such as bridges and even spirals. Watercourses must be avoided as far as possible. These pose especially serious problems if variations between drought and flood conditions are wide and unpredictable. Langbein and Hoyt (1959) estimated that no less than one-quarter of the capital expenditure on roads in the USA was for the construction of bridges, culverts, and other drainage works. On the other hand, the chosen route must be accessible to supplies of water and construction materials. Finally, soil conditions are important, although these are often not determined in detail until after the final decision on the route. The key attributes are soil stability and strength under the passage of heavy loads, deriving from such factors as depth to bedrock, texture, Atterberg limits, cut slope stability, compactibility, and shear strength, within the context of seasonal movements of the water-table and the general drainage situation.

Where the bedrock is deep, the most important characteristic is probably the soil texture. The engineer places more emphasis on the physical and less on the chemical properties of the soil than does the agriculturist, so that quite different systems of textural classification have been developed. Most engineering schemes are based on the work of Casagrande (1947), a widely used example being the Unified Soil Classification System or USCS for short (presented in different forms by, for instance, the USAEWES, 1963a, and Grant, 1973/74). It is summarized in Table 24.1. It differs from agricultural classifications of soil texture, such as that of the USDA Soil Survey Staff (1976) by including both gravel and organic materials and using the liquid limit as a criterion. The latter is also indicative of clay content.

Cut slope stability is measured in terms of the maximum angle of rest at which unconsolidated materials will remain stable, and is important in road cuttings. This angle generally increases with the coarseness of the materials, but steep slopes can occur in finer materials if their particles interlock, as in the case of loess.

Table 24.1 Unified soil classification system (USCS)

1	2		3 Group Symbols	4 Typical Names	5 Field identification procedures (excluding particles larger than 3 in and basing fractions on estimated weights)
Course-grained soils — More than half of material is *larger* than No. 200 sieve size (lest particle visible to the naked eye.)	Gravels — More than half of course fraction is larger than No. 4 sieve size	Clean gravels (little or no fines)	GW	Well-graded gravels, gravel-sand mixtures, little or no fines	Wide range is grain sizes and substantial amounts of all intermediate particle sizes
			GP	Poorly graded gravels or gravel-sand mixtures, little or no fines	Predominantly one size or a range of sizes with some intermediate sizes missing
		Gravels with fines (appreciable amount of fines)	GM	Silty gravels, gravel-sand-silt mixture	Nonplastic fines or fines with low plasticity
			GC	Clayey gravels, gravel-sand-clay mixtures	Plastic fines
	Sands — More than half of course fraction is smaller than No. 4 sieve size (for visual classification, the 1/4-in size may be used as equivalent to the No. 4 sieve size	Clean sands (little or no fines)	SW	Well-graded sands, gravelly sands, little or no fines	Wide range in grain size and substantial amounts of all intermediate particle sizes
			SP	Poorly graded sands or gravelly sands, little or no fines	Predominantly one size or a range of sizes with some intermediate sizes missing
		Sands with fines (appreciable amount of fines)	SM	Silty sands, sand-silt mixtures	Nonplastic fines or fines with low plasticity
			SC	Clayey sands, sand-clay mixtures	Plastic fines

| | | Identification procedures on fraction smaller than No. 40 sieve size | | |
		Dry strength (crushing characteristics)	Dilatancy (Reaction to shaking)	Toughness (consistency near PL)
Fine-grained soils — More than half of material is *smaller* than No. 200 sieve size. The No. 200 sieve size is about the smal[lest]	Silts and clays — Liquid limit is less than 50	ML — Inorganic silts and very fine sands, rock flour, silty or clayey fine sands or clayey silts with slight plasticity	None to slight	Quick to slow	None
		CL — Inorganic clays of low to medium plasticity, gravelly clays, sandy clays, silty clays, lean clays	Medium to high	None to very slow	Medium
		OL — Organic silts and organic silty clays of low plasticity	Slight to medium	Slow	Slight
	Silts and clays — Liquid limit is greater than 50	ME — Inorganic silts, micaceous or diatomaceous fine sandy or silty soils, eleastic silts	Slight to medium	Slow to none	Slight to medium
		CE — Inorganic clays of high plasticity, fat clays	High to very high	None	High
		OE — Organic clays of medium to high plasticity, organic silts	Medium to high	None to very slow	Slight to medium
Highly organic soils		Pt — Peat and other highly organic soils	Readily identified by colour, odour, spongy feel and frequently by fibrous texture		

Soil strength is an index of the suitability of the ground for the off-road mobility of vehicles. Tests can be made in field or laboratory, the former with penetrometers for vertical soil resistance or shear vanes for lateral resistance, and the latter with the 'triaxial' apparatus which quantifies the resistance of soil samples to multidimensional stress. The need to characterize the terrain in terms applicable to actual vehicle performance has led to a number of empirical tests. The 'going coefficient' indicates the sinkage and slippage of a wheel by comparing the number of its revolutions with the actual ground distance. The amount of diversion caused by obstacles can be measured in terms of the percentage excess distance required to negotiate a course between two points of known separation. The 'bumposity' of the ground can be measured with a spring-mounted continuously recording arm on a vehicle. Methods used in 'going coefficient' and 'bumposity' tests are outlined in section 13.7.

Earth processes affect engineering works, especially in areas of fluvial or coastal erosion or moving dunes. An example of the dangers of the first is provided by the Dharan–Dhankuta road, which runs across slopes up to 1000 m in length and inclined at 30–45° in the Himalayan foothills of eastern Nepal. A geomorphological survey led to a crucial decision to avoid unstable ground by incorporating a set of complex hairpin stacks. A severe storm in 1984 led to cut slope failures and blocked and scoured culverts, but the survey was amply justified by the fact that the road, including all the hairpin stacks, remained intact (Hearn and Jones, 1986).

24.3 Site surveys

Site surveys involve small areas of ground and require concentration on the details of land facets and their subdivisions. The type of information needed is that affecting the site, its area, and the engineering works in question. It will include information about stratigraphy, geological structures, earthquakes, mining activities, stability hazards, karst conditions, unstable ground, hydrogeology, geomorphological processes, and pollution. Sequential aerial photography is especially valuable for recording changes in ground conditions (Working Party, 1982).

24.4 Systems of engineering terrain evaluation

The most important terrain evaluation systems for engineering have been developed by road research organizations and are of the landscape type. Three are notable: those of the UK Transport and Road Research Laboratory (TRRL), the South African Department of Scientific and Industrial Research's National Institute of Road Research (NIRR), and the Australian CSIRO

Division of Applied Geomechanics (formerly Division of Soil Mechanics). Their views have converged, and since the Oxford symposium (Brink *et al.*, 1966), there has been a degree of standardization among them and also with the UK Military Vehicles Engineering Establishment (MVEE, formerly the Military Engineering Experimental Establishment, MEXE) and the Australian CSIRO WR, formerly the Division of Land Research and Regional Survey. The earlier work of all these organizations has been summarized by Beckett and Webster (1969).

24.4.1 *Transport and Road Research Laboratory, UK (TRRL)*

The Tropical Section of TRRL has studied the occurrence of road-building materials in relation to geology, climate, and topography since about 1960. The work started with regional surveys but progressed to more detailed examination of selected areas using aerial photographs to locate road lines, bridge sites, and suitable road-building materials such as laterite gravels. The first detailed study was made in Nigeria, and subsequent research has been done in Malaya and elsewhere. Since the Oxford conference (Brink *et al.*, 1966) TRRL have applied the land system approach in a wide variety of environments including Nepal, the Colombian Andes and, using satellite imagery for landscape interpretation, Niger (Beaumont, 1985). An example of the method of presentation of data is the diagram of the Doto land system, Nigeria, shown in Fig. 24.1.

24.4.2 *National Institute of Road Research, South Africa (NIRR)*

In order to design highways which achieved optimal route alignment consistent with accessibility to constructional materials, NIRR developed a system of 'geotechnical mapping' based on aerial photographic interpretation followed by field sampling. The earliest examples were for the Mariental–Asab route and the Etosha Pan area in South West Africa (now Namibia) (Kantey and Templer, 1959; Mountain, 1964). The maps showed lithological classes but did not incorporate surface geometry into the definition of units or provide for extrapolation outside the areas studied. Since the Oxford conference (Brink *et al.*, 1966) NIRR have used the land system method.

They have covered southern Africa with higher land units, starting with maps of considerable areas in Transvaal and South West Africa. An index map has been produced, showing the location of specific road-building projects and

311

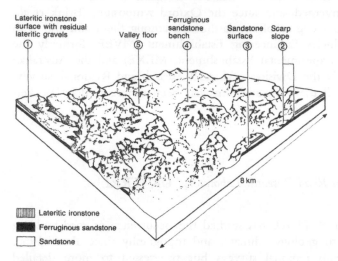

Lateritic ironstone
surface with residual
lateritic gravels ①
Valley floor ⑤
Ferruginous sandstone bench ④
Sandstone surface ③
Scarp slope ②

8 km

▦ Lateritic ironstone
■ Ferruginous sandstone
☐ Sandstone

Land facet	Form	Soils, materials and hydrology	Land cover
1	LATERITIC IRONSTONE SURFACE Flat to gently sloping surface at 600 m OD. Occurrences vary in size from 100 m² to 2 km². Occurs in flat interfluves and isolated mesa surfaces.	Lateritic ironstone 8 cm thick occurs as surface capping to sandstone, the upper part of which is often heavily ferruginized. Surface materials comprise bare ironstone and thin stony soils. Where the laterite has been stripped away, stony soils overlie sandstone.	Mixed *Detarium* woodland where sufficient soil occurs.
2	SCARP SLOPE Uneven steep topography generally becoming steeper on upper part of slope below margin of land facet 1. Concave lower slopes with uneven bouldery microrelief. Steep rocky ribs and pinnacles and incised gullies are common.	Lateritic ironstone and sandstone rock, talus and thin, stony soils subject to erosion by hill wash.	*Detarium* woodland where sufficient soil occurs.
3	SANDSTONE SURFACE Flat to moderately sloping ground corresponding to a structural bench formed by dissection of a flat bedded sandstone.	Rock and thin, stony soils.	As above.
4	FERRUGINOUS SANDSTONE BENCH As above, but more persistently developed and determined by the presence of a ferruginous sandstone horizon 0.5 m thick, midway between land facits 1 and 5.	Rock and thin stony soils. The indurated, ferruginous sandstone protects the underlying softer sandstone and forms a prominent bench, often bare rock, but sometimes overlain by rubbly soils.	
5	VALLEY FLOOR Flat but steepening slightly at margin with land facet 2. Vary in width from 10 m to 900 m. Length may be as much as 13 m.	Moderately deep, light grey to reddish brown sandy colluvial and alluvial soils. Permeable with no well-defined watercourses. Active downcutting of the gullies has now ceased.	Savanna 'parkland' heavily cultivated.

Fig. 24.1 Example of terrain evaluation by the Transport and Road Research Laboratory: evaluation of Doto land system, Nigeria. (Source: Dowling, 1968)

indicating the sources of data about them. Detailed land system maps have been completed of the Johannesburg–Pretoria district and the whole Karroo formation area. An example of a land system from this area is shown in Fig. 24.2.

The emphasis in the South African work is on the short-term engineering project level approach. This is because, for any given area, they produce not only a land system map but also a soil engineering map which translates the land facets into road engineering terms and enables the user to tap the NIRR data storage system in Pretoria (Brink *et al.*, 1968).

24.4.3 *Division of applied Geomechanics, CSIRO, Australia (AG)*

The AG division of CSIRO is separate from WLR, and is concerned with the engineering rather than the agricultural properties of terrain. Their mutual contacts in the field date from 1963 when the Division of Land Research and Regional Survey (the forerunner of WR) requested the Division of Soil Mechanics (the forerunner of AG) to undertake an engineering assessment of the Tipperary land system in the Katherine–Darwin region, Northern Territory, which the former had previously surveyed. The results of this assessment were contained in a report which made a quantitative evaluation of each unit in terms of its engineering characteristics (Grant and Aitchison, 1965).

The inadequacy of conventional land system surveys for engineering purposes thus revealed soon compelled AG to devise a modified system for their own needs. This had more narrowly defined land systems and land facets, which were proved by engineering experiment and defined entirely in terms of recognizable features before specific engineering data were attached to them.

The terrain evaluation process was seen as consisting of three phases: the establishment phase, the quantification phase, and the interpretation and application phase (Aitchison and Grant, 1968a). The establishment phase consisted of terrain classification and mapping rigorously aimed at engineering needs. It brought classifiers and users together so that the latter could, without training in geomorphology or soils, not only operate the system at any level of generalization but also build their requirements into the original classification.

The land classification was based entirely on land properties normally recognizable by engineers and avoided abstractions at all levels. There were four levels of generalization: 'province', 'terrain pattern', 'terrain unit', and 'terrain component' in descending order of size. Only the last three were considered as suitable storage units for engineering data. The method of relating such data to these was called the PUCE (pattern unit component evaluation) programme, and is described by Aitchison and Grant (1967, 1968a,

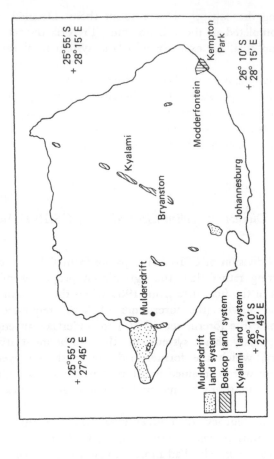

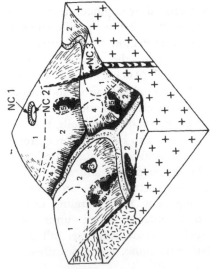

Fig. 24.2 Example of terrain evaluation by the South African NIRR. Part of the description of the Kyalami land system. (Source: Brink and Partridge, 1967): (SP) structural pattern; (NC) noncognate (facet)

314

TABLE 1 Kyalami Land System: Table of Land Facets

Land Facet	Form		Soils, materials and hydrology	Land cover
1	Hill crest. Slope <2 width <2,000 yd	Variant 1:	(Old erosion surface). Residual sandy clay with collapsing grain structure (<50 ft) on granite. Sometimes overlain by reworked soil (<10 ft) Above ground water influence except at depth.	Crops and grass.
		Variant 2:	Residual expansive clay (<20 ft) on schists and basic metamorphic rocks, with occasional low outcrop. Above ground water influence.	
		Variant 3:	Weathered granite (<10 ft) sometimes covered by thin vein quartz gravel (<3 ft) and/or reworked soil (<4 ft). Above ground water influence.	
2	Convex side slope <12 width <1,000 yd	Variant 1:	Hillwash of silty sand derived from granite (<3 ft) on granite, schists or basic metamorphic rocks. Occasionally saturated.	Crops, grass and low bush.
		Variant 2:	Hillwash of expansive silty clay derived from schists or basic metamorphic rocks (<3 ft), on granite, schists or basic metamorphic rocks. Occasionally saturated.	
3	Tor. Slope <90 diameter <100 yd		Fresh granite outcrop on side slope. Above ground water influence.	Trees and bush.
4	Whaleback. Slope <10 width <200 yd		Fresh granite outcrop on side slope. Above ground water influence.	Bush, grass and crops.
5	Gully, Slope <15 width <300 yd	Variant 1:	Sandy gully wash derived from granite (<4 ft) on granite. Periodically saturated and sometimes containing ferricrete with perched water table.	Bush, grass and crops.
		Variant 2:	Expansive clayey gully wash derived from schists and basic metamorphic rocks (<10 ft) on granite, schists and basic metamorphic rocks.	
6	Alluvial terrace. Slope <10 width <100 yd with steep bank over-looking stream height <25 ft slope <30		Sub-angular gravel and boulders of mixed origin in silty matric (<20 ft) on granite, schists and basic metamorphic rocks. Above ground water influence.	Grass and trees.
7	Alluvial floodplain. Slope <10 width <100 yd with incised stream course flanked by vertical banks of <15 ft		Expansive alluvial clays and sands (<20 ft) on granite schists or basic metamorphic rocks. High water table.	Grass.

non-cognate facet

1	Pan side slope (<20 width <100 yd)		Hillwash of silty sand derived from granite (<5 ft) on granite. Occasionally saturated near foot of slope.	Bush and grass
2	Pan floor. Slope <1 dia-meter <200 yd		Poorly drained black expansive clay (<3 ft) on granite. High water table.	Grass
3	Dyke ridge. Height <20 ft slope <30 width <150 yd	Variant 1:	Bouldery outcrop of diabase syenite or felsite. Usually above ground water influence.	Trees and bush
		Variant 2:	Residual expansive clay (<10 ft) on diabase, syenite or felsite. Usually above ground water influence.	
4	Pediment (below marginal escarpments) slope <9 width <2,000 yd.		Hillwash of silty sand derived from quartzite (<5 ft) on granite, schists or basic metamorphic rocks. Usually above ground water influence	Crops and grass.

Fig. 24.2 continued

315

TABLE 2 *Kyalami Land System: Elements of the Photographic Image*

Facet	Tone	Texture	Shape and internal pattern*		Stereoscopic appearance	Associative characteristics
			Shape and structural pattern	Drainage form*		
1	Light Grey.	Moderately fine.	Amorphous, showing occasional angular jointing pattern in igneous rocks, and foliation lines in schists.	Absent except on old erosion surface where dislocated, with occasional pans.	Gently sloping, convex.	Some agriculture.
1	Medium grey with occasional dark patches.	Medium, sometimes mottled.	Amorphous around hill crests, no S.P.	Sub-parallel/radial, low density, good integration.	Gently sloping, convex.	Patches of low bush and crops.
3	Light grey with heavy shadows.	Medium.	Circular, with angular jointing pattern.	—	Steep, with substantial relief.	Quarries.
4	Almost white.	Moderately fine.	Amorphous showing occasional angular jointing pattern.	—	Gently sloping, convex.	Absence of vegetation.
5	Medium grey, sometimes patchy.	Fine, except where rilled at head.	Hemi-lemniscate, no S.P.	Centripetal/sub-parallel at head. Dense but poorly integrated.	Gently sloping, concave.	Tall grass, crops, sand quarries.
6	Light grey.	Medium.	Linear/lenticular, no S.P.	—	Gently sloping, with steep side.	Trees.
7	Medium grey.	Fine, except where gullied.	Linear, sinuous, no S.P.	Sub-parallel where gullied. Moderately dense but poorly integrated.	Gently sloping, except where incised by stream.	Dams, tall grass.
N.C.						
1	Medium grey with occasional dark patches.	Medium, sometimes mottled.	Ring-shaped, no S.P.	—	Moderate slope, concave.	Patches of low bush, sand quarries.
2	Medium grey.	Very even, fine.	Circular, no S.P.	—	Virtually flat.	Standing water, beach line, tall grass.
3	Medium grey to black.	Mottled, moderately coarse.	Linear, with occasional angular jointing pattern.	—	Broken low ridge.	Trees.
4	Medium grey.	Medium.	Broad belt below marginal escarpments, no S.P.	Sub-parallel, low density, good integration.	Gently sloping, concave.	Crops.

*Includes pattern, density and degree of integration.

Fig. 24.2 continued

1968b) and Grant (1973/74). The procedure began with a regional terrain classification as a basis for feasibility or planning studies and concluded with a detailed classification of smaller units for detailed design and construction purposes.

The terrain pattern is recognized mainly on aerial photographs and is equivalent to the land system. It has constant geomorphology and associations of terrain units. It is mapped at 1:250 000 and illustrated by a schematic block diagram (see Fig. 24.3), being defined qualitatively in terms of its principal rock, soil, and vegetation characteristics, and quantitatively in terms of its horizontal and vertical dimensions. The terrain province, equivalent to the land region, is also mapped at 1:250 000, but relates only to areas of constant geology as revealed on aerial photographs. About 20 districts of Australia have now been covered by these reports. In order to provide a synthesis to allow the prediction of road-building costs throughout Australia, Grant *et al.* (1984) produced a *Geotechnical Landscape Map* of the country at a scale of 1:2 500 000.

The terrain unit is equivalent to the land facet and is generally the most important category. It has to be readily recognizable and is defined as 'an area occupied by a single physiographic feature formed of a characteristic association of earthen materials with a characteristic vegetation cover'. It is mapped at a scale of 1:50 000 and described qualitatively in terms of principal rock, soil, and vegetation characteristics, and quantitatively in terms of lateral dimensions and relief, using either aerial photographic interpretation or ground methods.

The terrain component is roughly equivalent to the land element. It has a constant rate of change of slope, consistent soil at primary profile level, and consistent vegetation associations. It is generally too small to be mapped or interpreted on aerial photographs, but where necessary is mapped *in situ* and its relative importance in the terrain unit recorded.

The second phase was the quantification of the units in order to simplify the nomenclature, refine their definitions, and increase the capability of matching conditions between different areas. The principle was to devise a numerical code for each scale of unit which assigned it to an exactly defined class range for each topographic, soil, vegetation, and land use attribute. The terrain provinces were identified in terms of their geology by two digits, the first for the major period from Archaean (1) to Cainozoic (5) and the second for the stratigraphic system. Thus Cambrian is 31, Jurassic 42, etc. The terrain patterns were characterized by two digits, the first for the 'greatest local relief amplitude' from '0–15 m' as 0 to >3600 m as 9, and the second for mean drainage density in terms of the number (from 0 to 9) of stream lines intersecting N–S and E–W lines per 1.6 km. The terrain units were described by four digits representing all combinations of topography, soil, and vegetation, and terrain components by eight digits representing slope profile, areal magnitude in two directions, soil profile, land use, and vegetation (Grant, 1973/74). Thus every terrain component in an area had its own unique numerical code. The quantification phase also included a system for information storage and retrieval.

Province No. 43.001 Rolling Downs Group

TERRAIN PATTERN No. 04

LITHOLOGY — Shale, claystone, siltstone, sandstone; often overlain by tertiary or quaternary sandstone, conglomerate, gravel, clay, silty clay, often gypsiferous

OCCURRENCE — Scattered occurrences mostly bordering drainage systems adjacent to Flinders and Willouran Ranges

TOPOGRAPHY — Moderately undulating dissected terrain

INCLUSIONS — Terrain patterns 01/3, 11, province 43.001, terrain pattern 01, province 50.007

NOTE - Parts of this terrain pattern may be covered by a veneer of aeolian sand

CHARACTERISTIC CROSS-SECTION SHOWING TYPICAL LOCATION OF TERRAIN UNITS

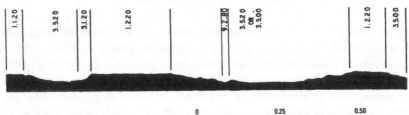

Vert. scale = twice horiz. scale

Miles

0   0.25   0.50

TERRAIN UNITS

Number	Terrain Pattern Area (%)	Occurrence	Description of Dominant		
			Topography	Soil	Surface Cover
1.1.20	5	Alternate to terrain unit 1.2.20	Flat surface	Stratified yellow brown medium to heavy-textured sandy clay with lenses of gravelly clay (CL-SC-GC), over variable gypsiferous sands and gravels (SC-GC) or occasionally over grey or purple decomposed and/or wholly or partly silicified rock of various ages	Sparse to mediumly dense rounded silcrete and quartz
1.1.80	5	Adjacent to drainage below terrain unit 3.5.20	Flat surface (floodplain)	Stratified medium-textured sandy or gravelly clay (CL-SC-GC) over red brown heavy-textured clay (CH) with lenses of sand and gravel	Mostly nil; occasional areas of sparse rounded silcrete
1.2.80	30	Continuous; extensive; included all other terrain units	Gently undulating surface	Stratified red brown medium to heavy-textured clay (CL-CH) over sandy or gravelly clay (SC-GC) over highly gypsiferous clay commonly with kopi lenses (ML), over grey, or purple decomposed and/or wholly or partly silicified rock of various ages	Sparse to mediumly dense rounded silcrete and quartz
3.1.20	<1	Discontinuous; mostly replacing terrain unit 3.5.00	Smooth steep slope	Grey gypsiferous silt over heavy-textured clay (decomposed rock) (ML/CH); some areas may have a capping of wholly or partly silicified purple sandstone and conglomerate	Mostly crystalline gypsum and kopi rubble, areas of rubble derived from rock outcrop

TYPICAL DRAINAGE NET OF TERRAIN PATTERN

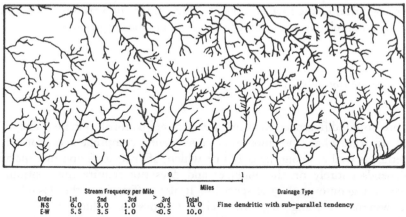

	0		1	
		Miles		

Order	1st	Stream Frequency per Mile 2nd	3rd	> 3rd	Total	Drainage Type
N-S	6.0	3.0	1.0	<0.5	10.0	Fine dendritic with sub-parallel tendency
E-W	5.5	3.5	1.0	<0.5	10.0	

Vegetation	Inclusions	Terrain Parameters			
		Terrain Unit No.	Max. Local Relief Amplitude (ft)	Length of Terrain Unit	Width of Terrain Unit
Mostly bare; seasonal grass and forbs	-	1.1.20	3	1 mile	1000 yards
		1.1.80	3	Extensive	500 yards
		1.2.80	10	Extensive	1000 yards
		3.1.20	50	2 miles	50 yards
		3.5.00	50	Extensive	1000 yards
		or			
		3.5.20			
		9.2.80	10	Extensive	10 yards
Mostly bare; seasonal grass and forbs; occasional shrubs	-				
Mostly bare; seasonal grass and forbs	Terrain pattern 01/3, 11, province 43.002, terrain pattern 01, province 50.007				
Mostly bare	-				

DIAGRAMMATIC REPRESENTATION OF TOPOGRAPHY AND
ARRANGEMENT OF TERRAIN UNITS WITHIN TERRAIN PATTERN

3.5.00
or
9.2.80   3.5.20   12.20   3.1.20

1.1.20

Shale

Sand
Silty clay

Fig. 24.3 Example of terrain evaluation by the Australian CSIRO Division of Soil Mechanics: terrain pattern no. 04. (Source: Grant, 1970)

319

The interpretation and application phase produces the information for the engineering user, employing the recognition criteria of the land units defined in the first phase to enter the store of quantified descriptions produced in the second. The user is provided with a map of the terrain patterns and guidance on the recognition of the smaller units. The quantified information about these is considered completely adequate for all normal engineering purposes.

24.4.4 *Comparison of systems*

There are some differences between these three national systems. The AG method is somewhat the most developed and comprehensive for civil engineering as well as being the only one which, after its initiation in a specific area, depends entirely on the engineer and does not require the continuing presence of a geomorphological specialist. It also differs from the TRRL and NIRR systems in using vegetation characteristics as definitive of classes, probably because of the more 'natural' vegetation patterns in most of the areas it covers. The TRRL scheme differs from that of the NIRR in being concerned with a wider range of engineering and agricultural land uses, and in including all scales from the broad national to the detailed local. Nevertheless, the three schemes are essentially similar in approach, using airphoto-recognizable terrain classes as the basis for storing environmental information for the engineer.

24.5 Synthesis

The systems discussed above derive their basic economies from the ease with which the layman can recognize the terrain units and the capability of extrapolating practical information between geomorphologically analogous areas.

Such systems can be used for operations ranging in scale from national or international appreciations of land resources and hazards to the provision of detailed data about specific sites. They serve the engineer at the broad scale by providing, at relatively small cost, information on which the planning of large schemes can be based, and can identify key areas which, because of their importance or problems, require special study. Also, because they permit the measurement of the total areas occupied by each land system and estimates of those occupied by each land facet, they can give a basis for national inventories of features significant to engineering. At the opposite extreme, they can provide the local worker with the site information required for small-scale projects. For these, it may be valuable to identify subdivisions of land facets.

Further reading

Beckett (1971), Beckett and Webster (1969), Bell (1987), Brink *et al.* (1966, 1968), Casagrande (1947), Fookes and Gray (1986), Grant (1973/74), Grant *et al.* (1984), Hearn and Jones (1986), Mitchell (1987), Unesco (1976), USDA (1976).

25

Systems for military purposes

25.1 Terrain in military activity

Knowledge of the terrain is basic to all military activity and its understanding can make the difference between victory and defeat. Its discriminating use has been the hallmark of great commanders, allowing 300 Spartans to hold the Persian army at bay in the narrow defile of Thermopylae, the Black Prince to gain success at Poitiers by pushing forward his light troops on to ground too soft for the French armour, and Wellington to screen his main dispositions at Waterloo. With the advent of gunpowder, questions of cover and the intervisibility of sites gained importance. Since 1914, the vastly increased mobility of armies, the expanded range and destructiveness of projectiles and, above all, the use of air power have transformed warfare. This has altered, but not removed, its dependence on the nature and configuration of the ground surface.

Military assessments of terrain have two aspects: strategic and tactical. At a strategic scale the concern is with the gross spatial distribution of economically important lowlands with their cities and associated transport lines, and also with the mountains, sea, and river barriers which divide them. Tactical assessments consider the landscape in more detail. They focus on five types of problem: position in relation to vantage and refuge, cross–country mobility, the reaction of terrain to deformation, the availability of water and construction materials, and natural hazards.

First, it is important to know the distribution of potential vantage points and areas of concealment which combine maximum visibility over an enemy with minimum exposure to oversight and vulnerability to projectiles. The dominating and inaccessible positions of ancient forts and medieval castles are obvious examples of this. Modern fortifications are likewise located to command traffic lanes with minimum exposure to ground–based artillery and aerial approach and observation. The same considerations govern route selection for motorized vehicles. Aircraft and cruise missiles, which fly low to foil hostile radar, follow the same principle. All mobile weapons require to process rapidly changing topographic information at high speed.

Second, terrain information is needed for the off-road movement, variously known as 'going' and 'trafficability', of vehicles and bodies of troops, and for the reception of parachute drops. These involve considerations of the slope, evenness, hardness, and slipperiness of the ground surface and the number and nature of the obstacles it contains.

Third, terrain must be judged in terms of its reaction to deformation. This includes the reception of tent pegs, excavation of trenches, caves, and dugouts which are stable against collapse, and resistance to explosives and projectiles.

Fourth, it is important to know the location of construction materials near to the site of use, and of potable water without the attendant evil of a high water-table which floods entrenchments and makes the surface soft and impassable.

Finally, a capacity is needed to foresee and protect against hazards deriving from the climate, such as gales, winds capable of moving poison gas, or storms which might cause flash floods and make the ground impassable. In arid climates such hazards include the overheating of surfaces and damage to vehicle brakes and engines from sand and dust. Because of the number and complexity of these factors, it is appropriate to begin by giving examples of the way in which terrain has affected specific military operations.

25.2 Case studies

Three twentieth-century examples: the Somme–Flanders, Sinai, and Falkland Islands battlefields, are illustrative.

25.2.1 *The Somme–Flanders battlefield*

The Somme–Flanders lowland was notable in both world wars in the way the military operations underlined the importance of even minor relief features. As Johnson (1921) has pointed out, this area is strategically vital not only in being the narrowest point on the great northern European plain but specifically in covering the gap between the Artois and Ardennes barriers which is the only route into France from Germany without formidable topographic obstacles. It also has strategic importance in its dense population, highly developed both agriculturally and industrially (Fig. 25.1).

The battles of both world wars centred around the lowland via command of the surrounding heights. Specifically, the Artois barrier is a chalk upland terminated on the north by the Vimy Ridge and cut by entrenched NW–SE streams with steep sides and marshy bottoms. The Ardennes barrier is a Palaeozoic upland reaching a height of about 600 m, cut only by the deep Meuse Trench.

Fig. 25.1   The Somme–Flanders area

The lowland itself is in two parts: the Flanders clay plain to the north of the uplands and the rolling chalk plain of the Somme between them, sometimes called the 'Seuil de Vermandois'.

The plain of Flanders, illustrated on Fig. 25.2, is one of the lowest and flattest tracts in Europe. The clay of which it is composed is especially fine grained and impermeable, and usually wet. When in this condition, all movements are slowed and troops are exposed for a longer time to hostile observation. Shells lose effect; equipment and weapons become clogged, and the wounded suffocate. Excavations are difficult. Trenches will not stand up, soon fill with water, and become unusable unless shored with timber and pumped out. Water, though abundant, is contaminated and natural building materials are absent. All these factors adversely affect morale. There are numerous instances of the obstacle posed by the mud of Flanders to military activities in 1914–18, and it was an important factor in delaying the German advance and in permitting the British evacuation from Dunkirk in 1940.

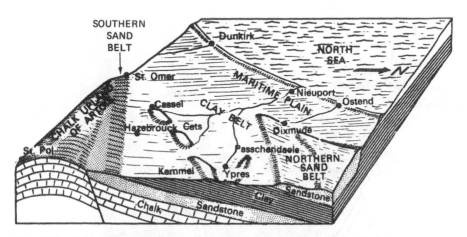

Fig. 25.2   Schematic block diagram showing the Flanders battlefield of the First World War (after Johnson, 1921)

Within this clay belt is an intercalated sandy bed which forms a cuesta with a south-facing escarpment (Fig. 25.2). This is the Mont Kassel–Mont des Cats–Mont Kemmel upland, whose eroded backslope remnant is the Passchendaele–Messines Ridge. Mont Kemmel, though barely 150 m high, has been the key position in Flanders since Roman times. From its summit a wide view may be obtained, stretching to the sea in the west, the Artois upland to the south, and almost to Brussels in the east. It forms the northern, as Vimy Ridge forms the southern, hinge of defensive positions for southwestern Flanders and the Pas de Calais. In the First World War no less than six major battles were fought for its control and hundreds of thousands of British and German lives were sacrificed in attempts to gain it. At no time did either side possess the whole.

The neighbouring Seuil de Vermandois is the main natural route from Flanders to Paris. It was the scene of all three Somme campaigns in 1914–18, and was briefly the battlefront in 1940 and 1944. Movement across it is relatively easy except where the surface is covered with sticky clay with flints or loam, or traversed by the swampy bottoms of the larger valleys such as the Scheldt, Sensée, and Somme, whose banks are sometimes steepened by undercutting meanders. These valleys, as well as the extensive caves in the chalk, were used defensively by both sides, and the dryness and relative ease of excavation of this material made the area suitable for trench warfare. Most of the campaigns centred around control of the river crossings and the heights commanding them.

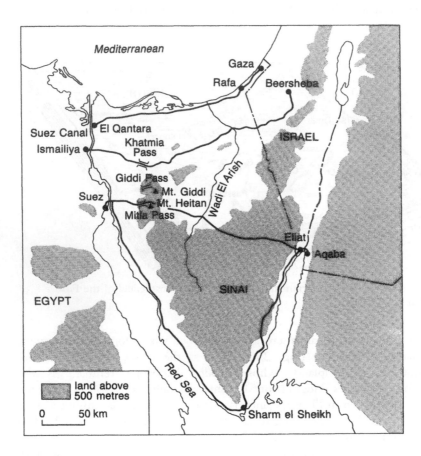

Fig. 25.3 The Sinai area

25.2.2 *The Sinai and Falklands battlefields*

More recent wars in different environments have not changed the basic dependence on terrain but have introduced some new factors. The Sinai Peninsula has been involved in war no less than four times in the present century: in 1916, 1956, 1967, and 1973, and witnessed subsidiary military movements in 1941–43 and 1948. Three distinct aspects of terrain can be identified as having had a critical effect on military operations.

First, the key feature in the frontier zone between Israel and Egypt is the Central Range of Sinai called Giddi Mountain (843 m) in its northern and Heitan Mountain (768 m) in its southern part (Fig. 25.3). This overlooks the coastal plain where it is narrowest and most encumbered with sand, and is crossed by the Giddi, Mitla, and Khatmia passes, which are thus the critical military positions in the whole peninsula. The Mitla Pass, commanding the approaches to Suez, is more important than the nearby Giddi, because although somewhat longer, it is much less sandy. Israeli paratroops seized it at the start of the 1956 Sinai campaign, and an armoured column of only four tanks took it in 1967. The restricted capacity of the tarmac and the surrounding impassable soft sand turned it into a trap leading on both occasions to the destruction, not only of the Egyptian reinforcements, but also of their retreating army, by aerial attack.

Second, earth materials determined tactics. The success of the initial Egyptian attack across the Suez Canal in 1973 was due not only to surprise but also to the use of high-powered hoses which made breaches in the sandy banks wide enough for the passage of tanks. These worked well in the sandy ground in the north but were less successful further south because the barrier was wider and the material became a mass of mud having to be bulldozed, causing delay.

Third, in the level country, small differences in topography were critical. After the successful Egyptian canal crossing, the main battle in northern Sinai was one which involved about 1600 tanks. An important factor in Israel's victory was that their tanks not only had cannon with a longer range, but also could point 10° below the horizontal compared with the 4° which was the maximum possible for the Egyptian tanks. This much increased the degree to which folds in the ground could be used for attack and concealment (Mitchell and Gavish, 1980).

The Falklands War underlined the need for the rapid availability of terrain intelligence about an unexpected and little-known area. This had to be both general, as in identifying potential landing sites for aircraft and paratroops, and detailed, in predicting site conditions at particular beaches, optimal traverse lines across boggy and exposed terrain, and eminences commanding points of tactical importance.

25.2.3 *Summary of case studies*

These examples illustrate the central importance of understanding terrain factors to obtain military success, both in providing protection in defence and opportunities for attack. In the strategic sphere the importance of terrain is seen in the location and dominating position of uplands, the rear cover provided by their possession, and the importance of passes. In the tactical sphere the significant characteristics are the reaction of terrain to excavations and other deformations under different moisture conditions (natural or induced), its effects on mobility and concealment, and in its provision of raw materials and water.

25.3 Physiographic systems of military terrain evaluation

25.3.1 *The MEXE system*

Military authorities require that detailed intelligence about the terrain of a battlefield be rapidly and continuously available at the planning stage and to the commander in the field. The pioneering research in this direction was that of the UK Military Vehicles and Engineering Establishment (MVEE, formerly MEXE) (Beckett and Webster, 1969). This was aimed to have application to all forms of land use in addition to the military. The basis was the use of land systems and land facets as 'pigeon-holes' for the storage of practical information which could rapidly be made available and extrapolated to unknown areas on the basis of physiographic analogies recognized on remotely sensed imagery.

Land systems are useful at the broader scale of strategic planning at theatre and army corps level, land facets over areas of a few tens of kilometres or less – the scale of operations of the division or the battalion. For smaller areas, the problem becomes more complex because of the difficulty of recognizing accurately the subtlest subdivisions of the landscape and, once recognized, of communicating relevant information about them to the men in the field rapidly and continuously. Since the field commander is faced with a kaleidoscope of problems concerning the terrain in his vicinity, he requires detailed data with a rapidity and accuracy at least equivalent, for instance, to those on weather provided for aircrews, and usually under hazardous and changing field conditions.

Research on this subject showed that one solution could be a terrain intelligence centre or central data store of the type discussed in Chapters 14 and 15. The basis would be the interlocking operation of two automated stores and an indexing system. The first store would be of 'items' of practical information relevant to military activities, such as 'going', sources of building materials, etc. indexed by content and by the land units to which they referred. The second would be a 'terrain store', containing land system maps and descriptions, supported by libraries of remotely sensed imagery and topographic maps. It would be indexed by location and physiographic type, to facilitate the matching of local with more distant terrain. The indexing system would be designed either to assemble information about particular land units or alternatively to find land units with particular characteristics with rapidity. The outstanding advantages of such a system are its simplicity and its capability for improving predictions as more data are acquired. Apart from the overriding need for speed, the most difficult task in such a system is probably the identification of the local land facet at the unvisited site, and this requires geomorphological expertise.

25.3.2 *Other physiographic systems*

The Indian army have adopted a physiographic system and were reported by Beckett as having by 1967 classified areas in the Punjab Plains, around Saugor in Madhya Pradesh (MP), and in Ganhati–Assam, and carried out some studies on the uniformity of terrain units.

The US Army Engineers have carried out terrain research on a largely parametric basis, which is considered in section 25.4, but other US army organizations, concerned with more specific needs or environments, have used a physiographic approach. The Air Force Cambridge Research Laboratories have sponsored two major types of terrain research: on tropical soils and on playas. The former showed how the military engineering problems of the major groups of soils could be related to their origins and character, and gave a key for their recognition on aerial photographs (Ta Liang, 1964).

Playas were studied by Neal (1965), Neal *et al.* (1968), Motts (1970, Krinsley (1970), and others. They are level, seasonally dry, lake beds, devoid of vegetation, with surfaces composed of clay or silt containing large amounts of soluble salts, and occur widely in all arid areas. They may be hard or soft, smooth or rough, depending on moisture conditions. The width and flatness of playa surfaces have important military implications. Apart from containing deposits of exploitable minerals, they can be used seasonally for the landing of aircraft without pretreatment, and they smooth themselves after rain. They provide possible recovery sites for spacecraft or locations for large arrays of antennae. They also provide an analogy with conditions on the moon and Mars, and for this reason have been used in designing space-landing equipment.

A large amount of data about over 200 North American, Australian, Iranian, and North African playas was acquired on their physical characteristics, especially those recognizable on aerial photographs, and their formative processes. Satellite photographs were used to monitor the seasonal variations in water cover and the giant desiccation polygons of some American examples. These data made it possible to classify them in terms of chemical composition and the surface hardness which affects their usability for aircraft landings. Further studies showed that terrain types could be rapidly distinguished on aerial photographs using microdensitometer techniques, but it was not possible to quantify their internal character or separate the image effects due to vegetation, soil, and cultural factors.

Special terrain conditions are required for constructing the sites for the movable advanced US intercontinental ballistic missile (ICBM) called MX.

The missile and its transporter has to move thousands of metres among a number of fixed shelters over unsurfaced roads in Nevada. In practice, sandy and gravelly deposits provided the best aggregates, fine-grained materials were the easiest excavated, while aeolian sands were the least suitable both as terrain and as aggregates. The survey procedure was to overprint the units on the geological map with composite symbols expressing the geomorphological origin and the grain size of the surface materials (Christenson *et al.*, 1982).

The Cornell Aeronautical Laboratory carried out a programme of research which evaluated broad areas of the earth's surface in climatic and physiographic terms in relation to general transport systems requirements. They also devised a quantitative system for determining from maps the degree to which topography limits lines of sight to surface targets from a source (Deitchman, *c.* 1966).

25.4 Parametric systems

The aim of the parametric approach is to provide a source of quantified terrain data based on selected attribute values, appropriate at a wide variety of scales, and rapidly available for any part of the world. It has been most fully developed by the US and Canadian armies, mainly by the Quartermaster Research and Engineering Centre at Natick, Massachusetts (QREC), the US Army Engineer Waterways Experiment Station at Vicksburg, Mississippi (USAEWES), and the Canadian Defence Research Board. These three organizations differ somewhat in objective, the first being concerned mainly with how to derive generalized relief data relevant to military activities from a sample of topographic maps, and the other two with producing, in map or digital form, a detailed data source on all those aspects of terrain which impinge on military activities, most specifically concerning cross-country mobility.

25.4.1 *The 'Natick' approach (QREC)*

The QREC produced terrain bibliographies of the Russian Arctic and western Africa, environmental handbooks on different parts of the world, and also evolved methods for analysing lines of sight and surface relief. In order to derive a predictive method for topographic analysis, 204 separate maps of parts of the USA at a scale of 1:62 500 were chosen at random and 10 sample areas selected in each by the expedient of drawing concentric circles around the centre of the maps to enclose respectively 0.81, 1.62, 3.24, 6.47, 12.95, 25.9, 51.8, 103.6, 207.2, and 414.4 kilometres^2. For each circled area the following measurements were made:

1. Highest elevation;
2. Lowest elevation;
3. Relief (obtained by subtracting (2) from (1);
4. Number of closed contours ('hilltops');
5. Number of crossings of contours (here at 6.1 metre intervals) in directions due N–S and E–W from the centre point ('contour count');
6. Number of valleys and divides on the same traverse as used in (5) ('slope direction changes').

This process yielded 6 items for each of the 10 sampling circles on each of the 204 map sheets, or 12 240 separate items of data. When it was seen that the variation was orderly, the number was reduced by ignoring every other sample area, i.e. those enclosing 0.81, 3.24, 12.95, 51.8, and 207.2 kilometres^2.

A ranking method of analysis was used. Correlations were found in all sized areas between relief, contour counts, and slope direction changes. The mean relief of a 0.81 kilometre^2 area was 73.2 metres and for a 414.4 kilometre^2 area was 432.8 metres. A successive doubling of areas from the smallest to the largest was accompanied by mean values for relief which increased in a regular progression. This permitted the determination of the approximate mean relief for areas in the USA of any size from 0.81 to 414.4 kilometres^2. Certain other values were of interest. The average elevation of the USA is about 701 metres. On a random traverse, a 6.1 metre contour was crossed every 88.4 metres along the ground, and three crests averaging 45.7 metres above valleys were encountered every 1.61 kilometres. Three hilltops sufficiently extensive to be represented by an enclosed contour occurred on average in every square mile. These values were means, and the medians were appreciably lower, indicating that the distribution of data was strongly skew (Wood and Snell 1957, 1959).

Six somewhat different quantitative indices were used in a further study of that part of Europe lying between 48° and 52° N and 7° and 16° E (Wood and Snell, 1960). These were 'grain', 'relief', 'average elevation–relief ratio', and 'slope direction changes'. They were measured on the 1:100 000 sheets of the US Army Map Service Central Europe Series, having a contour interval of 25 metres.

Grain was the spacing of major ridges and valleys. It was assessed by selecting a random point on a map, drawing a series of concentric circles with diameter increments of one kilometre, and determining the maximum relief difference within each circle. When these values were plotted against circle size on a graph, it was found that a 'knick point' occurred where relief ceased to increase appreciably. The sample area size equivalent to this knick point represented the grain of the area. Relief was the difference between the highest and lowest elevations in a unit area equivalent to the grain size.

Average elevation was derived from the mean of nine randomly chosen points within the unit area. Elevation–relief ratio, the relative proportion of upland and lowland, was derived by subtracting the lowest elevation from the average elevation within the area and dividing the remainder by the relief. The resulting value therefore always fell between 0 and 1 and was expressed as a decimal – the higher the value the higher the ratio of upland to lowland. Average slope was determined by counting the number of contours crossed by straight lines in directions NW–SE, N–S, NE–SW, and E–W across a circular unit area equivalent to the grain size, and computing the slope tangent from the equation

$$S \tan = \frac{I \times M}{6847} \qquad\qquad [25.1]$$

where $S \tan$ = slope tangent;
   I = contour interval in metres;
   M = number of contours crossed per kilometre of random traverse.

Slope direction changes denoted the dissection of an area and were another expression of topographic texture. They were assessed by counting the number of changes from rise to fall and vice versa along the same traverses as were used for counting contours.

   When the results of the study were analysed, it was found that groupings of the numerical data from the different indices gave 25 distinct regions which conformed well to the landform regions identified on previous physiographic maps, such as those by Lobeck (1923) and van Valkenburg and Huntington (1935) and had the double advantage of being based on a simple analysis and of resulting in groups which were quantitatively defined.

25.4.2 The 'Vicksburg' approach (USAEWES)

The USAEWES at Vicksburg, Mississippi, responsible for evaluating the effects of terrain on military activities, carried out a programme of research, inspired by Dr Paul Siple in 1953 and called MEGA (military evaluation of geographic areas). It began with the selection of the key terrain factors on which study should be concentrated. They had to be based on the following overriding needs: to limit the total number of factors to manageable proportions and to favour those basic parameters which were easily visualized, simple to map, significant to military uses, suitable in developing analogues between different areas, and in combination giving a complete picture of the terrain. Microrelief, i.e. terrain with less than 10 feet (3 m) of vertical amplitude, was excluded because the features were too small to map.

   Numerical subdivisions of the terrain factors were chosen for suitability as mapping units. This involved careful consideration of the availability of data, especially in the less well mapped parts of the world, of their military significance and of their amenability to quantitative measurement. Natural breaks in the landscape were used wherever possible. It was found, for instance, that alluvial fans almost always sloped less than 6° and the windward slopes of barkhans between 5° and 14°. This justified the use of 6° and 14° as critical values. In an initial study of deserts, terrain attributes were subdivided into classes and grouped into 'factor families'. These are discussed in detail in a handbook (USAEWES, 1959).

The first factor family was called 'aggregate or general factors', consisting of 'physiography', 'hypsometry', and 'landform and surface conditions'. Physiography included terms such as 'plateau', 'hill', and 'plain' because of the usefulness of the generalizations despite their imprecision of definition. Hypsometry showed altitude classes. As vehicles showed efficiency breaks at approximately 5000 and 9000 feet (1524 and 2743 m), these two limiting contours were included in the classification.

Landform and surface conditions included 'surface geometry and form' and 'ground and vegetation' factor families. The former were introduced by physiographic sketch maps by Raisz (1938, 1946) and others, and by outline maps of the main geomorphic surface types: depositional, erosional, tectonic, volcanic, intrusive, etc. Four quantitative geometry and form factors were assessed and mapped. 'Characteristic slope' was the range of slopes commonest in the mapped unit, derived from the spacing of 10 foot (3 m) contours. The class divisions used aimed at being natural and accorded with research which showed that the tangents of observed slope angles tended to cluster mainly around values of 0.05, 0.1, 0.2, 0.4, 0.6, and 0.8. 'Characteristic relief' was the maximum difference in elevation per unit area. This had to be assessed differently depending on whether the characteristic slope of an area was greater or less than 6° and whether the drainage lines were well or poorly developed. 'Occurrence of slopes greater than 50 per cent' was assessed by counting the frequency of such slopes on lines of random traverse.

'Characteristic plan–profile' really defined the areal relations of the three preceding factors. It attempted to express the 'peakedness', areal occupance, degree of elongation, and orientation of topographic highs in quantitative terms to form a legend. The classes derived are as shown on Fig. 25.4. They total 25 as each of the 4 kinds of plan arrangement can be combined with each of the 6 kinds of profile, and an extra class is included for areas without pronounced highs or lows which for this reason has no particular plan arrangement.

In order to secure precise definitions of the plan–profile classes, specific directions are given for quantifying the gradations from 'peakedness' to 'flat-toppedness', from 'linearity' to 'non-linearity', and from 'randomness' to 'parallelism'. These are defined respectively in terms of the 'peakedness index', the 'elongation number', and the 'parallelism number'. The 'peakedness index' is derived from random transects from hilltop to valley bottom. Each is graphed and the horizontal distance from the hilltop to the points which represent respectively 10, 50, and 90 per cent of the height difference to the valley bottom measured (see Fig. 25.5). The 'profile area' is then calculated from the formula

$$A = 0.05 + 0.25(d_{10} + d_{90})/d + 0.4d_{50}/d \qquad [25.2]$$

Where A is less than 0.5, the profile is 'peaked', where more, 'flat-topped'. The 'peakedness index' is defined as $0.1d/d_{90}$; the higher the value the more peaked the landscape.

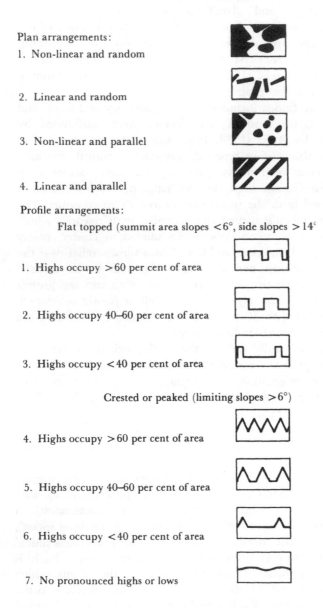

Plan arrangements:

1. Non-linear and random

2. Linear and random

3. Non-linear and parallel

4. Linear and parallel

Profile arrangements:

Flat topped (summit area slopes $<6°$, side slopes $>14°$

1. Highs occupy >60 per cent of area

2. Highs occupy $40-60$ per cent of area

3. Highs occupy <40 per cent of area

Crested or peaked (limiting slopes $>6°$)

4. Highs occupy >60 per cent of area

5. Highs occupy $40-60$ per cent of area

6. Highs occupy <40 per cent of area

7. No pronounced highs or lows

Fig. 25.4 Characteristic plan–profile: plan and profile arrangement of topographic highs and lows. (Source: USAEWES, 1959)

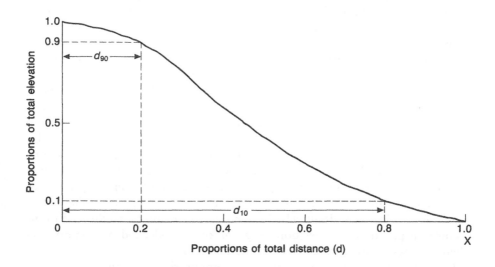

Fig. 25.5 Diagram to illustrate the method of calculating the peakedness index. d_{10} and d_{90} are at 10 and 90 per cent of the valley to crest height difference. They are measured and then calculated as proportions of the horizontal line OX. In the figure, $d_{10} = 0.8$; $d_{90} = 0.2$

In order to obtain the elongation number, it is first necessary to define the 'terrain unit'. This is the polygon circumscribed by the points at which nine randomly drawn lines from a central point on the map first reach valley bottoms. The elongation number (E) of this terrain unit is calculated by taking the compass direction (i.e. N–S, NE–SW, etc.) of the nearest contour line to each of a minimum of eight randomly chosen points and solving the formula

$$E = N/N_c \times F_h \qquad [25.3]$$

where N = number of sample points;
       N_c = number of direction classes;
       F_h = number of samples in the largest class.

Then, if $E = 0$–0.3, the area is characterized by ridges, 0.31–0.49 by ridges strongly to weakly developed, or 0.49–1 by hills.

The procedure to calculate the parallelism number (P) is an extension of the same principle. An adequate number of terrain units are constructed by the same method and the compass direction of the longest axis of each determined. These directions are then used in the same formula as for calculating the elongation number. Then, where P is 0–0.03, the landscape is strongly parallel, if 0.31–0.49 moderately to weakly parallel, and if 0.49–1 essentially random.

A map of the four surface geometry factors can be synthesized into another showing 'generalized landscape', although care must be taken to avoid a plethora of mapping units if too many different combinations of individual factors occur.

The 'ground and vegetation' factor family consisted of five land attributes. 'Soil type' mainly related to soil texture and the proportion of surface occupied by bare rock. 'Soil consistency' showed the degree of layering, cohesiveness, and crustiness of the soil. 'Surface rock' supplemented soil type by showing the lithology of the rock exposed at the surface, and 'vegetation' used a quantified physiognomic classification based on Dansereau (see Ch. 8). Although the importance of microrelief, arbitrarily defined as having a vertical amplitude of less than 10 feet (3 m), was recognized, it was not included in the scheme because of the necessity of imposing a lower limit on the scale of generalization and the fact that there was a lack of information on the earth's surface at this degree of detail. The 'ground' factors could be synthesized on to a single map, but the vegetation maps make an independent contribution when overlays are made to answer specific questions.

All factors were investigated and maps produced, in the first instance, of world deserts, at a scale of approximately 1:3 000 000. These maps aimed at being data sources which could be used to answer specific questions. For instance, a trafficability classification would demand overlaying the maps of soil strength, characteristic slope, slope occurrence, and possibly others, to obtain composite parametric units. A search for sources of building materials, on the other hand, might only require the use of the surface rock and soil type maps.

The Yuma Test Station, Arizona, was mapped with the same factors at a scale of 1:400 000. The purpose of this was to determine the degree of analogy between Yuma and other world deserts, to provide the basis for comparing them with each other and with Yuma, and to help evaluate the relations between terrain and such quantified military activities as the trafficability of given vehicles or the grading of naturally occurring materials required for construction. To achieve the first purpose 'analogue maps' were produced which assessed the similarity of Yuma to other world deserts.

The non-arid Fort Leonard Wood, Missouri, has since been mapped using the same parameters, and analyses made of the relations between vehicle mobility and soil strength, surface geometry, and tree stem spacings in test areas (USAEWES, 1963a).

The study also explored the question of recognizing quantified classes on aerial photographs, but concluded that only the physiographic landscape types could be clearly distinguished. Parry and Beswick (1973) used an area in eastern Canada to test such recognizability of both the QREC and the USAEWES parametric units. They concluded that among the former the index of change of slope direction could be interpreted if there were enough traverse lines and that grain could be determined from stereoscopic identification of maximum and minimum elevations. For the USAEWES system they concluded that the characteristic plan–profile and its definitive parameters – profile area, peakedness index, elongation number, and parallelism number – could be interpreted using a stereomicrometer or parallax converter if the relief was greater than 10 m. The process would, however, be virtually as laborious as making such interpretations from maps and would not lend itself to making predictions between different areas by physiographic analogy.

The Vicksburg work then moved in the direction of automation (Grabau, as quoted by Grant, 1968). Research is no longer devoted to terrain evaluation *per se* but rather to the development of mathematical models of all conceivable military activities involving terrain as a component. Engineer laboratories abstract from the literature any engineering formulations which might be adapted to military purposes, such as standard design for highways and airports, structural relations in bridges, etc. The resulting body of mathematical knowledge is used to identify the parameters needed by the military and to develop the instrumentation for acquiring them. Automation is used wherever possible. For example, a laser theodolite can automatically measure horizontal and vertical angles and put the information directly on to magnetic tape which a computer can be programmed to print out as a topographic profile.

25.4.3 *The Canadian army system*

The system employed by the Canadian army is mainly concerned with vehicle mobility. Parry *et al.* (1968) describe how Camp Petawawa, Ontario, was used as a sample area and surveyed in detail to determine the potential of large-scale photographs (1:5 000) in assessing the environmental factors affecting the cross-country mobility of military vehicles. The parameters considered were: surface composition, macromorphology, micromorphology, and surface cover. Map overlays were prepared for each, and a rather complex composite map produced, part of which is reproduced as Figure 25.6. Surface composition was based on the rigidity, elasticity, and viscosity of the surface materials. Macromorphology consisted of a subdivision of slope into steepness classes, distinguished according to whether they were convex, planar, or concave. Micromorphology was the most difficult factor to assess on aerial photographs because it consisted of a detailed evaluation of small surface features. Surface cover consisted in a quantification of vegetation and man–made features. Tests indicated that vehicle performance was similar for areas with similar arrays of terrain characteristics, and that successful predictions of speed could be made for test runs traversing a variety of terrain units.

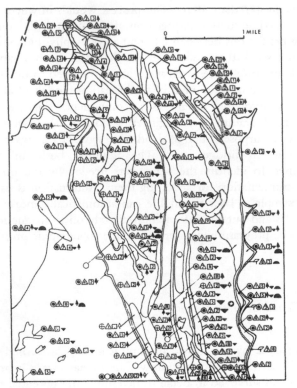

Composite terrain map of part of Camp Petawawa, Ontario

LEGEND FOR TERRAIN MAP

Surface Composition

☐   Consolidated rock — outcrops of granites and gneisses

⚲   Non-consolidated material

◉   Mineral soil — poorly graded sands and silty sands,
       SP-SM Unified Soil Classification System.

⊕   Organic soil — fine and coarse fibrous muskeg, types 9 and 12
       Radforth Classification System

○   Water — water bodies more than 3 ft deep and 1 acre in area

Fig. 25.6 Example of the system of terrain mapping employed by the Canadian army.
(Source: Parry *et al.*, 1968)

25.4.4 *Mathematical developments to parametric classifications*

It would be desirable to arrive at simple quantitative statements which could simplify the complexities of surface topography by mathematical means. Two

338

studies with this end are notable, one which seeks to reduce the variety of significant land attributes through factor analysis and another which quantifies topographic periodicity through constructing 'power spectra'.

The US Geological Survey, on contract to the US army, used direct factor analysis to derive principal components for nine geomorphological variables, such as characteristic slope, maximum and minimum elevation, etc. for each of 130 contiguous 1 × 1 km 'cells' in Fort Belvoir, Virginia (Cadigan *et al.*, 1972). These were used to make a second matrix showing Pearson's correlation coefficient for the comparison of each variable with every other. The factor analysis of this second matrix gave eigenvalues of 4.52 and 1.78 for the first two components, accounting for 50 and 20 per cent repectively of the variation. The first component seemed mainly associated with features deriving from geomorphic uplift, notably slope, maximum elevation, and relief, while the second mainly represented attributes associated with subsidence such as frequency of valley bottoms and closed water bodies. This suggests that in some areas a simplification of surface classification might be based on only two indices: relief and frequency of closed basin sites.

Pike and Rozema (1975) suggested a model which not only represented the topographic variations but also their periodicity. They transformed a longitudinal profile of terrain into a log–log graph or 'power spectrum'. The *x*-axis was the 'wavelength' of each of the main recognizable horizontal undulations (in metres), and the *y*-axis the variance of elevations within each. The value of the method is in finding the less obvious periodicities in the landscape. For instance, in the analysis on 1:24 000 scale maps, of five contrasting areas in the USA, periodicities were shown to occur in dendritic drainage systems in homogeneous rock with weak structural control with an approximately 700 m wavelength, corresponding to a third-order drainage basin diameter. This confirmed that there is spatial as well as statistical order in the seemingly random pattern of dendritic streams, probably according with the most efficient arrangement for sediment transport through a river system. More generally, such analyses and the resulting graphs could be used to partition wider topographic systems into smaller morphological regions, a parametric approach to the venerable taxonomic problem of obtaining 'optimally derived' subdivisions of topography distinguished by maximal internal homogeneity and mutual distinguishability.

25.5 Future developments

The main changes to military activity likely before the end of the century result from two different developments: advances in technology and changes in the nature of conflicts. The most significant directions for the former appear to be in aerospace technology, communications, and weapons systems. An example, combining features of all three, is the development of airborne and ground-based missiles which find their targets by ground sensing. The principle used is that of terrain contour matching, called 'Tercom', a guidance system based on a computer in the missile's nose which senses the landform over which it is travelling, converts it into digital form, searches its 'remembered landform' for a match, and then compiles position and course corrections from these data to control its itinerary (Craig, 1982). The cruise missile is the airborne version of this and has a potential range of up to 3200 km and an accuracy of 10 m. The same principle can also be used in ordinary short-range battlefield weapons with a simple explosive warhead. Both would require computer coding of accurate and detailed topographic information about potential areas of conflict. Digital elevation models, which are either static or mobile simulations of two-dimensional simulations of three-dimensional oblique images of a geographical matrix of height data, are used in the flight simulators in which pilots are trained in the guidance systems of nuclear missiles.

These developments are increasing the speed, range, and accuracy of weapons systems both on the ground and in the air, expanding the destructive powers in the hands of an attacker and enabling him to reach targets hitherto inaccessible through remoteness or depth of protection. They make possible larger and more varied forms of military engagement in a wider range of environments than hitherto. The result has been to expand the scale of military objectives and to decrease the dependence on local conditions and resources. The cost and sophistication of this technology involve the whole economic power of states and increase the dependence of smaller on larger.

At the same time, the ideological conflicts of the present age have changed the nature of military actions, widening the variety of potential targets and internalizing conflicts within states. The vital role of communications has made the media, such as newspapers and television, into weapons. Conflicts have become indifferent to the distinction between military and civil targets, blurring the roles of army and police, and involving whole populations.

The main effects of these changes have been, at all scales, to widen the range of militarily important data and force an acceleration of their communication. This increases their reliance on computers and remote sensing, which in turn adds to the pressure for the data to be in quantitative form. However, although some quantitative data are available, the experience of recent wars and the current state of technology make it likely that most armies will rely for the foreseeable future on a physiographic intelligence system of the MVEE type, but will gradually upgrade it by increasing the use of coded and automated parametric data.

Further reading

Beckett and Webster (1969), Cadigan *et al.* (1972), Craig (1982), Johnson (1921), Krinsley (1970), Mitchell (1973), Mitchell and Gavish (1980), Motts (1970), Parry *et al.* (1968), Parry and Beswick (1973), USAEWES (1959), Wood and Snell (1957, 1959, 1960).

26

Terrain in landscape resource evaluation

26.1 Principles and status of landscape resource evaluation

The importance of landscape quality has been increasingly recognized in recent years, due to expanded leisure and travel and the greater availability of environmental information. Outstanding areas such as the Alps and the Spanish coast owe their prosperity almost entirely to their scenic and recreational attractions. Even within relatively uniform landscapes, significant economic differences can be due to scenic factors. Residences along the Hudson Valley near New York, and in the North Downs and Chilterns near London, command higher prices than less favoured neighbouring areas. Spas such as Saratoga, Montreux, and Tunbridge Wells and of resorts such as Atlantic City, Brighton, and Marbella show the locational attractions of hilly and coastal sites respectively. On the other hand, bad landscape has both economic and social costs. Monotonous lowland sites and older mining and industrial areas (with the exception of some such as Ironbridge with special historical interest) attract few tourists, and their inhabitants often retire elsewhere. They also fail to attract mobile professional and self-employed workers, whose absence further compounds their economic and social disadvantages.

These concerns have stimulated scientific involvement. In the USA comprehensive and innovative studies by McHarg (1969), Lewis (1964), and Zube and Carlozzi (1967) received considerable attention. In Britain, the Landscape Research Group was formed in 1967, and government bodies such as the Forestry Commission and the Institute of Terrestrial Ecology developed programmes of research into landscape quality. The Unesco Man and Biosphere programme (1971, 1973b) selected the perception of environmental quality as one of its 13 projects. By the early 1970s there were dozens of studies contributing to the art and science of landscape assessment.

Landscape evaluation is now widely embodied in legislation. In the USA it is addressed in many public land use and management decisions and in some public decisions relating to private land uses as well (Zube, 1976). Since the Environmental Policy Act of 1969, environmental impact statements (EISs) are required for all major federal developments. Both the Act and its results have had wide repercussions in the private sector. The documentation relating to the EIS on the Alaska pipeline, for instance, was 3 m thick and cost $9 million to compile.

In England and Wales, all development must now pass a test of environmental suitability before receiving planning permission and an EEC directive has made EISs mandatory since July 1988. The degree of control on development varies with the type of area. Protection is afforded to 'national parks', 'areas of outstanding natural beauty' (AONBs), 'heritage coasts', and 'sites of special scientific interest' (SSSIs). Although the legal background varies between these categories, they identify areas, typified respectively by the Lake District, the Isles of Scilly, the Pembrokeshire coast, and parts of the Fens, where the justification for any building has to be especially strong. Most are graded into 'heritage sites' (about 10 per cent of the country) with almost total protection, and 'conservation zones' (about 20 per cent) with somewhat less. These categories do not include 'green belts' around large cities, or the higher grades of agricultural land, which also have special protection. Elsewhere, conservation objectives do not receive automatic priority.

26.2 Stages in landscape resource evaluation

There are three stages in developing a system for the aesthetic evaluation of landscape: (a) determination of human perceptions; (b) measurement of landscape attributes; and (c) coordination of the two. It is important that (a) precedes (b) because research has shown that the quantification of landscape attributes only has relevance where it reflects real environmental preferences. Stage (c) involves combining the results of (a) and (b) into a single system which is inherently credible, generally applicable, and capable of replication over a considerable area.

The resulting land evaluation system can then be used to make an inventory of the environmental resources of an area on the basis of their recognized public value. As Steinitz (1970) pointed out, these can then be linked into 'simulation models' which can help forecast future trends. Understanding these trends enables the planner to predict the changes likely to take place in the public demand for landscape, in the landscape itself, and in the transport networks which serve it.

26.3 The measurement of environmental perception

26.3.1 *Bases of human perception*

Human appreciation of landscape is a process of interpretation which derives more from an observer's social, educational, and cultural bases than from

343

absolutes in the landscape itself (Prince, 1980). Therefore, it is necessary to define biophysical categories on the basis of preference ratings (Kaplan, 1975), and the critical problem is to find predictors of such ratings. Appleton (1975) has suggested atavistic, Tuan (1974) cultural, and Laurie (1975) intuitive causes for preferences, but the complexity is such that no single explanation can be fully satisfactory (Whittow, 1979).

The West has witnessed a historical evolution of ideas about landscape, marked by fundamental changes such as those associated with the development of the medieval and Renaissance world-views, and the Romantic movement of the nineteenth century. The present century has witnessed a multiplicity of strands resulting from its complex philosophical climate. Two major themes are perhaps discernible: a search for new and abstract ways of looking at and representing landscape, and an 'internationalization' of perceptions deriving from expanded leisure, travel, and information flow. The future promises a continuation of these trends and of a search for unfamiliar and exotic environments.

There is evidence that landscape appreciation requires training as well as visual sensitivity (Laurie, 1975). Most planning decisions have tended to be made by small groups of highly trained and experienced professional planners or members of the environment and design professions. This has been criticized on the grounds that they do not adequately represent public attitudes and preferences. Recent research, however, suggests that it is still advantageous to leave decision-making to such groups (Robinson *et al.*, 1976; Whittow, 1979). The frequency of obvious mistakes, however, shows the importance of checking decisions by reference to public preferences.

26.3.2 *Levels of perception*

Human beings perceive landscape at two main levels of consciousness, often through noting details (Jakle, 1987). 'Perception' is the apprehension or awareness of something present in the environment. 'Cognition' is a wider and more general term which includes perception but adds thinking, problem-solving, and the organization of ideas about it. It also extends to a wider geographical area, and need not be linked to the observer's immediate reaction to his proximate environment (Downs and Stea, 1976).

Cognition leads to three types of behavioural response, which can be graded according to their strength. 'Preferences' define the observer's initial reactions to a landscape. 'Attitudes' are more global and enduring. They derive from perception and cognition of particular localities and lead to a predisposition to behave in certain ways, such as to opt for a holiday in a given environment or to work for its conservation. When a particular attitude pervades a wide variety of objects over a considerable period of time, it can become a 'personality trait', a recognizable stamp that particular landscapes can place on long residents. This is perhaps especially noticeable in mountains, deserts, or small islands. One illustration is the way Polynesian seafarers have developed a self-reliant character and wide outlook from a semi-migratory life spent travelling in small hand-made canoes over wide tracts of ocean, guided only by a knowledge of coasts, star patterns, and ocean swells (Lewis, 1974; Ward, 1989).

Attitudes, preferences, and traits can be assessed for whole populations. This is done on a local scale by testing and quantifying group reactions to selected scenes. The results may be presented as graphs or maps which grade the landscapes or viewpoints so tested. For wider areas, cognitive maps are a useful technique. These are based on the concept that an observer's evaluation of the different scenes in an area reflects his mental image of their distance, direction, and relative importance from his point of reference, generally his home. There is evidence that cognitive maps are more accurate and more rapidly related to experimental studies when derived from cartographic maps than from ground observations (Lloyd, 1989).

A person's cognitive map of an area will differ from an actual map through 'incompleteness', 'distortion', 'schematization', and 'augmentation'. It will be incomplete because he will not know all the area, distorted because it will overemphasize features in which he is interested: golf courses for golfers, rock exposures for geologists, etc., schematized because his arrangement of features of interest will be a simplified abstraction of the true map, and augmented because it will also show features, such as particular trees or viewpoints, which do not appear on it. This mental map continually changes by 'accretion', 'diminution', or 'reorganization', i.e. respectively by adding, forgetting, and regrading the importance of known features.

Since every observer has a different cognitive map from every other, it is necessary to use group reactions to arrive at an agreed evaluation of aesthetic qualities of an area. Such group reactions, especially when measured over a long period, are especially useful in focusing attention on sites to which reaction, either favourable or unfavourable, is strong, and by building up an increasingly refined measure of public views. The usual way of obtaining group reactions is by questionnaires which ask observers to grade photographs of different scenes. The latter give the most valuable results when they use a large number of scenes with very similar content (Kaplan, 1975).

26.3.3 *Environmental simulation*

The testing of group reactions can be facilitated by using environmental simulation. A number of urban authorities, including those of Jerusalem, Stockholm, the Bouwcentrum in the Netherlands, and the San Diego area of

California, have constructed scale models to show the whole environment in three dimensions. The San Diego model, for instance, is at a scale of 1:600. It is housed in a large shed containing a gantry from which a periscope is suspended which can be moved about at any level above the ground. It is possible also to 'drive' the periscope along streets and to record the results on film. This has been used to test reactions from three scales of audience: small groups around the model, public hearings of up to 200 people showing films taken through the periscope, and film presentations through the mass media. Feedback is invited and used in the planning process (Appleyard *et al.*, 1964, 1973). The model has also provided a useful approach to classifying observers by their reactions to the same scene (Craik, 1975).

26.4 Aesthetic landscape classification

26.4.1 *The observer's view*

Landscape classification from an aesthetic point of view consists in quantifying land attributes shown to be significant from an analysis of human preferences. Since landscape is mainly viewed from fixed points at ground level, its classification must emphasize the effects of scale change with distance.

The observer will tend to view the outdoor world in terms of distinct scenes, or perceptual cells, defined in terms of three limits: the floor or ground plane, the isolating near-vertical lateral limits imposed by buildings, trees, etc., and the psychological enclosure which separates the near from the distant. The textural finish of the nearby ground sets visual scale and can be a source of delight in itself, as with certain natural pavements or swards. The transition to the distance can be defined in terms of Herbert's 'texture gradient' (1982). An observer normally uses the ground plane as his means of judging the relative scale and distance of objects. His eye tends to see its components changing from a near 'three-dimensional mode' to a far 'backdrop mode'. In open views, this change comes at a distance beyond which he ceases to be able to distinguish the detail of familiar objects, such as pebbles or grass stems, but it can also come at an interruption such as a hedge, trees, or a wall. Uphill views appear foreshortened, downhill views elongated. It may be necessary to include an evaluation of features which, though not immediately visible to the observer, may yet provide a context which increases his or her appreciation. Enjoyment of an Alpine view is enhanced by knowledge that it is part of a much larger total. It is necessary to avoid two opposite dangers: too great a reliance either on quantification of factors which may be irrelevant to the landscape experience, or on intuitive insights unchecked by objective measurements (Kaplan, 1975). It is necessary to maintain a balance in which field measurements are within the parameters of viewer reaction.

Certain basic perceptions would be generally agreed. Views are best from an elevated point of vantage. To be attractive, they must have contrast. This can be between light and dark, vertical and horizontal, straight lines and curves, foreground and background, land and water, soil and sky, and other such comparisons. There should generally be an absence of stiff, formal angles or of undue repetition of the same elements. Bold, abrupt foregrounds have a softening effect on distance. Curves should generally be gentle and wide, except when leading up to notable features.

Colours should be in harmonious combination. As pointed out by Dennis (1835), if one arranges the colours of the spectrum in order from 1 to 9 thus:

1. White   4. Yellow   7. Indigo
2. Red   5. Green   8. Violet
3. Orange   6. Blue   9. Black

neighbouring colours generally blend easily, the most harmonious combinations are those between alternate colours, i.e. (1) with (3), (4) with (6), etc. Contrasts are usually strong enough to be unattractive when more than one intervening colour is jumped, except when a primary colour: red, blue, or yellow is juxtaposed with a combination of the other two, i.e red with green, purple with yellow, and so on.

Physiographic features are beneficial when they give contrasts in texture, orientation, and shape, as well as colour. Hills and mountains, especially where steep or snow-capped, add drama to a scene. Open water is always advantageous, particularly when it is in motion and includes pools, waterfalls, boulders, and other eye-catching features. Coasts are attractive and are improved by the presence of sandy beaches and rock pools. Woodland adds charm to all scenes, and is valued by the tourist when open enough to be generally penetrable to the rider and walker, and includes clearings.

Historical associations also influence our appraisals of landscape because they provide 'a living link between what we were and what we have become' (Drabble, 1979; Lowenthal; 1982), or as Lynch and Hack (1984) express it, 'the desirable image of a place is one that celebrates and enlarges the present while making connections with the past and future'. Modern change is often for the worse. Hoskins (1955) adjudged the English countryside of 1500 'infinitely more pleasant a place . . . than the mid-20th century's Nissen huts, arterial bypasses and murderous lorries'. Especially since 1914, 'every single change . . . has either uglified it or destroyed its meaning, or both'. The past has different types of attraction to the observer. Sheer age lends romance, and the further removed the times, the more magical they can appear. The past may be seen as associated with a slower tempo of existence when, for instance, simpler forms of agriculture allowed more time to enjoy the good things of life. It can also help us to distance ourselves from the stressful present, taking us back to a serener, simpler time with fewer responsibilities, such as one seems to find, for instance, in remote mountain villages or religious communities. Continuity, the sense of unbroken succession of long periods of history, is a valued aspect of this, which we can feel in Egypt or among the ancient earthworks of Salisbury Plain. The element of mystery is also important. It can invest antique landscapes with the aura of a romantic past seen through a vista of centuries. Landscape can also be made colourful and exciting by association with great historical events. Runnymede is scenically undistinguished but is rendered interesting by its association with the signing of Magna Carta. Scenery can also be historically informative, not only illuminating sites in classical lands, but also enhancing understanding of more recent events such as the landing of the *Mayflower* pilgrims on Plymouth Rock or the signing of the Armistice at Compiègne. Historical associations thus add both emotional and educational dimensions to the natural beauty of a landscape, and these may constitute the main part of its attraction to the observer. Changes to familiar scenes can be traumatic. 'We feel such profound and apparently disproportionate anguish when a loved landscape is altered out of recognition; we lose not only a place, but a part of ourselves, a continuity between the shifting phases of our life' (Drabble, 1979).

The classification of the physical attributes of the landscape varies between different environments and it is practical to use separate approaches for urban, rural, and transport land. In towns, the visual scale is smaller and more confined than in the country, so that aesthetic problems are mainly architectural, competition for the use of land is more severe, and aerial pollution is of greater relative importance. In rural areas, views are wider, recreational use less intensive, and conservation is more concerned with the protection of the landscape against extractive industry and unsightly buildings. Land used for transport and communications is a special case because of the necessity of harmonizing the interests of users and abutting owners.

26.4.2 *Requirements for urban land*

Urban land includes all built-up areas: residential, central, commercial, industrial, and transport, all of which generally require flat, well-drained sites with bedrock deep enough below the surface to avoid constructional and drainage problems. Residential property prefers somewhat steeper gradients than industrial plants and commercial centres. It can also accommodate itself to smaller and less compact blocks of land, while they can afford the extra drainage or infill costs of poorly drained land. Table 26.1 indicates slope standards appropriate to some common land uses.

Table 26.1 Influence of slope on land use (adapted from Lynch and Hack, 1984)

Slope Angle (°)	Remarks
<1/2	Poor drainage, liable to flooding if low lying.
<2 1/2	Usable for all kinds of intense human activity, e.g. playing fields, sports stadia etc.
2 1/2 – 6	Easy grades, suitable for informal movement and activity, e.g. shopping.
>6	Appears steep, make unfavourable roads in urban areas. Unsuitable for most sports except golf, riding etc.
8 1/2	Limit of slope that a normal loaded vehicle can climb for a sustained period.
14	Limit of surfaces that can be machine-mown in parks etc.
18	Limit of gradient (1 in 3) that a normal car can climb for a limited period.
>26	Liable to erosion in a humid climate.

26.4.3 Requirements for rural land

Rural or undeveloped terrain fulfils a considerable variety of practical, aesthetic, and recreational needs, which can be divided into three main groups. First, and most intensive, is the land which is used for team sports and children's group play. Individual areas may be small but are frequent where population is dense. The land should be relatively flat and level. In dry climates it must be neither too hard and stony nor too soft and dusty. In humid climates it needs to be well drained, so that it can have a good bearing strength and traction soon after heavy rain, and there should be sufficient depth of soil to sustain grass cover at drier times of year. It is, however, often of necessity placed on reclaimed derelict land on the urban fringe or on flood plains subject to seasonal inundation (Palmer and Jarvis, 1979).

Second, there is rural land which is suited to scientific and educational purposes. This includes nature reserves, arboreta, and centres for field studies. The dominant concern is not for aesthetics, but for the well-being of the plant and animal species involved. Sites are most suitable which are either representative of an area, or exceptional. An agricultural research station, for instance, needs to occupy a site typical of the area it serves, while a nature reserve seeks to protect some unusual conditions of rock, vegetation, or habitat. Marginal land, such as unused field headlands, rough corners, and road margins, is often of critical importance in retaining wildlife in an intensively used area, and it is desirable to take active steps to manage such areas with this in view (Thompson, 1979).

Local and national parks cater for more individual and wider ranging forms of sport and recreation which put a somewhat greater value on the visual and aesthetic qualities of landscape. Local parks include public gardens as well as cemeteries and memorial grounds devoted generally to leisurely exercise and calm enjoyment. National parks are larger and wilder, providing special terrain conditions in a rural setting. They have some or all of the following: footpaths for running, rambling, and picnicking, bridle paths, golf courses, mountain trails, and lakes with beaches for swimming, fishing, sailing, etc. They sometimes contain open sites aerodynamically suitable for sports such as hang-gliding, parascending, and ballooning. Land must usually be reserved for tent and caravan sites. Its surface must sustain heavy foot traffic and some wheeled traffic. It should be able to absorb rainfall and runoff from tents, caravans, and cars, and should otherwise have the characteristics noted above for sports fields (George and Jarvis, 1979).

26.4.4 *Requirements for transport land*

Land required for transport purposes, such as roads, railways, canals, and airports, and also for long–distance pipelines and cables, demands special treatment because of the divergent interests of users and abutting owners. Highways provide the means for bringing most people into contact with the countryside, but at the same time they bring danger, noise, and air pollution to their surroundings. Pipelines and cables may be less obtrusive but can include some dangers from leakage of fluids or electricity and are unsightly especially if carried on pylons or poles. Land near highways should reveal an attractive sequence of views to the moving observer. Economic and engineering considerations alone are an unsafe guide to the choice of routes, and the overriding principle must be to choose those which provide maximum social benefit at least social cost. They should go through areas of low value both in terms of land and building costs and also from the aesthetic point of view. An illustration of the absurdity of basing the location of transport facilities on purely economic considerations was an analysis that showed that on economic grounds alone the cheapest place for London's airport was in Hyde Park.

26.5 The identification and measurement of aesthetic parameters

A wide range of methods have been suggested to place the assessment of the aesthetic qualities of landscape on an objective and quantitative basis. They can be categorized as environment-specific or problem-specific.

26.5.1 *Environment-specific schemes of landscape evaluation*

Studies of the general landscape quality of particular areas range in scale from countries to local sites. An early example was Linton's (1968) assessment of Scottish scenery on the basis of 42 numerically graded combinations of landform and land use. The resulting map, though on very small scale, gave a reasonable overview of the country.

Some British local planning authorities have graded landscape from a grid of points and produced maps of the results. The Hampshire County Council, for instance, used sites 0.8–1.6 km apart, awarding good marks for favourable factors such as the amount of vertical relief, and bad for 'detractors' such as neglected land, rubbish tips, battery chicken houses, etc. A similar approach in East Sussex, resulting from the need to route a 400 kV power line, graded each site on a scale from 0 to 32 from 'unsightly' to 'spectacular'. This gave means for both the county and country of 5–6, a highest value for Britain of 18, and for the world (excluding atmospheric phenomena) of 24 (Fines, 1968). More comprehensive approaches including proposals for management and conservation were made in southeastern New England by Riotte *et al.* (1975) and in the Basle region by Plattner (1975).

For a local area Leopold (1969) proposed a system based on a numerical scale of 'uniqueness' of landscapes, which could be positive or negative depending on whether it was due to attractiveness or ugliness. The scale could be used for more specific evaluations, such as grading of the 'valley character' of a given site between extremes of 'ordinary and urban' and 'spectacular and wild'. Tandy's 'isovist' method (1967) aimed at mapping visually distinct units around 'focal centres', each commanding a 360° circle of vision. The area visible from each was called the 'visual zone' and its enclosing line the 'isovist line' in rough analogy with isobars, isotherms, etc. These zones vary in character but can be classified into types to which names are given, such as 'extroverted', 'introverted', 'linear', 'outward looking' (subdivisible into one-direction, open ended, etc.), as illustrated on Fig. 26.1. When two zones are separated by a ridge or other strongly limiting feature, the boundary is considered a 'visual watershed'. The overall appraisal of an area is based on a synthesis of the visual quality of the constituent zones.

Some studies have been designed to aid the conservation of particular rural environments. They include avifaunal habitat classifications by the British Trust for Ornithology and the Malta Museum, and a classification of the 'visual–cultural value' of the Massachusetts wetlands for planning purposes (Smardon, 1975).

26.5.2 *Problem-specific schemes of landscape evaluation*

Jacobs and Way (1969) used concepts of 'visual transparency' and 'visual complexity' to quantify the tolerance of landscapes to building encroachment. These parameters yielded indices of 'opaqueness' and 'complexity' positively correlated with the landscape's ability to absorb such intrusions. On a larger scale studies of the Californian mountains (Iverson, 1975) and of the rural

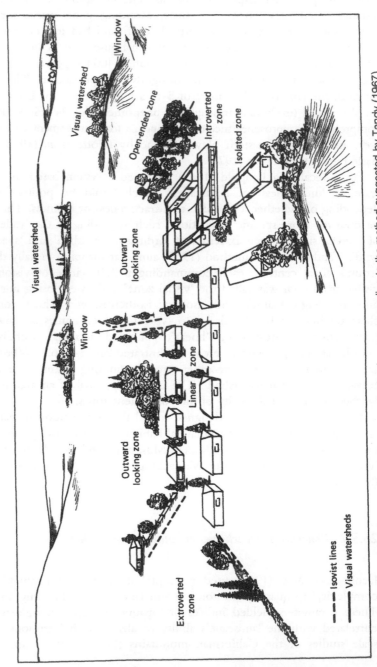

Fig. 26.1 Sketched example of an 'isovist' summary of a landscape, according to the method suggested by Tandy (1967)

margins of the Boston metropolitan area (Fabos *et al.*, 1975) used quantified measurements of 'visual attractiveness' and 'landscape resource value' respectively to quantify vulnerability to building intrusion. Dumansky *et al.* (1979) identified areas suitable for urban expansion around Ottawa by overlaying maps of nine 'land factors', emphasizing those due to wetness, stoniness, topography, and poor soils. McHarg (1969) underlined the importance of seeing urban growth as part of overall environmental improvement, based on a multidisciplinary assessment of the ability of a region to support it.

A particular problem occurs with power stations, often large eyesores in country areas. Murray (1967) devised a method for evaluating their general appearance in relation to the numbers and types of observers who view them, the circumstances of viewing such as weather, season, and time of day, and the activities of the observers.

This evaluation was based on five numerical dimensionless coefficients for each potential viewpoint:

1. *The silhouette factor*: the visible area of silhouette from a given viewpoint according to an arbitrarily chosen scale. This can be presented by building a scale model or by superimposing a silhouette of the proposed station on to a view of the site.
2. *The distance coefficient*: the ratio between 25° ('the normal effective vision cone') and the greatest angle subtended by the silhouette.
3. *The visibility coefficient*: the proportion of days in the year in which the visibility is adequate to see the station.
4. *The displacement coefficient*: the angular distance of the station from an observer's preferred line of sight. When the viewpoint is one used by someone driving a car, the displacement coefficient must be modified to take account of the decreasing cone of vision which results from increasing driving speed. This modification is known as the 'speed coefficient'.

The author does not specify exactly how these coefficients are to be applied. Their value lies in their comprehensiveness in quantifying the visual aspects of power stations in relation to the locations and conditions of viewing. They make it possible to compare sites from one viewpoint or viewpoints to one site.

It is sometimes necessary to evaluate specific areas for recreation purposes. Goodall and Whittow (1973) studied all British forests, and evolved the concept of the 'recreational potential index' (P_R). This was based on the calculation of three factors: P_1 the topographic factor, P_2 the mantle factor, and P_3 the access/uniqueness factor, for each of 20 forest reserves chosen as typical of the country as a whole. P_R was calculated from the formula

$$P_R = P_1 + P_2 - P_3 \qquad [26.1]$$

Factors P_1 and P_2 were based on calculations which gave maximum values for

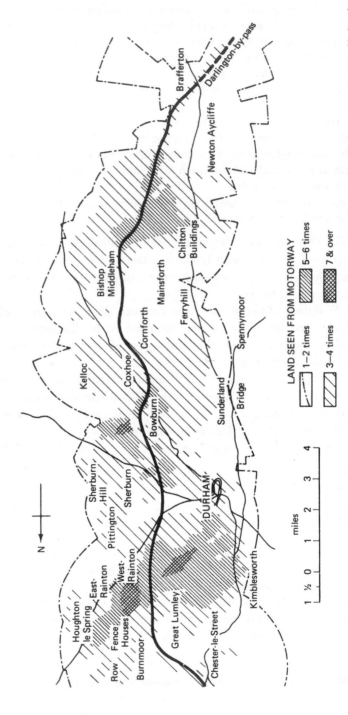

Fig. 26.2 The use of visual criteria in road planning. In a study for the Durham motorway land was classified in accordance with its degree of visibility from the proposed route line with a 'visual corridor' and areas of 'visual frequency'. (Source: Clouston, 1967)

attractive features, P_3 for detractors. For P_1, these were slope steepness, relief amplitude, and the number of water bodies (including the sea), for P_2, tree spacing, tree height, and the proportion of broad-leaf woodland, and for P_3, the degrees of impenetrability, inaccessibility, and elongation of forest plots. The resulting values for P_R made it possible to grade and compare the forest plots from a recreational point of view. A special feature of the method was its provision for the assessment of the relevant parameters from map evidence alone without field visits.

Some attempts have been made to assess landscape quality from the point of view of the traveller. An example was the semi-quantitative scheme for judging the proposed route of the Durham motorway from Darlington to Chester-le-Street described by J. B. Clouston (Landscape Research Group, 1967). A composite map (part of which is reproduced as Fig. 26.2) gave a simple landscape assessment of the 'visual corridor' along which a passing driver would travel, modified by the 'vision frequency', i.e. the number of times he would see any part of this corridor. Its width was in part determined by the frequency of poor visibility in the area. The method allowed an optimal route to be chosen both from the point of view of the motorist and of abutting land users.

Lewis (1964) identified the four main types of area along the Wisconsin heritage trail attractive to the tourist: those with 'significant' water, topography, wetlands, or landscape personality. The major roads and trails of the state are superimposed on this base (Fig. 26.3). Polakowski (1975) used a two-scale approach for locating a scenic highway around parts of the Great Lakes: a 'macro-geomorphic analysis' for defining the major areas and a 'micro-composition analysis', based on earlier evaluative schemes, for subdividing them. The least intrusive route for an electricity line across a $770\,000\ km^2$ area of northwestern USA was selected by overlaying digital maps of critical environmental attributes (Murray and Niemann, 1975).

26.6 Applying landscape evaluations

The result of combining observer testing with environmental measurement should be to provide lasting aesthetic evaluations of geographical areas for planning purposes. Since many studies use categories of land system and land facet type, it is reasonable to use this basis as a sampling framework for aesthetic as well as for agricultural or engineering evaluations. If so used, group ratings of landscape quality could be added to other data pigeon-holed by land units.

To assess the visual qualities of an area such as the land region partly shown in Fig. 26.4, for instance, one would first assess it as a whole from different viewpoints by group ratings and then separately analyse the five sub-parallel land systems: 'White Peak', Hope Valley, Mam Tor Ridge, Edale Valley, and

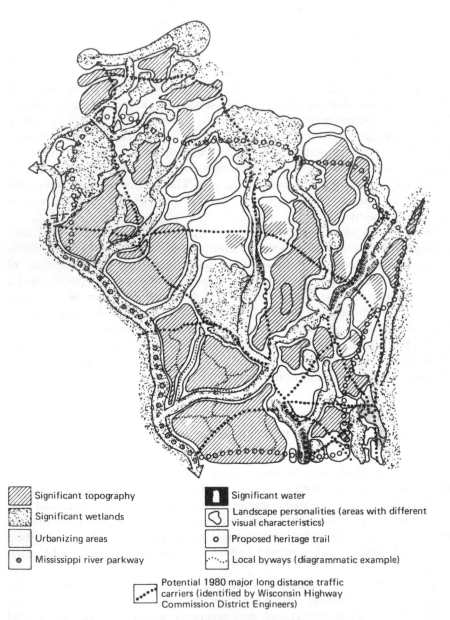

Significant topography

Significant wetlands

Urbanizing areas

Mississippi river parkway

Significant water

Landscape personalities (areas with different visual characteristics)

Proposed heritage trail

Local byways (diagrammatic example)

Potential 1980 major long distance traffic carriers (identified by Wisconsin Highway Commission District Engineers)

Fig. 26.3 The Wisconsin 'heritage trails' proposal (Source: Lewis, 1964)

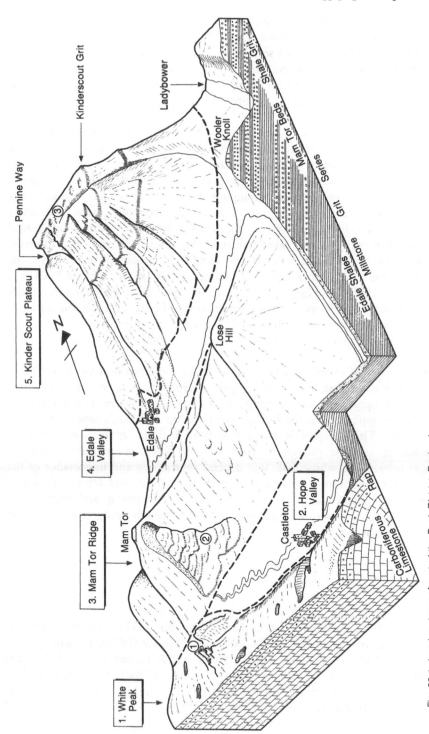

Fig. 26 4 Land systems of part of the Peak District, England

Fig. 26.4 Land systems of part of the Peak District, England

'Dark Peak', making specific evaluations of the visual properties of individual land facets such as the limestone gorge, the Mam Tor landslide, and the summit plateau of Kinder Scout seen from a range of viewpoints.

Once an area has been evaluated, it is often necessary to take special measures both to present it to the public and to carry out improvements. This involves careful site planning, publicity, improvements to facilities, and landscape beautification.

Urban landscaping aims to make the best use of, and where necessary to modify, the components of the landscape: rock, soil, water, and plants, so as to optimize the arrangement of buildings at a site or groups of buildings in a wider area. This would include giving a sense of balance through rhythmic repetitions and strong contrasts, arranging shapes, colours, and textures into harmonious but interesting groups, maximizing desirable vistas and axes, optimizing views, hiding blemishes, dramatizing desirable features, and saving areas such as small woods or lakes as habitats for fish, birds, and other animals. At the same time, human residences would be secluded with 'cove' or 'harbour' type approaches and 'defensible' front space (Simonds, 1961; Rubenstein, 1969; McHarg, 1969). An example of a decision in which environmental considerations were paramount was that of the Netherlands 'Delta Enclosing Project', which left a partly opened dam to retain an ecological link between the enclosed fresh water and the North Sea (Vink, 1983).

Appreciation of country parks is enhanced by judicious presentation. Sharpe (1976) emphasizes that 'park and outdoor interpretation must embody a total environment approach' and has suggested a comprehensive programme for the park planner applicable to the USA, comparable to Aldridge's proposals for the UK (1975). Their object is to give visitors a fuller appreciation of their environment and the processes which form it. Such programmes involve a series of operations, whose scale will depend on the size and importance of the site, ranging from simple provision of parking facilities and litter bins at one extreme to marked nature trails, 'park centres', camping and recreational facilities, shops, museums, educational and research programmes, media publicity, etc. at the other.

Further reading

Aldridge (1975), Appleton (1975, 1976), Drabble (1979), Downs and Stea (1976), Dumansky *et al.* (1979), Goodall and Whittow (1973), Hoskins (1955), Jacobs and Way (1969), Jakle (1987), Jarvis and Mackney (1979), Landscape Research Group (1967), Leopold (1969), Linton (1968), Lowenthal (1982), McHarg (1969), Rubenstein (1969), Sharpe (1976), Unesco (1971, 1973b), Whittow (1979), Zube, Brush, and Fabos (1975).

27

The future

27.1 The requirement

Growing world population is leading to an increasing pressure on land resources. In industrialized countries, agriculture and other rural activities are becoming more efficient and thus more selective in their choice of land and more intensive in its use. This has led to overproduction and a trend to the release of agricultural land for other purposes. At the same time, non-agricultural uses of land have increased. Urban developments with high space standards expand into the countryside. Extended leisure multiplies tourism, so that in many areas recreation has replaced agriculture as the main land use. The transport revolution vastly increases the accessibility of hitherto remote rural areas. The result is intense competition for the use of land. Areas previously regarded as natural wilderness rapidly acquire considerable economic importance. A wider public understands the challenge of the present situation: the extent to which thoughtless exploitation of natural resources and pollution threaten the quality of the environment, but at the same time the new opportunities for its conservation and improvement.

The same trends have been apparent in Third World countries but more because of rapidly growing and urbanizing populations than because of the rise of living standards. An FAO study (1978a; Higgins and Kassam, 1981; Higgins et al., 1984) of the 117 developing countries of Central and South America, Africa, and southern Asia concluded that although they contain much undeveloped land in these areas and great potential for the development of both rainfed and irrigated agriculture, the rapidity of population growth will leave an increasing number of them unable to grow their own food or biofuels between now and AD 2000. Assuming improved but non-mechanized levels of agricultural management, the number of such countries will be 37; assuming unimproved traditional methods, 65. It is further estimated that 29 per cent of the total rainfed crop potential in these countries will be lost through erosion in the same period if conservation measures are not introduced.

As emphasized in Chapter 23, a vital aspect to the solution of these problems is the availability of information, especially about natural resources and their management. There are today two opposite problems concerning environmental data. First, in spite of the rapid growth of knowledge in the past few years, for much of the world there is still only a very sparse amount of information about the land and about the processes affecting it. We, for instance, have little quantitative data about rates of erosion, siltation, pollution, or over-exploitation of natural resources such as the destruction of tropical forests. This is especially true of Africa, Asia, Latin America, and the polar regions; a fact which must not be obscured by the significant advances that have taken place. The gap in terrain intelligence between developed and developing parts of the world is widening.

Nevertheless, the growth of information has everywhere been rapid in recent years, so that the stock is impressive when old and new sources are considered together. Libraries of books, maps, and computerized data have multiplied and become accessible to more people. Black and white photography from aircraft and multispectral imagery from satellites have now become available for most of the world and there are now large areas covered by radar and thermal imagery.

So large is the amount of data on the environment currently in existence, that it has led to another problem: that of data management. Apart from published literature and computer files, there is a bewildering amount of information in unpublished and uncatalogued form: government and commercial files, unindexed publications, and in the memory and experience of individuals. We cannot effectively use much of this information, which has often been acquired at considerable cost and will before long be lost beyond recall. Thus, there is a need for expanding 'banks' of terrain intelligence. In many areas data management has replaced data acquisition as the chief bottleneck in resources evaluation and this problem is becoming more general.

These problems are compounded by an older one: that of fragmentation of effort. Coordination of operations between government departments and still more between nations is slow and often ineffective. Commercial competition imposes secretiveness. There is thus considerable overlap and waste in gathering and managing terrain intelligence.

27.2 The capability

27.2.1 *Terrain classification*

Integrated survey based on terrain classification is the basis of much environmental planning today. Since terrain is unchanging and its natural components are simple and readily observed, worldwide cover could be achieved more easily than in most other thematic surveys. Terrain units provide an integrating framework for a wide range of land resources, notably surface materials, soils, water, and vegetation, and thus for an equally wide range of users, including disciplines as diverse as forestry, archaeology, or dam construction. They can be linked into a recognizable hierarchy useful for planning at all scales from the international to the local. Finally, terrain units, unlike data referenced only by grid coordinates, provide a basis for extrapolation, so that data from one area can be used, within measurable limits, to predict site conditions in analogous areas.

Although the general validity of the approach to landscape planning through terrain classification has been tested in temperate and arid areas, however, it requires further verification especially in the polar, tundra, equatorial, and tropical savanna zones, It is necessary, too, to gain a more complete world cover of terrain classification than exists at the moment, especially where geology or soils are not mapped in detail. Land system or geomorphological maps at scales useful for local planning today only cover relatively small parts of industrialized countries and a few other areas such as Uganda, Swaziland, Papua New Guinea, Nigeria, Argentina, and Jordan. The greater part of the earth's land surface as yet lacks such mapping or evaluation at any scale.

Moreover, improvements continue to be needed in the concepts and methods of terrain classification. First, definitions should be made more quantitative. This is especially important because of the need for statistical analysis and computer manipulation of data. Second, there should be better integration between different land users so that each will have a wider appreciation of concerns and capabilities of others. The main emphasis will remain on agricultural, engineering, and conservation aspects, but it is necessary to include the microclimatic view of altitude, exposure, and aspect, and the hydrological emphasis on river catchments and terrain factors which determine the water balance in any locality. Even more important in some areas are the aesthetic aspects of landscape which affect urban and recreational development. These demand the incorporation of boundaries of the isovist or visual watershed type, with the aesthetic quality of mapped areas assessed in terms of quantitative visual criteria and also sometimes, in these days of pollution, according to olfactory and audial criteria as well.

27.2.2 GIS and remote sensing technology

While terrain classification provides the cross-disciplinary or 'horizontal' basis for integrated survey, GIS and remote sensing now provide the technology for a 'longitudinal' integration of the survey process. In essence the combination of the two provides a foundation for the systemization of natural resource information from initial survey up to the point at which it can be issued for making planning decisions.

The technology of GIS is advancing so rapidly that progress is measured in months rather than years, and forecasts are risky. Nevertheless, certain capabilities and trends appear likely to dominate the foreseeable future. The essence of modern GIS technology is the ability to store and manipulate a database consisting of a number of data planes of different types, and to present the processed data in required form. The manipulation capability has moved furthest in two directions: in developing small, very powerful intelligence systems for particular topics such as weather forecasting and census analysis, and in merging diverse data sets. The latter includes a capability to make summaries, selective searches, measurements, comparisons of data within and between planes, generate new data, and present processed data in textual, graphic, or map form. Ultimately, it can automatically monitor environmental changes revealed by the data, and even trigger remedial action. Hitherto, storage and manipulation has been on the basis of grid coordinates, but there is an increasing need for the use of 'envelopes' of effectively homogeneous terrain which can, for instance, be identified in terms of parametrically defined landscape polygons.

Remote sensing can provide one or more of the data planes within a GIS database. Because of the increasing flood of multiscale, multispectral, and multitemporal data there is a trend for it to increase in importance relative to other sources of data. This is because of technological advances which make it possible to refine for remotely sensed imagery into precision geocoded form accurate enough both to replace conventional topographic surveys as a database, and to provide a wealth of thematic information. There remains, however, a need for an effective means of indexing and storing imagery, especially that from satellites (Howard and van Dijk, 1980).

Future trends can be discerned which will further enhance the contribution of remote sensing. Hitherto, it has been mainly oriented towards the single instrument 'ride of opportunity' approach where a satellite is orbited for one purpose only. But more and more we are entering the age of the 'space transportation system' with larger and more sophisticated platforms carrying a number of different sensors. The future will probably see a move to exploit this trend with integrated earth science missions carrying a battery of sensors. Research by the US Government, however, has shown that satellite-based earth science research falls into three broad areas or 'plateaux', each of which could be covered by a single multi-sensor mission, but which it would be more difficult to combine until after the remote sensing stage (Broome, 1983). These plateaux are: (1) water cycle, geology, biomass, and land use; (2) biochemical cycles other than water together with atmospheric chemistry; and (3) climate and the circulations of atmosphere and oceans. Each would require a user-friendly technology and multidisciplinary databases.

27.3 Meeting the future need

Terrain classification optimizes the definition of landscape polygons, and provides a framework for the multidisciplinary integration of monodisciplinary data. It is particularly suited to the use of remote sensing, which today forms the basis of almost all topographic and thematic mapping, especially in areas where other sources of data are sparse. The technology for manipulating a wide variety of geographically referenced data sources is provided by GISs.

The integration of these three technologies: terrain classification, remote sensing, and GIS, offers far-reaching prospects for advancing our understanding of the world environment in the future. They will provide the essential basis for planning decisions, and become an input in the direction of operational analysis, whose potentiality was first stressed by Grabau (Grant, 1968). This is essentially the analysis of human activities in a spatial context. Since this context is largely one of terrain, the operations cannot be characterized or their course predicted unless, in some prior form, the terrain has also been specified. The provision of information by which this can be done is the ultimate objective of terrain evaluation.

Further reading

Broome (1983), Grant (1968), Higgins *et al.* (1984), Howard and Mitchell (1985), Howard and van Dijk (1980), Jackson and Mason (1986), Tomlinson (1984).

Bibliography

The literature bearing on terrain evaluation is extensive and comes from a variety of disciplines, but few books are wholly devoted to it. The works mentioned below are given in full in the list of references.

There is a basic three-way distinction between works on: (1) the geomorphic analysis and classification of terrain and its relation to soils and vegetation; (2) practical applications of terrain information to land uses; and (3) survey technologies. Selection is difficult, but among the first group could be mentioned Lobeck's graphic work (1939), Strahler (1969), Thompson et al. (1986), and the more purely systems approach of Dury (1981) for the geomorphological aspects, Buol et al. (1980) for soil geography, and Leser (1978) and Anderson (1981) for ecology, with Fairbridge (1968) and Whittow (1984) being useful works of reference. On the specific question of terrain classification, important contributions are Linton (1951), Savigear (1965), and Verstappen and Van Zuidam (1968). A summary is given in Howard and Mitchell (1985).

Texts on the applications of terrain study to practical land use problems include Stewart (1968), which reviews research to that time, Dawson and Doornkamp (1973), Townshend (1981), and Dent and Young (1981). Within a more restricted compass, summaries by organizations involved in terrain evaluation include Beckett and Webster (1969, MEXE Report 1123) Christian (1983), and Grant's reports on the PUCE programme (1973–74).

Critical path analysis is given by Baboulene (1969), remote sensing by Curran (1985), Townshend (1981) adding coverage of sampling and interpretation. Field-work is given by the USDA (1962) and Hodgson (1976). These should be practised in line with the FAO (1976a, 1985), whose application in the tropics is summarized by Young (1976). Snedecor and Cochran (1978) is a useful text on statistical methods and the British Standards Institution (1975) a basic reference on laboratory methods. Maguire (1989) is a comprehensive guide to this subject. Tomlinson (1968) gave an early introduction to geographical information systems, which can be updated by reference to Jackson and Mason (1986) and Burrough (1986).

Appendix A

Example of a method of sampling the terrain over a large area: the hot deserts of the world

Arid land shows a wide range of geological and physiographic conditions whose character and distribution have been much less studied than have those of more settled regions. The total area is so large that it is virtually impossible to study its surface conditions as a whole. A method of obtaining a representative sample of physiography is therefore generally desirable. Such a sample would make it possible to estimate the character and relative proportions of the surface covered by different landform types. This would be of assistance to many specialists concerned with terrain: to agriculturists attempting to estimate soil and water resources, to designers of cross–country vehicles, to engineers calculating requirements of local materials for engineering projects over large areas, and to anyone seeking physiographic analogies between one part of the world and another.

A comprehensive definition and classification of the arid areas of the world based directly on climatic criteria was devised by P. Meigs for Unesco in 1957. It included all areas with 'aridity indices' (Thorntwaite, 1948) between –20 and a theoretical maximum of –60. Under this scheme Joly (1957) calculated that arid lands totalled approximately 48 million km^2 of which 19 million km^2 were semi–arid (indices of –20 to –40), and 29 million km^2 were arid (indices below –40). Of the latter category 7 million km^2 were separated into a special extremely arid class if consecutive months had been recorded without rainfall and there was no clear seasonal pattern.

They are distributed by continents as shown in Table A.1.

The sampling scheme is outlined below. It does not, however, quite cover the total area given in the table. The colder deserts of North America and Eurasia (i.e. Soviet and Chinese central Asia) were excluded because of the belief that they would have appreciably different ecosystems which would give them a different physiography from that found in hotter arid areas. The remaining 24.9 million km^2 were further somewhat arbitrarily reduced by excluding all Latin American deserts (2 572 000 km^2) and the small hot arid areas in southwestern Madagascar and central India (23 000 and 16 000 km^2

respectively) because of limited time and resources, the unavailability of published information and difficulties of translation. The residual area thus totalled 22 294 000 km² and included, broadly speaking, the Sahara, the Kalahari, southwestern USA, and the Australian and Middle Eastern deserts. It is shown in Fig. A.1.

Table A.1 Distribution of world arid lands by continents (after Joly, 1957). Values are given in thousands of square kilometres

Region	With no month averaging below 0 °C	With at least one month below 0 °C
Americas		
USA and Canada	682	437
South and Central America	2 572	0
Africa		
Northern hemisphere	10 807	0
Southern hemisphere	1 022	0
Madagascar	23	0
Eurasia	5 870	3 650
Australasia	3 929	0
Total	24 905	4 087

The sampling scheme was devised as part of the MEXE–Cambridge desert terrain evaluation project and is described by Perrin and Mitchell (1970). Briefly, physiographic units were outlined on the basis of all available geological and topographic information on 1:4 000 000 base maps. The resulting total of 238 regions are shown in Figs. A.2 to A.10 and listed in Table A.2, together with a representative 1° square of latitude and longitude located near to the centre of each and regarded as being representative of it. These regions were classified into the 57 lithomorphological types given in Table A.2. It was assumed that a 1° square from a single region in each of the 57 groups would represent an adequate sample of the physiography of the arid zone. Assuming a median latitude of 25°, a square of this type represents approximately 11 000 km², so that the method represents the 22 million km² of the arid areas by 57 × 11 000, or about 600 000 km², i.e. about a 2.5 per cent sample.

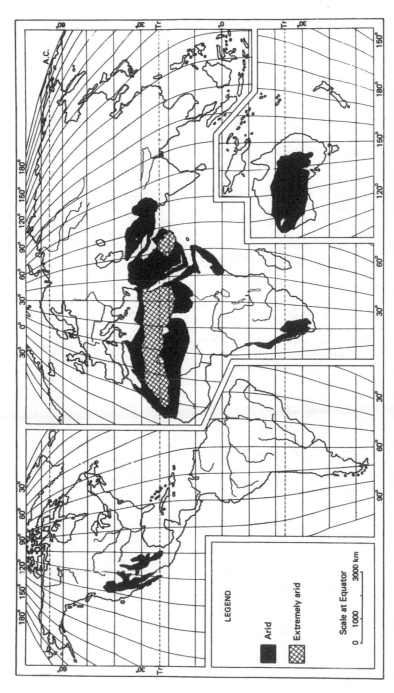

Fig. A.1 Arid and extremely arid areas of the world used for the terrain sample (Source: Perrin and Mitchell, 1970, © British Crown copyright 1970/MOD)

Table A.2. Classified list of regions (after Joly, 1957)

Group No.		Class and name	Limiting coordinates of representative 1°squares		No on Map
		I. *Mountains*			
1	A	Crystalline and metamorphic			
		Central ranges, Australia	23–23° S	133–134° E	218
		Mountains of S. Sinai, Egypt	28–29° S	34– 35° E	124
		Red Sea hills, Egypt and Sudan	26–27° N	33– 34° E	117
		Namaqualand Highlands, SW Africa	24–25° S	16– 17° E	207
		Adrar des Ifoghas, Algeria	19–20° N	1– 3° E	63
		Qain-Birjand Highlands, Iran	33–34° N	59– 60° E	163
		Socotran Archipelago*			173
		Ethiopian Coast Range			237
2	B	Crystalline, metamorphic and volcanic mixed			
		Baja California hills, Mexico	33–34° N	109–110° W	7
		Mts of Midian, Saudi Arabia	28–29° N	35– 36° E	125
		Ahaggar foothills, Algeria	22–23° N	4– 8° E	64
		Asir-Yemen Highlands, Saudi Arabia and Yemen	19–20° N	42– 44° E	127
		Ougarta mountains, Algeria	29–30° N	2– 3° W	39
3	C	Volcanic			
		Jebel Harug, Libya	27–28° N	17– 18° E	83
		Jebel Es Soda, Libya	28–29° N	15– 16° E	82
		Mountain areas of French Somaliland	11–12° N	42– 43° E	177
		Hanich islets in Red Sea	13–14° N	42– 43° E	157
		Jebel Marra, Darfur, Sudan			118
4	D	Sandstone			
		Jebel Uweinat–Archenu, Egypt–Sudan–Libya	21–22° N	25– 26° E	110
5	E	Limestone and dolomite			
		Mekran, W Pakistan	25–26° N	64– 65° E	198
		N Rangers, Baluchistan, W Pakistan	29–30° N	68– 69° E	199
		NW hills, W Pakistan	33–34° N	72– 73° E	200
		Saharan Atlas, Algeria	33–34° N	1– 2° E	21
		Jebel Nefusa–Ksour, Libya and Tunisia	31–32° N	11– 12° E	72
		II. *Hills*			
6	A	Foothills: Limestone			
		Ogaden Fringe, Somalia Ethiopia	7– 8° N	47– 48° E	183
		Foothills of Somalia ('Ogo')	10–11° N	45– 46° E	182
		Shebeli-Juba Plateau, Somalia–Kenya	3– 4° N	41– 42° E	184
7	B	Undulating country 1. Crystalline with volcanics			
		Air Massif, Niger	17–18° N	8– 9° E	70

Group No.		Class and name	Limiting coordinates of representative 1°squares		No on Map
8		2. Foliated:			
		Barrier–Grey Ranges, Australia	28–29° S	143–144° E	219
9		3. Limestone:			
		Hamad Plateau. Arabia–Jordan–Iraq–			
		Syria	32–33° N	39– 40° E	139
		Upper Tigris–Euphrates plains, Iraq	33–34° N	42– 43° E	140
		III. *Piedmonts*			
10	A	Erosional			
		1. Crystalline			
		NW Slopes of NW Tableland,			
		Australia	23–24° S	115–116° E	220
		Uplands of NW SW Africa	19–20° S	13– 14° E	208
11		2. Volcanic			
		Central zone of French Somaliland	11–12° N	42– 43° E	178
12		3. Limestone			
		Persian Gulf Coastal plain, Iran	29–30° N	50– 51° E	164
13	B	Depositional			
		1. Mainly calcareous			
		Batina coastal plain, Muscat	23–24° N	57– 58° E	159
		Djeffara coastal plain, Libya–Tunisia	32–33° N	12– 13° E	71
		Coastal strips of S Somalia			192
		Egyptian coastal plain W of Delta	31–32° N	27– 28° E	113
		Red–Sea coastal plain Egypt–Sudan	26–27° N	34– 35° E	112
		S slopes of Saharan Atlas, Algeria			22
14		2. Mainly acidic			
		Aden coastal plain	12°30'–	44°30'–	
			13°30' N	45°30' E	160
		Coastal plain of W Australia	23–24° S	113–115° E	221
		San Joachim valley basin, California,			
		USA	36–37° N	119–120° W	1
		Baja California Gulf coastal plain,			
		Mexico	33–34° N	109–110° W	6
		Baja California Pacific coastal plain,			
		Mexico	33–34° N	110–111° W	8
		Tihama coastal plain, Saudi Arabia–			
		Yemen	10–20° N	41– 42° E	128
		Ethiopian coastal plain	15–16° N	39– 40° E	174
		Namib coastal plain, Angola,			
		SW Africa and S Africa	19–20° S	12– 13° E	209
		IV. *Plateaux*			
15	A	Slightly dissected			
		1. Crystalline or metamorphic			
		Salt Lake Division, Australia	28–29° S	120–123° E	222
		Bur Region, Kenya	1– 2° N	40– 41° E	185
		Hasa Coastal Desert Saudi Arabia	26–27° N	49– 50° 30'E	150
		Eglab, Algeria	26–27° N	3– 4° W	34
		Central crystalline plateau, Saudi			
		Arabia	24–25° N	44– 45° E	129

Group No.	Class and name	Limiting coordinates of representative 1°squares		No on Map
16	2. Volcanic 'harras'			
	Northern Harras Area, Saudi Arabia	31–32° N	38– 39° E	130
	Central Harras Area, Saudi Arabia	22–23° N	41– 42° E	131
	Plateaux of French Somaliland	11–12° N	42– 43° E	175
	Nasratabad–Taftan Highlands, Iran	28–29° N	61– 62° E	166
17	3. Limestone: with some vegetation			
	High plateau of Chotts, Algeria	33–34° N	1– 2° W	20
18	4. Limestone: bare			
	Nullarbor Plain, Australia	30–31° S	130–131° E	221
	Altoplanice, Mexico, USA	32–33° N	107–108° W	9
	Limestone plateau of SE Jordan	30–31° N	37– 38° E	133
	Hamada of Murzuch, Libya	26–27° N	12– 13° E	74
	Hamada of Tademait, Algeria	28–29° N	1– 2° E	52
	Hamada of Dra, Algeria	28–29° N	7– 8° W	14
	Hamada of Daoura, Algeria	30–31° N	3– 4° W	18
	Hamada of El Gantara, Algeria	30–31° N	3– 4° E	51
	Hamada of Bou Denib, Algeria Morocco	32–33° N	3– 4° W	16
	Hamada of Bou Laouaiche, Morocco	31–32° N	4– 5° W	17
	Hamada of El Haricha, Mali	22–23° N	3– 4° W	40
	Hamadas of NE Spanish Sahara			12
	Hamadas of S Morocco			11
	Hamadas of Krenachich–Mahia, Mali			42
	Hamada El Homra, Libya	29–30° N	12– 13° E	73
	Interior plateau, Baluchistan, W Pakistan	28–29° N	63– 64° E	197
	Central plateau, Egypt	26–27° N	30– 31° E	108
	NW plateau, Egypt	30–31° N	26– 27° E	107
19	5. Sandstone			
	South African Highveld	31–32° S	22– 23° E	215
	Cape Middleveld, South Africa			216
	Gilf Kebir, Egypt	23–24° N	26– 27° E	111
	Hamada of Guir, Algeria–Morocco	30–31° N	3– 4° W	19
	Tindoui,	27–28° N	7– 8° W	25
	Araouane,	25–26° N	3– 4° W	35
	Sandstone plateau of N Hejaz, Saudi Arabia	28–29° N	37– 38° E	34
	Mauritanian interior plateau	23–24° N	9– 10° W	26
	Jefjief Plateau, Tchad–Libya	20–21° N	20– 21° E	86
	Coastal plain of S Morocco and Spanish Sahara	30–31° N	9– 10° W	13
20 B	Moderately dissected			
	1. Crystalline or metamorphic			
	NW tableland of Australia	23–24° S	116–118° E	223
	Cutch, India	23–24° N	69– 80° E	206
	Sol Plateau, Somalia			191
21	2. Limestone			
	Al Wadiyab Area, Saudi Arabia–Iraq	32–33° N	11– 13° E	142
	Dibdibba–Hasa Plains, Saudi Arabia	29–30° N	45– 46° E	149
	Hajara Plain, Saudi Arabia	29–30° N	42– 43° E	238

Group No.	Class and name	Limiting coordinates of representative 1°squares		No on Map
22	3. Sandstone			
	Great Karroo, S Africa	32–33° S	23– 24° E	214
	Erdi Plateau, Tchad–Libya	19–20° N	22– 23° E	87
	Sandstone plateau of SE of Aswan			116
C	Much dissected			
23	1. Crystalline or foliated			
	Central Hejaz Uplands, Saudi Arabia	24–25° N	38– 39° E	126
	Jabrin Plateau, Saudi Arabia	23–24° N	48– 49° E	151
	Tuweiq Plateau, Saudi Arabia	25–26° N	45– 46° E	152
	Biyadh Plateau, Saudi Arabia	23–24° N	47– 48° E	153
	Arma Plateau, Saudi Arabia	25–26° N	46– 47° E	154
	Summan Plateau, Saudi Arabia	25–26° N	48– 49° E	155
24	2. Limestone (including kem–kems)			
	Chela-Otair Highlands, SW Africa	19–20° S	14– 15° E	210
	Jol and Kathiri-Mahra Plateaux, Aden	15–16° N	49– 50° E	162
	Kem-Kem, Algeria	30–31° N	4– 5° W	15
	Mzab Plateau, Algeria	32–33° N	3– 4° E	24
	Tinghert Plateau, Algeria	28–29° N	6– 7° E	69
	Central Sinai–Negev, Egypt–Israel	30–31° N	34– 35° E	123
	Maaza Plateau, Egypt	26–27° N	32– 33° E	109
	Terecht–Timetrim, Mali	21–22° N	0– 1° E	56
	Ain Sefra Plateau, Algeria			23
25	3. Sandstone ('tassili')			
	Tibesti outliers, Libya	21–22° N	19– 20° E	89
	Tassili, Algeria	25–26° N	8– 9° E	59
	Ahnet, Algeria	24–25° N	2– 3° E	60
	Mouydir, Algeria	25–26° N	4– 5° E	61
	Mekran, Iran	26–27° N	58– 59° E	172
	Ennedi, Tehad	17–18° N	22– 23° E	88
	V. *Desert plains*			
A	With inselbergs			
	1. Gravelly			
26	(a) with crystalline inselbergs			
	Plains of E and Air Massif, Niger	18–19° N	11– 12° E	96
27	(b) with volcanic inselbergs			
	Farah Lowlands, SW Afghanistan	33–34° N	61– 62° E	94
28	(c) with limestone inselbergs			
	Qatar and Bahrein	25–26° N	51– 52° E	156
	2. Sandy			
29	(a) with crystalline inselbergs			
	Great Basin, USA	37–38° N	117–118° W	2
	Sonoran Desert, USA	32–34° N	113–116° W	4
	W plains of NSW and Queensland, Australia	28–29° S	144–145° E	225
	E Namaqualand, SW Africa			217
	Little Namaqualand–Bushmanland plain, SW Africa	29–30° S	19– 20° E	211
	S Kalahari, SW Africa	26–27° N	20– 21° E	212
	Mortcha, Tchad	16–17° N	20– 21° E	93

Group No.		Class and name	Limiting coordinates of representative 1°squares		No on Map
30		(b) with volcanic inselbergs			
		Lake Rudolf Basin, Kenya	3– 4° N	35– 36° E	180
		E lowlands of Ethiopia	12–13° N	41– 42° E	176
31		(c) with limestone inselbergs			
		Trucial Coast, Arabia	24–25° N	54– 55° E	158
		Sand and gravel plains of Oman Dhofar	19–20° N	57– 58° E	161
		Lowlands W of Air, Niger	18–19° N	5– 6° E	67
		Guban, Somalia	10–11° N	45– 46° E	181
32		(d) with sandstone inselbergs			
		SE desert plains, Libya	22–23° N	22– 23° E	85
		Central N desert of Sudan	21–22° N	30– 31° E	114
		Cent desert plains, Libya	25–26° N	17– 18° E	84
		S plateau, Egypt	21–22° N	28– 29° E	115
		Borkou, Tchad	18–19° N	20– 21° E	95
		Thar Desert, W Pakistan, India	26–27° N	70– 71° E	204
		Ténéré, Algeria	22–23° N	9– 10° E	58
33		3. Argillaceous			
		Low Plains, Somalia★			190
		Mauritanian coastal plain	19–20° N	16– 17° W	29
	B	Without inselbergs			
34		1. Gravelly			
		Serir of Calansho, Libya	27–28° N	22– 23° E	81
		Marmarica Lowlands, Libya	31–32° N	24– 25° E	80
		Tanezrouft, Algeria	23–24° N	0– 1° E	57
		Dasht-i-Margo Plain, Afghanistan	31–32° N	63– 64° E	195
		Tagama Plain, Niger	16—17° N	8– 9° E	68
		SE plains of Mali★			66
35		2. Sandy★			
		E central lowland of Australia	24–25° S	139–141° E	226
		Sirte Lowlands, Libya	30–31° N	18– 19° E	79
		Sinai coastal plain, Egypt	28–29° N	33– 34° E	122
		Djouf, Mauritania★			31
		Plains of Aouker and Hodh, Mali★			32
		Majabat Al Koubra, Mali★			44
		Isthmus of Suez, Egypt	30–31° N	32– 33° E	121
		Kanem, Tchad	14–15° N	15– 16° E	92
		Azaouad, Mali	18–19° N	2– 3° W	43
		Manga, Niger	14–15° N	13– 14° E	90
		Registan, Afghanistan	30–31° N	65– 66° E	196
36		3. Clayey			
		Sudan clay plains			120
37		4. Lateritic			
		Brakna-Douaich Plains, Mauritania	17–18° N	13– 14° W	30
		VI. *Main valleys*			
38	A	Large seasonal washes			
		Wadi Hadhramaut, Aden	15–16° N	48– 49° E	163
		IghargharRhir Basin, Algeria	32–33° N	6– 7° E	47
		Tabelbala Basin, Algeria	29–30° N	3– 4° W	37

Group No.		Class and name	Limiting coordinates of representative 1°squares		No on Map
		Wadi Saoura, Algeria	29–30° N	1– 2° W	49
		Wadi Tlemsi, Mali	17–18° N	0– 1° E	50
39	B	Upper eroding areas of perennial rivers 1. Crystalline			
		Orange River Gorge tract, SW Africa	28–29° S	17– 18° E	213
		Shabluka Gorge, Sudan	16–17° N	32– 33° E	98
		Aswan Cataract, Egypt	24–25° N	32– 33° E	99
40		2. Sandstone			
		Grand Canyon, USA	36–37° N	112–113° W	3
		Pecos Valley, Texas, USA	31–32° N	103–104° W	10
41		3. Limestone			
		Submontane Indus tract, W Pakistan	33–34° N	71– 72° E	201
		Oxus Plains, Afghanistan	37–38° N	66– 67° E	193
42	C	Middle alluvial areas 1. Siliccous			
		Nile Valley, Egypt and Sudan	26–27° N	31– 32° E	100
		Sind Plain, W Pakistan	27–28° N	68– 69° E	203
		Punjab Plain, W Pakistan	30–31° N	72– 73° E	202
43		2. Calcarcous			
		Tigris–Euphrates Lowlands, Iraq	32–33° N	35– 46° E	143
		Nogal Valley system, Somalia	8–9° N	49– 50° E	186
		Shebeli-Juba Lowlands, Somalia	1–2° N	43– 44° E	187
44	D	Lower swampy courses 1. Siliceous			
		Wadi Sirhan, Jordan	31–32° N	37– 38° E	135
		Niger Valley, Mali	16–17° N	2– 3° W	65
45		2. Calcareous			
		Jordan Valley, Jordan	31–32° N	35– 36° E	136

VII. *Eolian areas: ergs*

Group No.	Class and name	Limiting coordinates of representative 1°squares		No on Map
46	Fixed dune areas (*qoz*) of Kordofan, Sudan			119
	Sturt Desert, Australia	28–29° S	141–143° E	227
	Simpson Desert, Australia	24–25° S	136–137° E	228
	Gibson Desert, Australia	24–25° S	126–127° E	229
	Gt Victorian Desert, Australia	28–29° S	130–131° E	230
	NT Sand Desert, Australia	20–21° S	132–133° E	231
	Gt Sandy Desert, Australia	21–22° S	123–124° E	232
	Rebiana sand sea, Libya	24–25° N	21– 22° E	77
	Calansho sand sea, Libya			78
	Gt Egyptian sand sea, Egypt	26–27° N	26– 27° E	97
	Abu Muharik dunes, Egypt	26–27° N	30– 31° E	235
	Edeien Murzuch, Libya	24–25° N	13– 14° E	76
	Edeien Ubari, Libya	28–28° N	12– 13° E	75
	Gt E Erg, Algeria	30–31° N	7– 8° E	46
	Gt W Erg, Algeria	30–31° N	0– 1° E	45
	Erg Er Raoui	28–29° N	1– 2° W	38
	Erg Issaouane, Algeria			62
	Erg Chech, Algeria			36

Group No.		Class and name	Limiting coordinates of representative 1°squares		No on Map
		Erg Iguidi, Algeria			33
		Erg Hamami, Algeria			27
		Erg Makteir, Algeria			28
		Rub Al Khali, Saudi Arabia	19–20° N	50– 51° E	148
		Gt Nafud, Saudi Arabia	28–29° N	40– 41° E	143
		Nafud Dahi, Saudi Arabia	22–23° N	45– 46° E	147
		Jafura, Saudi Arabia	24–25° N	50– 51° E	145
		Inner Girdle Sand Desert, Saudi Arabia	25–26° N	44– 45° E	146
		Outer Girdle Sand Desert, Saudi Arabia or	25–26° N 23–24° N	47– 48° E 48– 49° E	144
		Coastal consolidated dunes, Somalia	2– 3° N	45– 46° E	188

VIII. *Large enclosed depressions*

Group No.		Class and name	Limiting coordinates of representative 1°squares		No on Map
47	A	Tectonic or fold			
		1. Internal drainage and salt lakes			
		Lake Eyre Basin, Australia	28–29° S	137–138° E	233
		Danakil Depression, Ethiopia	13–14° N	40– 41° E	179
		Chotts, Tunisia	33–34° N	8– 9° E	48
		Great Kavir, Iran	34–35° N	53– 54° E	167
		Jaz Murian, Iran	27–28° N	58– 59° E	168
		S Lut. Iran	30–31° N	58– 59° E	169
		Isfahan Siran, Iran	32–33° N	52– 53° E	170
		Sistan, Iran★			171
		Bodelé ('pays-bas'), Tchad	16–17° N	16– 17° E	91
48		2. with external drainage			
		Salton Trough, California, USA	33–34° N	115–116° W	5
		Arabah Graben, Jordan	30–31° N	35– 36° E	137
		Murdi Tchad	18–19° N	21– 22° E	94
		Al Medina Basin, Saudi Arabia	25–26° N	38– 39° E	138
	B	Large solution and deflation hollows			
49		1. Sandstone areas			
		Touat, Algeria	27–28° N	0– 1° E	54
50		2. Limestone areas			
		Jefr Depression, Jordan			132
		Kharga Oasis, Egypt	25–26° N	30– 31° E	106
		Dakhla Oasis, Egypt	25–26° N	29– 30° E	105
		Bahariya Oasis, Egypt	28–29° N	38– 29° E	103
		Farafra Oasis, Egypt	27–28° N	28– 29° E	104
		Fayum and Wadi Rayan, Egypt★			236
		Siwa Oasis, Egypt	29–30° N	25– 26° E	101
		Qattara Depression, Egypt	29–30° N	27– 28° E	102
		Wadi Natrun, Egypt	30–31° N	30– 31° E	234
		Tidikelt Oasis, Algeria	27–28° N	2– 3° E	55
		Gourara Oasis, Algeria	28–29° N	0– 1° E	53
		Taoudeni Oasis, Mali	22–23° N	3– 4° E	41

Group No.	Class and name	Limiting coordinates of representative 1°squares		No on Map
51 C	Interior basins with external drainage Plains of French Somaliland	11–12° N	42–43° E	189
52	IX. *Intertidal features: marsh*			
	Rann of Cutch, India	23–24° N	68–69° E	205

* Regions without representative 1° squares were not on the list originally used. They are included for the sake of completeness.

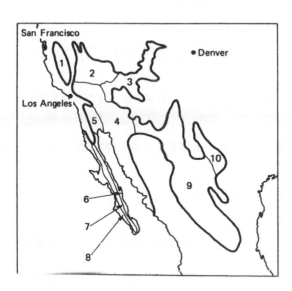

Fig. A.2 Physiographic regions of North America. (Source: Perrin and Mitchell, 1970, © British Crown copyright 1970/MOD)

Table A.3. Key to positions of regions in classified list (after Joly, 1957)

No on map	Group	No on map	Group	No on map	Group	No on map	Group	No on map	Group	No on map	Group	No on map	Group
1	14	35	19	69	24	103	50	137	48	171	47	205	52
2	29	36	46	70	7	104	50	138	48	172	25	206	20
3	40	37	38	71	14	105	50	139	9	173	1	207	1
4	29	38	46	72	5	106	50	140	9	174	14	208	10
5	48	39	2	73	18	107	18	141	43	175	16	209	14
6	14	40	18	74	18	108	18	142	21	176	30	210	24
7	2	41	50	75	46	109	24	143	46	177	3	211	29
8	14	42	18	76	46	110	4	144	46	178	11	212	29
9	18	43	35	77	46	111	19	145	46	179	47	213	39
10	40	44	35	78	46	112	13	146	46	180	30	214	22
11	18	45	46	79	35	113	13	147	46	181	31	215	19
12	18	46	46	80	34	114	32	148	46	182	6	216	19
13	19	47	38	81	34	115	32	149	21	183	6	217	29
14	18	48	47	82	3	116	22	150	15	184	6	218	1
15	24	49	38	83	3	117	1	151	23	185	15	219	8
16	18	50	38	84	32	118	3	152	23	186	43	220	10
17	18	51	18	85	32	119	46	153	23	187	43	221	14
18	18	52	18	86	19	120	36	154	23	188	46	222	15
19	19	53	50	87	22	121	35	155	23	189	51	223	20
20	17	54	49	88	25	122	35	156	28	190	33	224	18
21	5	55	50	89	25	123	24	157	3	191	20	225	29
22	13	56	24	90	35	124	1	158	31	192	13	226	35
23	24	57	34	91	47	125	2	159	13	193	41	227	46
24	24	58	32	92	35	126	23	160	14	194	27	228	46
25	19	59	25	93	29	127	2	161	31	195	34	229	46
26	19	60	25	94	48	128	14	162	24	196	35	230	46
27	46	61	25	95	32	129	15	163	38	197	18	231	46
28	46	62	46	96	26	130	16	164	12	198	12	232	46
29	33	63	1	97	46	131	16	165	1	199	5	233	47
30	37	64	2	98	39	132	50	166	16	200	5	234	50
31	35	65	44	99	39	133	18	167	47	201	41	235	46
32	35	66	34	100	42	134	19	168	47	202	42	236	50
33	46	67	31	101	50	135	44	169	47	203	42	237	1
34	15	68	34	102	50	136	45	170	47	204	32	238	21

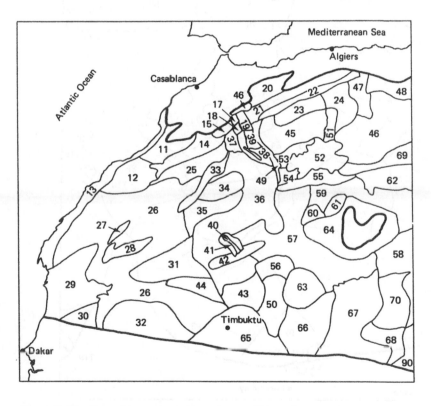

Fig. A.3 Physiographic regions of northwest Africa. (Source: Perrin and Mitchell, 1970, © British Crown copyright 1970/MOD)

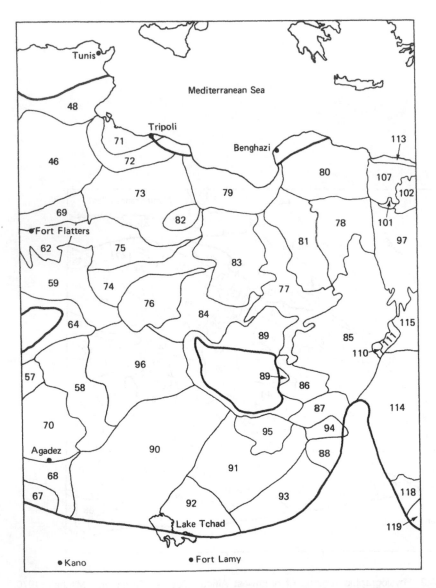

Fig. A.4 Physiographic regions of North Africa. (Source: Perrin and Mitchell, 1970, © British Crown copyright 1970/MOD)

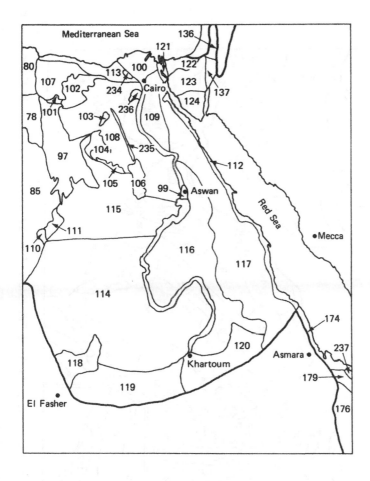

Fig. A.5 Physiographic regions of northeast Africa. (Source: Perrin and Mitchell, 1970,
© British Crown copyright 1970/MOD)

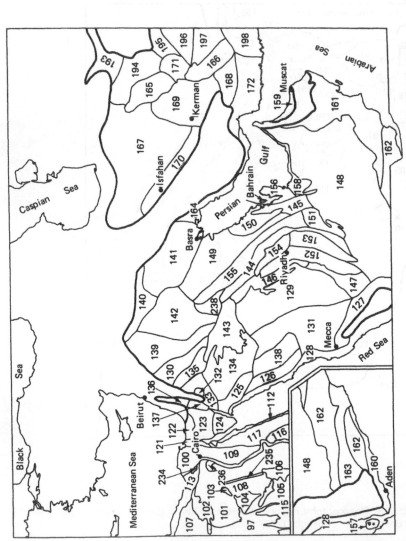

Fig. A.6 Physiographic regions of the Middle East. (Source: Perring and Mitchell, 1970, © British Crown copyright 1970/MOD)

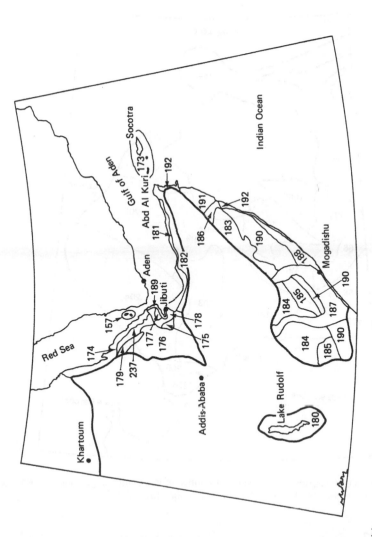

Fig. A.7 Physiographic regions of the arid lands near the Gulf of Aden. (Source: Perrin and Mitchell, 1970, © British Crown copyright 1970/MOD)

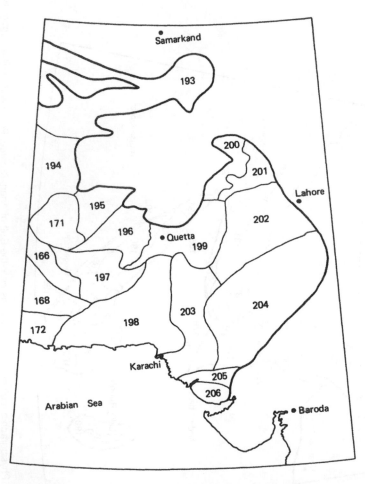

Fig. A.8 Physiographic regions of northwest India and neighbouring areas. (Source: Perrin and Mitchell, 1970, © British Crown copyright 1970/MOD)

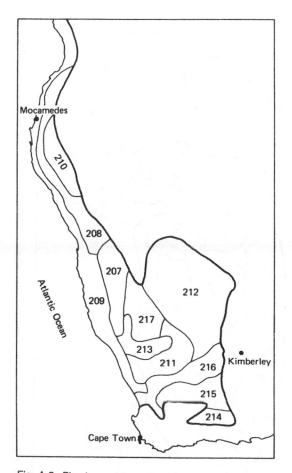

Fig. A.9 Physiographic regions of arid southern Africa. (Source: Perrin and Mitchell, 1970, © British Crown copyright 1970/MOD)

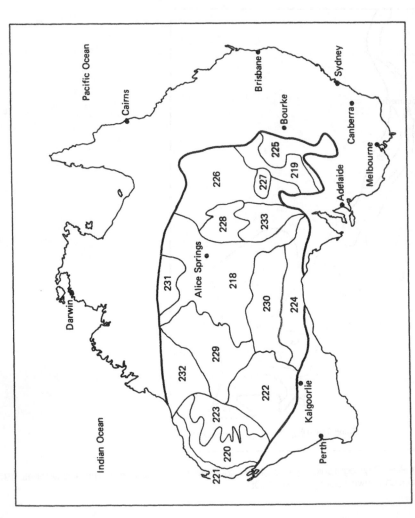

Fig. A.10 Physiographic regions of arid Australia. (Source: Perrin and Mitchell, 1970, © British Crown copyright 1970/MOD)

Appendix B

Field measurements of hydraulic conductivity and infiltration rate

1   The pump-out auger hole test

This test is described by van Beers (1958) and FAO (1976b). The equipment (Fig. B.1) includes a post-hole auger with sufficient extension rods to penetrate well below water-table depth, a baler, a stopwatch, and an electrical probe fitted to a galvanometer. Where the soil is likely to collapse, it is necessary to carry perforated metal tubing to case the holes. The baler is a section of metal tubing with a bottom opening on a hinge to admit the water and the other end threaded to take auger rods. The electrical probe (designed by Hunting Technical Services Limited) consists of an electrical two-pin plug on a wire connected to a galvanometer by a battery. The galvanometer needle gives a flick as soon as the plug is touched by the water surface as it rises. The field procedure is rapidly to bale out the auger hole and measure the exact time required for the water-table to rise in the hole. This is done by raising the probe regular amounts measured along a scale of marks at 1 cm intervals and waiting for the galvanometer to flick. The method is illustrated by Fig. B.2, and the hydraulic conductivity is calculated from the following formula:

$$k = \frac{4000r^2}{(H+20r)(2 - \frac{y}{H})y} \times \frac{\Delta y}{\Delta t} \qquad \text{[B.1]}$$

This is accurate to ±20 per cent where the following conditions are fulfilled: r = 3–7 cm, H = 20 –200 cm, $y > 0.2H$, $S > H$, Δy = or < 1/4 y_0. A similar method using a piezometer hole is described by FAO (1976b).

2   The pour-in auger hole method

This method is described by Hunting Technical Services (Sudan, 1963). A dry auger hole is bored through the horizon to be measured, which should be at

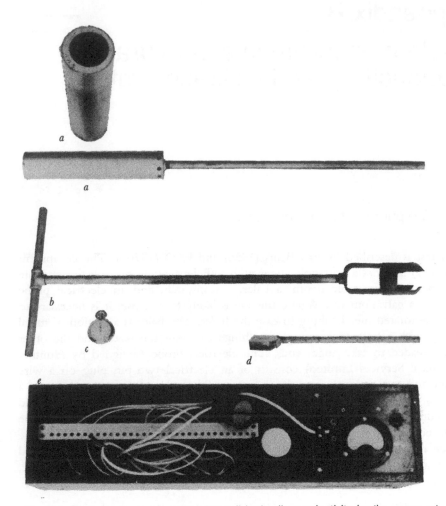

Fig. B.1 Equipment used in measuring soil hydraulic conductivity by the pump-out auger hole method: (a) baler; (b) 4-inch Jarrett post-hole auger; (c) stopwatch; (d) tape measure; (e) 'Δy box'. The electric plug drops though the white hole down into the auger hole and records the depth of the water-table by a flick on the galvanometer. The small clip fits into the small holes allowing the wire to be pulled up in exact increments of 1 cm. (Photograph: H. Walkland)

least 25 cm thick, making sure that there are no cracks. The hole is packed with gravel and water is run in and maintained at the level of its top. When inflow rate has stabilized after the soil has become saturated, it is measured by temporarily diverting this flow into a graduated cylinder over a timed period. Infiltration rate, in terms of the head of water used ($h/2$), is calculated from the following formula:

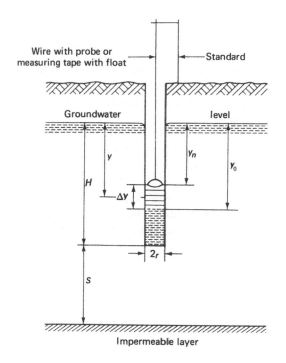

H = depth of the hole below groundwater table (cm)

y_0 = distance between the groundwater level and the elevation of the water surface in the hole after removal of water at the time of the first reading (cm)

y_n = the same at the end of the measurement. Usually about five readings are taken (cm)

Δy = $\Sigma \Delta y_t = y_n - y_0$, the rise of water level in the hole during the time of measurement (cm)

y = distance between the groundwater level and the average level of the water on the hole during the time of measurement (cm)

$$y = \frac{y_n - y_0}{2} = y_0 - \tfrac{1}{2}\Delta y$$

r = radius of the hole (cm)

S = depth of the impermeable layer below the bottom of the hole. It is assumed that this is greater than 2 m

Note: The measurements should be completed before $y_n < 3/4\ y_0$, or $\Delta y > 1/4\ y_0$

Fig. B.2 Diagram and formulae illustrating the method for measuring soil hydraulic conductivity by the pump-out auger hole method (Source: W.F.J. vanBeers, 1958)

$$I = 864Q/Cu\ r\ h \qquad\qquad [B.2]$$

where

l = *infiltration rate (m day^{-1})*; Q = rate of steady flow (ml s^{-1});

r = radius of bore (cm); h = height of water column (cm);

Cu = an empirical coefficient of conductivity which must be assessed from h/r by interpolation between 32 when $h/r = 10$ and 48.5 when $h/r = 20$.

3   Double cylinder pour-in method

This method also reflects gravity percolation in saturated soil under a preselected head, and is shown on Fig. B.3. It uses two lengths of open-ended cylindrical steel pipe, sharpened at one end and with diameters of at least 15 and 25 cm respectively. These are driven a short way into the ground concentrically and both are filled with water to the same predetermined level. Measurement is made in the inner one only, it being assumed that the water it contains penetrates vertically downwards into the underlying soil column. This assumption is not strictly correct but does not introduce serious error. A constant head is maintained in the inner cylinder by arranging that the supply is from an outlet tube under an inverted bottle. The infiltration rate is calculated by dividing the quantity of water fed from the bottle in a given time by the cross-sectional area of the inner cylinder.

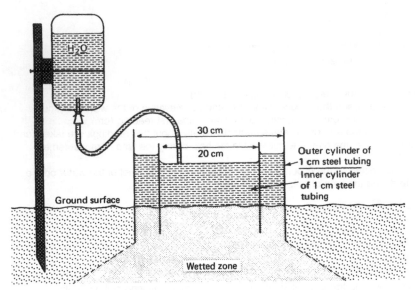

Fig. B.3 Concentric cylinder method for measuring soil infiltration rates

4   Single cylinder pour-in method

This method follows the same general principle as the last, but dispenses with the outer cylinder. The rate of water infiltration is measured by graduations on the feeder bottle (Hills, undated). Because the method has less control over the lateral movement of water than the double cylinder method, it represents the likely infiltration rate of water over a field plot less accurately. On the other hand, it is claimed that, because the restricting effect of the outer cylinder cannot be assessed accurately, this method is more satisfactory provided results for infiltrometer cylinders of different diameters are calibrated against the hydraulic properties under field conditions. Best results are obtained with cylinders with a diameter of over 150 mm (Youngs, 1987).

In order to measure infiltration rates under the pre-ponding conditions characteristic of the early stages of rainfall events, special infiltrometers have been designed to operate at water pressures which are negative to atmospheric pressure (Jarvis *et al.*, 1987).

References

Addor, E. E. (1963) Vegetation description for military purposes, in USAEWES, *Military Evaluation of Geographic Areas, Reports on Activities to April, 1963*, Miscellaneous Paper no. 3–610.

Afanasiev, J. N. (1927) The classification problem in Russian soil science, *Russian Pedology* **5**, 1–51 & 11 tables. Academy of Sciences of the USSR, Leningrad; quoted in USDA (1960) *Soil Classification: a comprehensive system*, seventh approximation, Washington, DC.

AGARD (North Atlantic Treaty Organization Advisory Group for Aerospace Research and Development) (1970) *Information Analysis Centres*, AGARD Conference Proceedings, No. 78, Schiphol, Amsterdam.

Aitchison, G. D. and **Grant, K.** (1967) The PUCE programme of terrain description, evaluation and interpretation for engineering purposes, *Proceedings of the Fourth Regional Conference for Africa on Soil Mechanics and Foundation Engineering*, Cape Town, pp. 1–8.

Aitchison, G. D. and **Grant, K.** (1968a) proposals for the application of the PUCE programme of terrain classification and evaluation to some engineering problems. Paper No. 452T, *Symposium on Terrain Evaluation for Engineering*, August 1968, convened by the Division of Soil Mechanics in association with the Fourth Conference of the Australian Road Research Board, CSIRO Division of Soil Mechanics Research Paper No. 119, Melbourne.

Aitchison, G. D. and **Grant, K.** (1968b) Terrain evaluation for engineering, in Stewart, G.A. (ed.) *Land Evaluation*, Macmillan of Australia, Melbourne, pp. 125–146

Aldrich, J. W. (1966) *Life Areas of North America*, US Fish and Wildlife Service BSFW Poster 102, Washington, DC.

Aldrick, J. M. and **Robinson, C. S.** (1972) *Report on the Land Units of the Katherine–Douglas area, N.T. (1970)*, Land Conservation Series No. 1, Northern Territory Administration, Darwin, Northern Territory, Australia.

Aldridge, D. (1975) *Principles of Countryside Interpretation*, Part One, Guide to Countryside Interpretation, Countryside Commission, Scotland, HMSO, Edinburgh.

Allan, J. A. and **Richards, T. S.** (1983) Use of satellite imagery in archaeological surveys, *Journal of Libyan Studies*, **14**, 4–8.

Allen, J. A. (1892) 'The geographical distribution of North American mammals', *Bulletin of the American Museum of Natural History*, **4**, 199–243.

390

Allen, J. R. L. (1970) *Physical Processes in Sedimentation*, Unwin University Books, George Allen & Unwin, London.

Allum, J. A. E. (1966) *Photogeology and Regional Mapping*, Faber and Faber, London.

American Geological Institute *Bibliography and Index of Geology* (monthly), Alexandria, Virginia 22302.

American Society of Photogrammetry (1980) *Manual of Photogrammetry* (2 vols), 4th edn, Falls Church, Virginia.

American Society of Photogrammetry (1983) *Manual of Remote Sensing* (2 vols), 2nd edn, Falls Church, Virginia.

Anderson, J. M. (1981) *Ecology for Environmental Sciences: biosphere, ecosystems, and man*, Edward Arnold, London.

Anderson, J. R., Hardy, E. E., Roach, J. T. and **Witmer, R. E.** (1976) *A Land Use and Land Cover Classification System for Use with Remote Sensor Data*, US Geological Survey Professional Paper 964, Washington, DC.

Anstey, F. (1960) *Digitized Environmental Data Processing,*, Headquarters Quartermaster Research and Engineering Command, Research Study Report RER-31, US army, Natick, Massachusetts.

Appleton, J. (1975) *The Experience of Landscape*, John Wiley & Sons, London.

Appleton, J. (ed.) (1976) *The Aesthetics of Landscape*, Rural Planning Services Ltd Publication No. 7, Department of Geography, University of Hull.

Appleyard, D., Craik, K. H., Klapp, M. and **Kreimer, A.** (1973) *The Berkeley Environmental Simulation Laboratory: Its Use in Environmental Impact Assessment*, Berkeley California Institute of Urban and Regional Development, University of California.

Appleyard, D., Lynch, K. and **Meyer, J. R.** (1964) *The View from the Road*, MIT Press, Cambridge, Massachusetts.

Arno, S. F. (1979) *Forest Regions of Montana*, US Forest Service Research Paper INT-218, Oden, Utah, 51.

Atkinson, R. J. C. (1970) The principles of archaeological air photography, in Fagan, B. M. (ed.) *Introductory Readings in Archaeology*, Little Brown & Company, Boston, Chapter 2, pp. 31–9.

Atkinson, T. C. (1971) Hydrology and erosion in a limestone terrain, Ph.D. Thesis, University of Bristol.

Austin, M. E. (1963) *Land Resource Regions and Major Land Resource Areas of the United States:Map at 1:10,000,000*, Soil Conservation Service, USDA, Beltsville, Maryland.

Australia, CSIRO (1967) *Lands of Bougainville and Buka Islands, Territory of Papua New Guinea*, Land Research Series No. 20, Melbourne.

Australia, CSIRO (1978) *Land Use on the South Coast of New South Wales: a Study in Methods of Acquiring and Using Information to Analyse Regional Land Use Options*, General editors: Austin, M. P. and Cocks, K. D., 4 vols, Melbourne.

Australia, CSIRO Division of Soils (1983) *Soils: An Australian Viewpoint*, CSIRO/Academic Press, Melbourne/London.

Avery, B. W. and **Bascomb, C. L.** (eds) (1974) *Soil Survey Laboratory Methods*, Soil Survey Technical Monograph No. 6, Harpenden.

Avery, B. W., Findlay, D. C. and **Mackney, D.** (1974) *Soil Map of England and Wales at 1:1 000 000*, Soil Survey of England and Wales, Harpenden.

Baboulene, B. (1969) *Critical Path Made Easy*, Duckworth, London.

Bagnold, R. A. (1941, 1965) *The Physics of Blown Sand and Desert Dunes*, Methuen, London.

Bagnold, R. A. (1966) *An Approach to the Sediment Transport Problem from General Physics* US Geological Survey Professional Paper 422-I, Washington, DC.

Bagwell, C., Sharma, G. C. and **Downs, S. W.** (1976) Ground truth study of a computer generated land use map of North Alabama. In Shahrokhi, F. (ed.) *Remote Sensing of Earth Resource*', vol. 5, F. Shahrokhi (ed.) University of Tennessee, Tullahoma, Tennessee, 139–50.

Bailey, R. G. (1980) *Description of the Ecoregions of the United States*, USDA Miscellaneous Publication 1391, Washington, DC.

Bailey, R. G. (1983) Delineation of ecosystem regions, *Environmental Management*, **7**, 365–73.

Bailey, R. G., Zoltai, S. C. and **Wiken, E. B.** (1985) Ecological regionalization in Canada and the United States, *Geoforum*, **16** (3), 265–75.

Ball, D. F. and **Barr, C. J.** (1986) *The Distribution of Sites of the Monitoring of Change Landscape Project in Relation to Classes in ITE Land Stratifications*, Hunting Technical Services Limited/NERC Contract f6/92/133, Institute of Terrestrial Ecology Project 987, Bangor, N. Wales and Merlewood, Cumbria.

Barnes, R. M. (ed.) (1979) *Applications of Plasma Emission Spectrometry*, Heyden, Philadelphia.

Battarbee, R. W., Cooke, R. U., Metcalfe, S. E. and **Derwent, R. G.** (1989) Geographical research on acid rain, *Geographical Journal*, **155**(3), 353–77.

Batty, M. (1987) *Microcomputer Graphics: Art Design and Creative Modelling*, Chapman & Hall, London.

Baylis, P. (1976) Photograph in article by Davis-Jones, R. P., Buress, D. W., and Lemon, L. R. An atypical tornado-producing cumulonimbus, *Weather*, **31**, 337–47.

Beaumont, T. E. (1985) An application of satellite imagery for highway maintenance and rehabilitation in Niger, *International Journal of Remote Sensing* **6** (7), 1263–7.

Beckett, P. H. T. (1962) Punched cards for terrain intelligence, *Royal Engineers Journal*, **76** (2), 185–93.

Beckett, P. H. T. (1967) *Report of a Visit to the Terrain Evaluation Cell of the R and D Organization of the Indian Army*, MEXE Report No. 1020, Christchurch.

Beckett, P. H. T. (1971) *Output for a Terrain Data Store*, MVEE (Military Vehicles and Engineering Establishment) MoD (army) Report No. 71506, Christchurch.

Beckett, P. H. T. and **Furley, P. A.** (1968) Soil formation and slope development, *Zeitschrift für Geomorphologie*, **12**, 1– 42.

Beckett, P. H. T. and **Webster, R.** (1962) *The Storage and Collation of Information on Terrain*, MEXE, Christchurch (Part 1 by Beckett, Part 2 by Webster).

Beckett, P. H. T. and **Webster, R.** (1965a) *A Classification System for Terrain*, MEXE Report 872, Christchurch.

Beckett, P. H. T. and **Webster, R.** (1965b) *Field Trials of a Terrain Classification System: Organization and Methods*, MEXE Report 873, Christchurch.

Beckett, P. H. T. and **Webster, R.** (1965c) *Field Trials of a Terrain Clasification System: Statistical Procedure*, MEXE Report 874, Christchurch.

Beckett, P. H. T. and **Webster, R.** (1965d) *Minor Statistical Studies on Terrain Evaluation*, MEXE Report 877, Christchurch.

Beckett, P. H. T. and **Webster, R.** (1969) *A Review of Studies on Terrain Evaluation by the Oxford–MEXE–Cambridge Group, 1960–1969*, MEXE Report 1123, Christchurch.

Beckett, P. H. T., Webster, R., McNeil, G. M. and **Mitchell, C. W.** (1972) Terrain evaluation by means of a data bank, *Geographical Journal*, **138** (4), 430–56.

Bell, F. G. (1987) *Ground Engineers Reference Book*, Butterworths, London.

Bell, J. P. (1976) *Neutron Probe Practice*, Report No. 19, Institute of Hydrology, Wallingford, Oxon.

Bendelow, V. C. and **Hartnup, R.** (1980) *Climatic Classification of England and Wales*. Soil Survey Technical Monograph No. 15, Harpenden, Herts.

Bibby, J. S. and **Mackney, D.** (1969) *Land Use Capability Classification*, Soil Survey Technical Monograph No. 1, Rothamsted Experimental Station and Macaulay Institute for Soil Research, Harpenden and Aberdeen.

Bie, S. W. (ed.) (1975) Soil information systems: *Proceedings of the Meeting of the ISSS Working Group on Soil Information Systems*, Centre for Agricultural Publishing and Documentation, Wageningen, Netherlands.

Bonn, F. (1976) Some problems and solutions related to round truth measurements for thermal infra-red remote sensing, *Proceedings of the American Society of Photogrammetry*, **42**, 1–11.

Bouille, F. (1984) Architecture of a geographic structured expert system, *Proceedings of the International Symposium on Spatial Data Handling*, Zurich, vol. 2, 520–43.

Bourne, R. (1931) *Regional Survey and its Relation to Stocktaking of the Agricultural Resources of the British Empire*, Oxford Forestry Memoirs No. 13.

Bowman, I. (1911) *Forest Physiography, Physiography of the US and Principal Soils in Relation to Forestry*, John Wiley & Sons, New York.

Bowman, I. (1916) *The Andes of Southern Peru: Geographical Reconnaissance Along the Seventy Third Meridian*, American Geographical Society of New York, Henry Holt & Company, New York.

Bradford, J. (1957) *Ancient Landscapes: Studies in Field Archeology*, G. Bell & Sons, London.

Brady, N. C. (1984) *The Nature and Properties of Soils*, 9th edn, Macmillan, New York.

Braun, E. L. (1964) *Deciduous Forest of Eastern North America*, Hofner, New York.

Braun–Blanquet, J. (1964) *Pflanzensozioloie: Grundzüge der Vegetationskunde*, 3rd edn, Springer–Verlag, Vienna–New York.

Bridges, E. M. and **Doornkamp, J. C.** (1963) Morphological mapping and the study of soil patterns, *Geography*, **48** (2), 175–81.

Brink, A. B. A., Mabbutt, J. A., Webster, R. and **Beckett, P. H. T.** (1966) *Report of the Working Group on Land Classification and Data Storage*, MEXE Report 940, Christchurch.

Brink, A. B. A. and **Partridge, T. C.** (1967) Kyalami land system: an example of physiographic classification for the storage of terrain data, *Proceedings of the Fourth Regional Conference for Africa on Soil Mechanics and Foundation Engineering*, vol. 1, p. 9.

Brink, A. B. A., Partridge, T. C., Webster, R. and **Williams, A. A. B.** (1968) Land classification and data storage for the engineering usage of natural materials, *Proceedings of the Symposium on Terrain Evaluation for Engineering*, Australian Road Research Board Fourth Conference, vol. 4, pp. 1624–47.

British Standards Institution (1975) *Methods of Testing Soils for Civil Engineering Purposes*, British Standard 1377, London.

Brooks, R. R. (1983) *Biological Methods of Prospecting for Minerals*, John Wiley & Sons, New York.

Broome, D. R. (1983) Future remote sensing instruments and systems, *Proceedings of 17th International Symposium on Remote Sensing of Environment*, ERIM, Ann Arbor, Michigan, vol. 1, pp. 65–73.

Brown, G. (ed.) (1962) *The X–ray Identification and Crystal Structure of Clay Minerals*, Mineralogical Society Clay Minerals Group, London.

Brubaker, S. (1972) *To Live on Earth*, Resources for the Future, Baltimore and London.

Bryan, K. (1919) Classification of springs, *Journal of Geology*, **27**, 522–61.

Bryan, M. L. and **Larson, R. W.** (1973) Application of dielectric constant measurements to radar imagery interpretation, in Shahrokhi, F. (ed.) *Remote Sensing of Earth Resources*, University of Tennessee, Tullahoma, Tennessee, vol. 2, pp. 529–48.

Büdel, J. (1963) Klima–genetische Geomorphologie, *Georaphische Rundschau*, **15**, 269–85.

Bunce, R. G. H., Barr, C. J. and **Whittaker, H.** (1983). A stratification system for ecological sampling, in Fuller, R. M. (ed.) *Ecological Mapping from Ground, Air and Space*, Institute of Terrestrial Ecology (ITE) Symposium No. 10, Monks Wood Experimental Station, Abbots Ripton, Huntingdon.

Buol, S. W., Hole, F. D. and **McCracken, R. J.** (1980) *Soil Genesis and Classification*, 2nd edn, Iowa State University Press, Ames.

Burrough, P. A. (1986) *Principles of Geographical Information Systems for Land Resources Assessment*, Monographs on Soil and Resources Survey 12, Clarendon Press, Oxford.

Burrough, P. A. (1987) Mapping and map analysis: new tools for land evaluation, *Soil Use and Management*, **3**(1), 20–5.

Burrough, P. A. and **Bie, S. W.** (eds) (1984) *Soil Information System Technology*, Pudoc, Wageningen.

Butler, B. E. (1959) *Periodic Phenomena in Landscapes as a Basis for Soil Studies*, CSIRO Australian Soil Publication No. 14, Melbourne.

Butler, B. E. (1967) Soil periodicity in relation to landform development in southeastern Australia. In Jennings, J. N. and Mabbutt, J. A. (eds) *Landform Studies in Australia and New Guinea*, Australian National University Press, Canberra, pp. 231–66.

Cadigan, R. A., Ormsbee, L. R., Palmer, R. A. and **Voegeli, P. T.** (1972) *Terrain Classification: A Multivariate Approach*, US Geological Survey Report USGS–GD–72–04; summarized in US Government Reports Announcements PB–211 090, Washington.

Campbell, R. C. (1967) *Statistics for Biologists*, Cambridge University Press.

Canada Land Inventory (1965) *Soil Capability Classification for Agriculture*, Report No. 2, Ottawa.

Canby, T. (1983) Satellites that serve us, *National Geographic Magazine*, **164**(3), 281–334.

Cannell, G. H. and **Weeks, L. G.** (1979) Erosion and its control in semi-arid regions, in Hall, A.E. Cannell, G.H. and Lawton, H.W. (eds) *Agriculture in Semi–arid Environments*, Springer, pp. 238–56.

Carol, H. (1957) Grundsätzliches zum Landschaftsbegriff, *Petermanns Geographische Mitteilungen*, **101**, pp. 93–7 and in Paffen, K.H. (ed.) (1973) *Das Wesen der Landschaft*, Darmstadt, pp. 142–55.

Carter, J. R. (1984) *Computer Mapping: Progress in the 80s*, Resource Publications on Geography, Association of American Geographers, Washington, DC.

Casagrande, A. (1947) Classification and identification of soils, *Proceedings of the American Society of Civil Engineers*, **73**(6), 783–810.

Casey, R. S., Perry, J. W., Berry, M. M. and **Kent, A. A.** (eds) (1958) *Punched Cards: Their Applications to Science and Industry*, Reinhold, New York.

Chang, J–H. (1968) *Climate and Agriculture: an Ecological Survey*, Aldine Publishing Company, Chicago.

Chang, L. and **Burrough, P. A.** (1987) Fuzzy reasoning, a new quantitative aid for land evaluation, *Soil Survey and Land Evaluation*, **7**, 69–80.

Chapman, P., Quade, J., Brennan, P. and **Blinn, J. C.** (1970) Ground truth/sensor correlation. *2nd Annual Earth Resources Review*, NASA, Houston, Texas, 2, Section 31.

Chartres, C. J. (1975) *Soil development on the terraces of the River Kennet*, Ph.D. Thesis, University of Reading.

Chepil, W. S. (1945) Dynamics of wind erosion: III. Transport capacity of the wind, *Soil Science*, **60**, 475–80.

Childs, D. R. (1967) The anatomy of the human environment, in *Landscape Research Group Symposium*, pp. 43–46.

Chorley, R. J. (ed.) (1969) *Water, Earth, and Man*, Methuen, London.

Chorley, R. J. (1978) The hillslope hydrological cycle. In Kirkby, M.J. (ed.) *Hillslope Hydrology*, John Wiley & Sons, Chichester, pp. 1–42.

Chow, Ven Te (ed.) (1964) *Handbook of Applied Hydrology*, McGraw-Hill, New York.

Christaller, W. (1933) *Die Zentralen Orte in Süddeutschland*, Jena.

Christenson, G. E., Miller, J. R. and **Pieratti, D. D.** (1982) Prediction of engineering properties and construction conditions from geomorphic mapping in regional siting studies, in Craig, R.G. and Craft, J.L. (eds) *Applied Geomorphology*, George Allen & Unwin, London, pp. 94–107.

Christian, C. S. (1958) The concept of land units and land systems, *Proceedings of the 9th Pacific Science Congress, 1957*, vol. 20, pp. 74–81.

Christian, C. S. (1983) *The Australian Approach to Environmental Mapping: A Reprint*, CSIRO Institute of Biological Resources, Division of Water and Land Resources, Technical Memorandum 83/5, Canberra. Also appeared as part of US Geological Survey Professional Paper 1193, edited by F.C. Whitmore and M.E. Williams (1982) *Resources for the Twenty-first Century*, United States Government Printing Office.

Christian, C. S. and **Stewart, G. A.** (1968) Methodology of integrated surveys, in *Unesco Conference on Principles and Methods of Integrating Aerial Survey Studies of Natural Resources for Potential Development*, Toulouse, Unesco, Paris, pp. 233–80.

Clarke, G. (1964) *Archaeology and Society*, 3rd edn, Methuen, London.

Clements, R. E. and **Shelford, V. E.** (1939) *Bioecology*, John Wiley & Sons, New York.

Clifton–Taylor, A. (1982) *The Pattern of English Building*, Faber & Faber, London and Boston.

Clouston, J. B. (1967) The Durham motorway landscape study, in *Landscape Research Group Symposium*, pp. 11–19.

Coates, D. R, (1987) Engineering aspects of geomorphology, in Bell, F.G. (ed.) *Ground Engineers Handbook*, Butterworths, London, 2/1–2/37.

Cochran, W.G. (1963) *Sampling Techniques*, 2nd edn, John Wiley & Sons, New York.

Cochrane, G. R. and **Brown, G. H.** (1981) Geomorphic mapping from Landsat 3 return beam videcon (RBV) imagery, *Photogrammetric Engineering and Remote Sensing*, **47**, 1305–14.

Cole, J. P. and **King, C. A. M.** (1968) *Quantitative Geography*, John Wiley & Sons, New York.

Cole, M. M. (1980) Geobotanical expression of ore bodies, *Transactions of the Institute of Mining and Metallurgy*, **B.89**, 73–91.

Coleman, A. and **Maggs, K. R. A.** (1965) *Land Use Survey Handbook, an Explanation of the Second Land Use Survey of Great Britain on the Scale of 1:250,000*, Isle of Thanet Geographical Association, Stanfords, Long Acre, London.

Colinvaux, P. (1973) *Introduction to Ecology*, John Wiley & Sons, New York.

Commission de Pédologie et de Cartographie des Sols (1967) *Classification des Sols*, Laboratoire de Géologie et Pédologie, Ecole Nationale Supérieure d'Agronomie, Grignon, France.

Conacher, A. J. and **Dalrymple, J. B.** (1977) The nine–unit land surface model: an approach to pedo–geomorphic research, *Geoderma*, **18** (1/2), 1–154.

Condon, R. W. (1968) Estimation of grazing capacity of arid grazing lands, in Stewart, G.A. (ed.) Land Evaluation, Macmillan of Australia, pp. 112–124.

Conklin, H. E. (1959) The Cornell system of economic land classification, *Journal of Farm Economics*, **41**(1), 548–57.

Cooke, H. J. and **Shaw, P. A.** (1986) Geomorphology in development, in Gardiner, V. (ed.) *International Geomorphology*, Part II, John Wiley & Sons Limited, Chichester, pp. 411–17.

Cooke, R. U. and **Harris, D. R.** (1970) Remote sensing of the terrestrial environment – principles and progress, *Transactions of the Institute of British Geographers*, **50**, 1–23.

Cornell Aeronautical Laboratory Inc. (1963) *Matrix Methods for Terrain Profile Simulation* (Project CATVAR), Ithaca, New York. See also Deitchman (c. 1966).

Cotton, C. A. (1944) *Volcanoes as Landscape Forms*, Whitcomb & Tombs, New Zealand.

Cousins, L. B. (1970) The Harwell heat transfer and fluid flow information analysis centre, in *AGARD Information Analysis Centres*, North Atlantic Treaty Organization Conference Proceedings, No. 78, London, **4**, 1–6.

Covell, R. R. and **Shaffer, M. E.** (1966) *Soil Interpretations for Town and Country Planning and Development. De Soto County, Mississippi*, USDA Soil Conservation Service/Mississippi Agricultural Experiment Station.

Cowan, D. J., Bayne, J. N. and **Fairey, D. A.** (1976) *Development and Applications of the South Carolina Computerized Land Use Information System*, South Carolina Land Resources Commission.

Cowan, W. L. (1956) Estimating hydraulic roughness coefficients, *Agricultural Engineering*, **37**, 473–75.

Cowardin, L. M., Varter, V., Golet, F. C. and **Laroe, E. T.** (1979) *Classification of Wetlands and Deepwater Habitats of the United States*, US Fish and Wildlife Service FWS/OBS – 79/31, Washington DC.

Craig, R. G. (1982) Criteria for constructing optimal digital terrain models in Craig, R. G. and Craft, J. L. (eds) *Applied Geomorphology*, George Allen & Unwin, London, pp. 108–30.

Craik, K. H. (1975) Individual variations in landscape description, in Zube, E. H., Brush, R. O. and Fabos, G. J. (eds) *Landscape Assessments: Values, Perceptions, and Resources*, Dowden, Hutchinson & Ross, Stroudsburg, Pennsylvania, pp. 130–50.

Cruickshank, J. D. and **Heidenreich, C. E.** (1969) Pedological investigations in the Huron Indian village of Cahiague, *Canadian Geographer*, **13**(1), 34–46.

Curran, P. (1985) *Principles of Remote Sensing*, Longman, London and New York.

Curtis, L. F., Doornkamp, J. C. and **Gregory, J. K.** (1965) The description of relief in field studies of soils, *Journal of Soil Science*, **16**(1), 16–30.

Dale, E. and **Michelon, L. C.** (1966) *Modern Management Methods*, Penguin Books, Harmondsworth.

Dalrymple, J. R., Blong, R. J. and **Conacher, R. J.** (1968) A hypothetical nine unit landsurface model, *Zeitschrift für Geomorphologie*, **12**, 67–76.

Dangermond, J. (1988) GIS trends and comments, *ARC News*, Summer/Fall, Environmental Systems Research Institute, Redlands, California, pp. 13–17.

Daniels, R. B., Gamble, E. E. and **Cady, J. G.** (1971) A relation between geomorphology and soil morphology and genesis, *Advances in Agronomy*, **23**, 51–88.

Dansereau, P. (1951) Description and recording of vegetation on a structural basis, *Ecology*, **32**, 172–229.

Dansereau, P. (1958) *A Universal System for Recording Vegetation*, Institut Botanique de l'Université de Montréal, Canada

Davis, J. R. and **Nanninga, P. M.** (1984) *Geomycin: Towards a Geographic Expert System for Resource Management*, CSIRO Division of Water and Land Resources Report, Canberra, ACT 2601, Australia.

Davis, W. M. (1909) *Geographical Essays*, edited by D.W. Johnson, Dover Publications, New York. Republished 1954.

Dawson, J. A. and **Doornkamp, J. C.** (1973) *Evaluating the Human Environment*, Edward Arnold, London.

Dawson, J. A. and **Unwin, D. J.** (1976) *Computing for Geographers*, David & Charles, Newton Abbot.

de Boodt, M. and **Gabriels, D.** (eds) (1980) *Assessment of Erosion*, Wiley–Interscience, Chichester.

de Candolle, A. P. (1856) *Géographie Botanique Raisonnée*, Paris.

Deitchman, S. J. (*c.*1966) *Classification and Quantitative Description of Large Geographic Areas to Define Transport System Requirements*, Cornell Aeronautical Laboratory, Ithaca, New York.

de Jong, G. (1962) *Chorological Differentiation as a Fundamental Principle of Geography*, J. B. Wolters, Groningen.

de Martonne, E. (1948) *Géographie Physique*, vol. 2, Le Relief du Sol, 8th Edn, Libraire Armand Colin, Paris.

Demek, J., Embleton, C., Gellert, J. F. and **Verstappen, H. Th.** (1972) *Manual of Detailed Geomorphological Mapping*, Academia, Prague.

Denmark: Ministry of Agriculture (1982) *Land Data Systems at Ministry of Agriculture Bureau of Land Data*, Vejle.

Dennis, J. (1835) *The Landscape Gardener*, James Ridgeway, London.

Dent, D. and **Young, A.** (1981) *Soil Survey and Land Evaluation*, George Allen & Unwin, London.

Department of Primary Industries (1974) *Western Arid Region Land Use Study*, Part I, Technical Bulletin No. 12, Queensland Division of Land Utilisation, Brisbane.

Department of Primary Industries (1978) *Western Arid Region Land Use Study*, Part IV, Technical Bulletin No. 23, Queensland Division of Land Utilisation, Brisbane.

Desaunettes, O. R. (1977) *Catalogue of Landforms for Indonesia*, prepared for the Land Capability Appraisal Project, Trust Fund for the Government of Indonesia and FAO, Working Paper No. 13, AGL/TFINS/44, Soil Research Institute, Bogor, Indonesia.

Dice, L. R. (1943) *The Biotic Provinces of North America*, University of Michigan Press, Ann Arbor, Michigan.

Dickinson, R. E. (1930) The regional functions and zones of influence of Leeds and Bradford, *Geography*, **15**, 548–59.

DoE (UK Department of the Environment) (1987) *Handling Geographic Information*, HMSO, London.

Dokuchaiev, V. V. (1899) On the Theory of Natural Zones. Horizontal and Vertical Zones, St. Petersburg; reprinted from the newspaper 'The Caucasus' nos 253 & 254 with small additions. In *Socheniya* (Collected Works), vol. 6, Moscow–Leningrad, 1951.

References

Douglas, I. (1977) *Humid Landforms*, An Introduction to Systematic Geomorphology, Vol. 1, MIT Press and Australian National University Press, Canberra.

Dowling, J. (1968) The classification of terrain for road engineering purposes, *Conference on Civil Engineering Problems Overseas*, Institution of Civil Engineers, Session V, Paper 1, 33–58, London.

Downs, R. M. and **Stea, D.** (1976) *Image and Environment*, Aldine Publishing Co., Chicago.

Doyle, F. J. (1978) Digital terrain models: an overview, *Photogrammetric Engineering and Remote Sensing*, **XLIV** (12), 1481–6.

Drabble, M. (1979) *A Writer's Britain: Landscape in Literature*, Thames & Hudson, London.

Dresch, J., d'Hollander, R. and **Verger, F.** (eds) (1985) *New Atlas of Relief Forms*, Nathan, Paris.

Driggers, W. G., Downs, J. M., Hickman, J. R. and **Packard, R. L.** (1978) Data base design for a worldwide multicrop information system. In *NASA, The LACIE Symposium: Proceedings of Technical Sessions*, NASA Johnson Space Center, Houston, Texas, pp. 1085–96.

Dubief, J. (1952) Le vent et le déplacement du sable au Sahara, *Travaux de l'Instutitut de Recherches Sahariennes*, **8**, 123–62.

Dufton, A. F. (1940–41) Heat transmission coefficients, *Journal of the Institute of Heating and Ventilating Engineers*, **8**.

Dumansky, J., Marshall, I. B. and **Huffman, E. C.** (1979) Soil capability analysis for regional land use planning – a study of the Ottawa urban fringe, *Canadian Journal of Soil Science*, **59**, 363–79.

Dunne, T. (1978) Field studies of hillslope flow processes. in Kirkby, M.J. (ed.) *Hillslope Hydrology*, John Wiley & Sons, Chichester, Chapter 7, pp. 227–324.

Du Rietz, G. E. (1936) Classification and nomenclature of vegetation units, *Svensk Botanisk Tidskrift*, **30**, 580.

Dury, G. H. (1981) *An Introduction to Environmental Systems*, Heinemann, London and Exeter, New Hampshire.

Ecological Land Survey Task Force (1980) *Ecological Land Survey Guidelines for Environmental Impact Analysis*, Environment Canada, Ottawa.

Elsevier Geoabstracts (1990–) Elsevier Science Publishers, Amsterdam. This was previously published by the University of East Anglia, Norwich, under the following titles; Geomorphological Abstracts 1960-5, Geographical Abstracts 1966–71, Geoabstracts 1972–86, and Geographical Abstracts 1986–9.

Elwell, H. A. (1981) A soil loss technique for southern Africa, in Morgan, R.P.C. (ed.) *Soil Conservation: Problems and Prospects*, John Wiley & Sons, Chichester, New York, Brisbane and Toronto, pp. 281–92.

Erol, Oguz (1983) *Die Naturräumliche Gliederung der Turkei*, Beihefte zum Tubinger Atlas des Vorderen Orients, Reihe A (Naturwissenschaften) Nr. 13, Dr Ludwig Reichert, Wiesbaden.

ESRI (Environmental Systems Research Institute) (1979) *Bernardino Forest Wildland Recreation Study*, Redlands, California.

Fabos, J. G., Hendrix, W. G. and **Greene, C. M.** (1975) Visual and cultural components of the landscape resource assessment model of the Metland study, in Zube, E.H., Brush, R.O., and Fabos, J.G. (eds) *Landscape Assessment: Values, Perceptions, and Resources*, Dowden, Hutchinson & Ross, Stroudsburg, Pennsylvania, pp. 319–43.

Fairbridge, R. W. (1946–47) Notes on the geomorphology of the Houtmans Abrolhos Islands, *Journal of the Royal Society of Western Australia*, **33**, 1–43.

Fairbridge, R. W. (ed.) (1968) *Encyclopedia of Geomorphology*, Reinhold, New York.

FAO (Food and Agriculture Organization of the United Nations) (1971) *Land Degradation*, Soils Bulletin No. 13, Rome.

FAO (1973) *Soil Survey Interpretation for Engineering Purposes*, Soils Bulletin No. 19, Rome.

FAO (1976a) *A Framework for Land Evaluation*, Soils Bulletin No. 32, Rome.

FAO (1976b) *Drainage Testing*, Irrigation and Drainage Paper No. 28, Rome.

FAO (1977) *Guidelines for Soil Profile Description*, Rome.

FAO (1978a) *Report of the Agro–Ecological Zones Project*, World Resource Report 48, Vol. 1. *Methodology and Results for Africa*, vol.2 *Results for Southwest Asia*, Project Coordinator: G. M. Higgins, Rome.

FAO (1978b) *Guidelines for Prognosis and Monitoring of Salinity and Sodicity*, Soils Bulletin No. 39, Rome.

FAO (1983) *Guidelines: Land Evaluation for Rainfed Agriculture*, Soils Bulletin 52, Rome.

FAO (1984) *Land Evaluation for Forestry*, FAO Forestry Paper 48, Rome.

FAO (1985), *Guidelines: Land Evaluation for Irrigated Agriculture* Soils Bulletin 55, Rome.

FAO/Unesco (1974–) *Soil Map of the World at 1:5,000,000* (17 sheets), Rome.

Farquharson, F. A. K., Mackney, D., Newson, M. D. and **Thomasson, A. J.** (1978) *Estimation of Run–off Potential of River Catchments from Soil Surveys*, Soil Survey of Great Britain (England and Wales) Special Publication No. 11, Harpenden.

Fenelon, P. (ed.)(1975) *Phénomènes Karstiques*, Mémoires et Documents, Série de Documentation et de Cartographie Géographique, CNRS, 15, Paris.

Fenneman, N. M. (1916) Physiographic divisions of the United States, *Annals of the Association of American Geographers*, **6**, 19–98.

Fenneman, N. M. (1928) Physiographic divisions of the United States, *Annals of the Association of American Geographers*, **18**, 261–353.

Fenton, T. E. (1982) Estimating soil erosion by remote sensing techniques. In Johannsen, C.L. and Sanders, J.L. (eds) *Remote Sensing for Resource Mapping*, Soil Conservation Society of America, pp. 217–23.

Fines, K. D. (1968) Landscape evaluation: a research project in East Sussex, *Regional Studies* 2(1), 41–5.

Finkel, H. J. (1959) The barchans of southern Peru, *Journal of Geology*, 67, 614–47.

Fisher, H. T. (1978) *Thematic Cartography – What it is and what is different about it*, Harvard Papers in Theoretical Cartography, Laboratory for Computer Graphics and Spatial Analysis, Harvard University.

Fogarty, P. and **Wood, B.** (1978) *The Land Resources of the Larapinta Valley and Undoolya area, Alice Springs*, Land Conservation Section, Darwin, Northern Territory, Australia.

Fookes, P. G. and **Gray, J. M.** (1987) Geomorphology and civil engineering, In Gardiner, V. (ed.) *International Geomorphology*, vol. 1, John Wiley & Sons, Chichester, pp. 83–105.

Ford, D. C. and **Williams, P. W.** (1989) *Karst Geomorphology and Hydrology*, Unwin Hyman, London.

Forer, P. (1984) *Applied Apple Graphics*, Prentice-Hall, Englewood Cliffs, New Jersey.

Foster, G. R., Meyer, L. D. and **Onstad, C. A.** (1973) Erosion equations derived from modelling principles, *American Society of Agricultural Engineers* Paper No. 73–2550, American Society of Agricultural Engineering, St Joseph, Michigan.

Fournier, F. (1960) *Climat et Erosion: la Relation entre l'Erosion du Sol par l'Eau et les Précipitations Atmosphériques*, Presses Universitaires de France, Paris.

Fox, J. W. (1956) *Land Use Survey: General Principles and a New Zealand Example*, Auckland University College Bulletin No. 49.

Franklin, J. R. and **Dyrness, C. T.** (1973) *Natural Vegetation of Oregon and Washington,* US Forest Service General Technical Report PNW–8, Portland, Oregon.

Gardner, N. (1985) *The Oxford Project on Microcomputers in Geographical Education,* The Geographic Information System Project Module, School of Geography, Oxford.

Garnett, A. (1935) Insolation, topography and settlement in the Alps, *Geographical Review,* **25,** 601–17.

Garnett, A. (1939) Diffused light and sunlight in relation to relief and settlement in high latitudes, *Scottish Geographical Magazine,* **55**(5), 271–84.

Geiger, R. (1965) *The Climate Near the Ground,* 3rd edn, Harvard University Press, Massachusetts.

Gellert, J. F. (1959–60) Die naturräumliche Gliederung des Landes Brandenburg und Altmark. *Wissenschaftliche Zeitschrift der Pädagogische Hochschule Potsdam* **5,** 3–22.

Geoffroy, J–L. (1978) Normes utilisées actuellement au Maroc pour l'établissement des cartes de classement des sols en vue de la culture sous irrigation, *ORSTOM, Ser. Pédol.,* **XVI**(2) 177–91.

Geological Society of America (annual) *Bibliography and Index of Geology.* Before 1969, separate volumes were published on North America and areas exclusive of North America.

Geomap (1968) *Topomorphic Map of Northern Afar, Ethiopia,* Geomap Cartographic Centre, Florence, now Lausanne.

George, H. and **Jarvis, M. G.** (1979) Land for camping and caravan sites, picnic sites, and footpaths, in Jarvis, M.G. and Mackney (eds) *Soil Survey Applications,* Soil Survey Memoir No. 13, Harpenden, pp. 166–82.

Gerrard, J. (1982) The use of hand–operated soil penetrometers, *Area,* **14**(3), 227–34.

Gibbons, F. R. (1983) Soil mapping in Australia, CSIRO Division of Soils, *Soils an Australian Viewpoint,* CSIRO/Academic Press, Melbourne/London, pp. 267–76.

Gloyne, R. W. (1955) Some effects of shelterbelts and windbreaks, *Meteorological Magazine,* **84,** 272–81.

Godfrey, A. E. (1977) A physiographic approach to land use planning, *Environmental Geology,* **2,** 43–50.

Goodall, B. and **Kirby, A.** (eds) (1979) *Resources and Planning,* Pergamon Press, Oxford.

Goodall, B. and **Whittow, J. B.** (1973) *The Recreational Potential of Forestry Commission Holdings,* a Report to the (UK) Forestry Commission, University of Reading.

Goodall, D. W. (1952) Quantitative aspects of plant distribution, *Biological Review,* **27,** 194–245.

Goudie, A. (1973) *Duricrusts in Tropical and Subtropical Landscapes,* Clarendon Press, Oxford.

Grabau, W. E. and **Rushing, W. N.** (1968) A computer–compatible system for quantitatively describing the physiognomy of vegetation assemblages, in Stewart, G.A. (ed.) *Land Evaluation,* Macmillan of Australia, Melbourne, pp. 263–75.

Grant, K. (ed.) (1968) *Proceedings of Study Tour and Symposium on Terrain Evaluation for Engineering,* Division of Applied Geomechanics, CSIRO, Australia.

Grant, K. (1970) *Terrain Classification for Engineering Purposes in the Maree Area, South Australia,* CSIRO Division of Soil Mechanics Technical Paper No. 4, Melbourne.

Grant, K. (1973–74) *The PUCE Programme for Terrain Evaluation for Engineering Purposes,* Vol. I: *Principles,* and II: *Procedures for Terrain Classification,* CSIRO Division of Applied Geomechanics, Technical Papers Nos 15 (1973) and 19 (1974) respectively, Melbourne.

Grant, K. and **Aitchison, G. D.** (1965) *An Engineering Assessment of the Tipperary Area, Northern Territory, Australia,* Soil Mechanics Section, CSIRO, Australia.

Grant, K., Davis, J. R. and De Visser, C. (1984) *A Geotechnical Landscape Map of Australia* Scale 1:250 000, Divisional Report 84/1, CSIRO Institute of Biological Resources, Division of Water and Land Resources, Canberra.

Great Britain LRDC (Land Resources Development Centre) (1958) *Land Use Map of Gambia,* scale 1:25 000, Directorate of Overseas Surveys 3001, HMSO, London.

Great Britain, MAFF (Ministry of Agriculture, Fisheries and Food) (1964) *Agricultural Land Classification Maps, Overall Agricultural Significance,* 1:250 000, with accompanying explanatory note (1968), HMSO, London.

Great Britain, MAFF (1967–73) *Types of Farm Maps* of the 10 regions of England and Wales at 1:250 000, London.

Great Britain, MAFF (1965–77), Agricultural Land Service, *Maps of Agricultural Land Classification of England and Wales* at 1:63 360 and 1:250 000, and Explanatory Note (1968). The classification is according to the method described in *Farm Classification in England and Wales,* (1965), HMSO, London.

Great Britain, Ordnance Survey (1939) *Types of Farming, England and Wales,* Map at 1:625 000 compiled by MAFF, London.

Great Britain, Soil Survey of Scotland, (1972) *Land Use Capability Maps,* scale 1:63 360, Macaulay Institute, Aberdeen. An example of a soil map with an attached land capability assessment is the *Kirkmaiden, Whithorn, Stranraer, and Wigtown* sheet (1, 2, 3, 4, and 7) dated 1972 by Bown, C.J. and Herslop, R.E.F.

Greenland, D. J. and **Lal, R.** (eds) (1977) *Soil Conservation and Management in the Humid Tropics,* John Wiley & Sons, New York.

Gregory, K. J. and **Walling, D. E.** (eds) (1974) *Fluvial Processes in Instrumented Watersheds,* Institute of British Geographers Special Publication No. 6, London.

Gregory, S. (1978) *Statistical Methods and the Geographer,* 4th edn, Longman, London.

Greig–Smith, P. (1964) *Quantitative Plant Ecology,* Butterworths, London.

Grigg, D. (1967) Regions, models, and classes, in Chorley, R.J. and Haggett, P. (eds) *Models in Geography,* Methuen, London, pp. 461–509.

Grim, R. E. (1968) *Clay Mineralogy,* McGraw–Hill, New York.

Gunn, R. H. and **Nix, H. A.** (1977) *Land Units of the Fitzroy Region, Queensland,* Land Resource Series No. 39, CSIRO, Canberra.

Gvozdetskiy, N. A. (1962) An attempt to classify the landscapes of the USSR, *Soviet Geography,* III(6), 30–9.

Haans, J. C. F. M., Steur, G. G. L. and **Heide, G. E.** (eds) (1984) *Progress in Land Evaluation,* Proceedings of a Seminar on Soil Survey and Land Evaluation, Wageningen, Netherlands, A A Balkema, Rotterdam and Boston.

Haantjens, H. A. (1968) Practical aspects of land system surveys in New Guinea, in Unesco *Aerial Surveys and Integrated Studies,* Proceedings of the Toulouse Conference, pp. 455–60.

Haase, G. (1964) Landschaftsökologische Detailuntersuchung und naturräumliche Gliederung, *Petermanns Geographische Mitteilungen,* **108,** 8–30.

Hack, J. T. (1957) *Studies in Longitudinal Stream Profiles in Virginia and Maryland,* US Geological Survey Professional Paper 294B, pp. 45–97.

Haeckel. E. (1866) *Generale Morphologie der Organismen* vol. 1, Berlin.

Haggett, P. and **Chorley, R. J.** (1969) *Network Analysis in Geography*, Edward Arnold, London.

Haggett, P., Chorley, R. J. and **Stoddart, D. R.** (1965) Scale standards in geographical research. A new measure of areal magnitude, *Nature*, **205**, 844–7.

Hagood, M. J., Damlevsky, M. D. and **Beum, C. O.** (1941) An examination of the use of factor analysis in the problem of sub–regional delineation, *Rural Sociology*, **6**, 216–33.

Hails, J. R. (1977) Applied geomorphology in coastal zone planning and management, in Hails, J.R. (ed.) *Applied Geomorphology*, Elsevier, Amsterdam, pp. 317–98.

Hammond, E. H. (1954) Small-scale continental landform maps, *Annals of the Association of American Geographers*, **44** (1), 33–42

Hammond, E. H. (1964) Analysis of properties in land form geography: application to broad scale land form mapping, *Annals of the Association of American Geographers*, **54**(1), 11–19, and Classes of land-surface form in the forty eight states. USA. Map Supplement 4 (1:5 000 000) (end pocket of same volume).

Hare, F. K. (1959) *A Photo–reconaissance Survey of Labrador–Ungava*, Canadian Department of Mines and Technical Surveys, Geographical Branch Memoir No. 6, Ottawa.

Harrop, J. F. (1974) *Design and Evaluation of Land Development Units for Indonesia*, Land Capability Appraisal, Indonesia, FAO Working Paper No. 9, AGL/INS/72/011, Bogor, Indonesia.

Hearn, G. and **Jones, D. K. C.** (1986) Geomorphology and mountain highway design: some lessons from the Dharan–Dhankuta highway in east Nepal, in Gardiner, V. (ed.) *International Geomorphology*, John Wiley & Sons, Chichester, Part 1, pp. 203–19.

Heath, G. R. (1956) A comparison of two basic theories of land classification and their adaptability to regional photo–interpretation key techniques, *Photogrammetric Engineering*, **22**, 144–68.

Heilman, J. L. and **Moore, D. G.** (1981) HCMM detection of high soil moisture areas, *Remote Sensing of Environment*, **5**, 137–45.

Henderson–Sellers, A. and **Robinson, P. J.** (1986) *Contemporary Climatology*, Longman, London.

Herbert, J. (1982) Lecture at Geography Department, Reading University.

Herbertson, A. J. (1905) The major natural regions, an essay in systematic geography, *Geographical Journal*, **25**, 300–12.

Herz, K. (1973) Beitrag zur Theorie der landschaftsanalytischen Massstabsbereiche, *Petermanns Geographische Mitteilungen*, **117**, 91–6.

Hettner, A. (1934) *Vergleichende Landerkunde*, vol III, Teubner, Leipzig.

Heyligers, P. C. (1968) Quantification of vegetation structure on vertical aerial photographs, in Stewart, G.A. (ed.) *Land Evaluation*, Macmillan of Australia, Melbourne, pp. 251–62.

Hicks, J. P. (1977) *Managing Natural Resources Data: Minnesota Land Management Information System*, Council of State Governments, Lexington, Kentucky.

Higgins, G. M. and **Kassam, A. H.** (1981) FAO's agro–ecological approach to determination of land potential, *Pédologie*, **XXXI**, 147–68.

Higgins, G. M., Kassam, A. H. and **Shah, M.** (1984) Land, food, and population in the developing world, *Nature and Resources*, Unesco, Paris, **XX**(3), 2–10.

Higginson, F. R. (1973) Soil erosion of land systems within the Hunter Valley, *Journal of the Soil Conservation Service of New South Wales*, **29**, 103–10.

Highland Regional Council (Scotland) (1985) *HRC/ITE Land Classification System*, Planning Department Information Paper No. 5, Inverness.

Hill, I. D. (1969) *An Assessment of the Possibilities of Oil Palm Cultivation in Western Division, The Gambia*, Directorate of Overseas Surveys Land Resource Study 6, Tolworth.

Hills, G. A. (1942) An approach to land settlement problems in northern Ontario, *Scientific Agriculture*, **23**, 212–16.

Hills, G. A. (1949) *The Classification of Northern Ontario Lands According to Their Potential for Agricultural Production*, Soil Report No. 1, Ontario Department of Lands and Forests, Research Division, Toronto.

Hills, G. A. (1950) The use of aerial photographs in mapping soil sites, *Forestry Chronicle*, **4**, 37.

Hills, G. A. (1960) Regional site research, *Forestry Chronicle*, **36**, 401–23.

Hills, G. A. and **Portelance, R.** (1960) *The Glackmeyer Report on Multiple Land Use Planning*, Ontario Department of Lands and Forests, Toronto.

Hills, R. C. (undated) *The Determination of the Infiltration Capacity of Field Soils Using the Cylinder Infiltrometer*, British Geomorphological Research Group Technical Bulletin, University of East Anglia, Norwich.

Hodgkins, E. J. (1965) *Southeastern Forest Habitat Regions Based on Physiography*, Auburn University Forestry Department, Series 2, Auburn, Alabama.

Hodgson, J. M. (1976) *Soil Survey Field Handbook*, Technical Memorandum No. 5, Soil Survey of Great Britain, Harpenden, Herts.

Hoefs, J. (1980) *Stable Isotope Geochemistry*, Springer–Verlag, Berlin, Heidelberg, New York.

Hoffman, G. J. and **van Genuchten, M. Th.** (1983) Soil properties and efficient water use: water management for salinity control, in Taylor, H.M. *et al.* (eds) *Limitations to Efficient Water Use in Crop Production*, American Society of Agronomy, Crop Science Society of America, and Soil Science Society of America, Madison, Wisconsin, pp. 73–85.

Holdridge, L. R. (1947) Determination of world plant formations from simple climatic data, *Science*, **105** (2727), 367–8.

Holdridge, L. R. (1966) The life zone system, *Adsonia*, **6**, 199–203.

Holdridge, L. R. and **Toshi, J. A.** (1972) *The World life Zone Classification System and Forestry Research*, Seventh World Forestry Congress, 7CFM/C:V2G, FAO, Rome.

Holmes, R. A. (1970) Field spectroscopy. in Committee on Remote Sensing for Agricultural Purposes *Remote Sensing with Special Reference to Agriculture and Forestry*, National Academy of Sciences, Washington, DC. pp. 298–323.

Horton, R. E. (1945) Erosional development of streams and their drainage basins: hydrophysical approach to quantitative morphology, *Bulletin of the Geological Society of America*, **56**, 275–370.

Hoskins, W. G. (1955) *The Making of the English Landscape*, Penguin, London.

Howard, A. D. and **Spock, L. E.** (1940) Classification of landforms, *Journal of Geomorphology*, **3**, 332–45.

Howard, J. A. (1970a) Multi–band concepts of forested land units, *Symposium on Photo–Interpretation*, International Society of Photogrammetry, Commission 7, Dresden, pp. 281–316.

Howard, J. A. (1970b) Stereoscopic profiling of land units from aerial photographs, *Australian Geographer*, **11**(3), 359–72.

Howard, J. A. (1970c) Multiband concepts of forested land units, in *International Symposium of Photo–Interpretation*, vol. 1, International Archives of Photogrammetry, Dresden, pp. 281–316.

Howard, J. A. and **Mitchell, C. W.** (1980) Phyto–geomorphic classification of the landscape, *Geoforum*, **11**, 85–106.

Howard, J. A. and **Mitchell, C. W.** (1985) *Phytogeomorphology*, John Wiley & Sons, New York.

Howard, J. A. and **van Dijk, A.** (1980) Towards a world index of space imagery, paper presented to *International Society of Photogrammetry*, Commission VII, Rome.

Hudson, N. (1981) *Soil Conservation*, Batsford, London.

Hunting Technical Services Limited (1954–) *Reports on Soil Surveys* of Makhmour, Ishaqi, Nahrwan, Diyala, and Middle Tigris areas in Iraq; Jebel Marra and Roseires in Sudan, Sind in Pakistan; *Range Classification* in Jordan; *Land Classification* in Ghana.

Hunting Technical Services Limited (1956) *Report on the Range Classification of the Hashemite Kingdom of Jordan*, Joint Jordan–US Fund for Special Economic Assistance, Department of Range and Water Resources, Amman.

Hutchings, G. E. (1960) *Landscape Drawing*, Methuen, London.

ICI (Imperial Chemical Industries) (1965) *A Farming Type Map of England and Wales*, based on agricultural census data for 1965, Kynoch Press, Birmingham. Classes defined in *Outlook on Agriculture*, **5**(5), 191–6.

ICSU (International Council of Scientific Unions) (1979) *Fourth Consolidated Guide to International Data Exchange Through the World Data Centers*, Secretariat of the ICSU Panel on World Data Centers, Washington, DC.

Ignatyev, G. M. (1968) Classification of cultural and natural vegetation sites as a basis for land evaluation, in Stewart, G.A. (ed.) *Land Evaluation*, Macmillan of Australia, Melbourne, pp. 104–11.

IRSIA (Institut pour l'Encouragement de la Recherche Scientifique dans l'Industrie et l'Agriculture, Belgium), (various dates) *Carte des Sols de la Belgique*, at 1:20 000, for example Sheet 22W, Bredene, by J. B. Ameryckx under the direction of R. Tavernier and F. R. Moorman.

Irving, E. G. (1962) Coastal cliffs: report of a symposium, *Geographical Journal*, **128**, 303–20.

Isacenko, A. G. (1965) *Probleme der Landschaftsforschung und physisch–geographischen Gliederung*, Moscow.

Isard, W. (1975) *Introduction to Regional Science*, Prentice–Hall, Englewood Cliffs, New Jersey.

Iverson, W. D. (1975) Assessing landscape resources: a proposed model, in Zube, E.H., Brush, R.O., and Fabos, J.G. (eds) *Landscape Assessment: Values, Perceptions, and Resources*, Dowden, Hutchinson & Ross, Stroudsburg, Pennsylvania, pp. 274–88.

Jackson, M. J. (1985) Workshop notes for RSS–CERMA workshop on geographic information systems, *Remote Sensing Society–CERMA International Conference on Advanced Technology for Monitoring and Processing Global Environmental Data*, School of Oriental and African Studies, London.

Jackson, M. J. and **Mason, D. C.** (1986) The development of integrated geo–information systems, *International Journal of Remote Sensing*, **7**(6), 723–40.

Jacobs, P. and **Way, D.** (1969) *Visual Analysis of Landscape Development*, Graduate School of Design, Harvard University, Cambridge, Massachusetts.

Jakle, J. A. (1987) *The Visual Elements of Landscape*, University of Massachusetts Press, Amherst.

Japan (since *c*. 1980) *Coloured Slope Classification Maps at a scale of 1:25 000*, Geological Survey, Tokyo.

Jarvis, M. G., Hazelden, J. and **Mackney, D.** (1979) *Soils of Berkshire*, Soil Survey of England and Wales, Bulletin No. 8, Harpenden.

Jarvis, M. G. and **Mackney, D.** (eds) (1979) *Soil Survey Applications,* Soil Survey of Great Britain (England and Wales), Technical Monograph No. 13, Harpenden.

Jarvis, N. J., Leeds–Harrison, P. B. and **Dosser, J. M.** (1987) The use of tension infiltrometers to assess routes and rates of infiltration in a clay soil, *Journal of Soil Science* **38**, 633–40.

Jenkin, R. N. and **Foale, M. A.** (1968) *An Investigation of the Coconut Growing Potential of Christmas Island* (vols 1 and 2), Directorate of Overseas Surveys, Land Resources Study 4, Tolworth, UK.

Jenny, H. (1941) *Factors of Soil Formation*, McGraw-Hill, New York.

Jenny, H. (1958) Role of the plant factor in pedogenic functions, *Ecology*, **39**, 5–16.

Jensen, M. L. and **Bateman, A. M.** (1979) *Economic Mineral Deposits*, 3rd edn, John Wiley & Sons, New York.

Jesty, C. and **Wainwright, A.** (1978) *A Guide to the View from Scafell Pike*, Cook & Jesty, Dolgelley, printed by Cook, Hammond, and Kell, Mitcham, England.

Jewitt, T. N. (1955) *Gezira Soil*, Sudan Government Ministry of Agriculture Bulletin 12, Khartoum.

Joerg, W. L. G. (1914) Natural regions of northern America, *Annals of the Association of American Geographers*, **4**, 55–83.

Johnson, D. W. (1921) *Battlefields of the World War*, American Geographical Society Series No. 3, New York.

Joly, F. (1957) Les milieux arides. définition, extension, *Notes Marocaines*, **8**, 15–30, Rabat.

Joly, F. (1962) Etudes sur le relief du sud–est marocain', *Travaux de L'Institut Scientifique Chérifien*, Série de Géologie et Géographie Physique, Rabat.

Jones, R. J. A. and **Thomasson, A. J.** (1985) *An Agricultural Data Bank for England and Wales*, Soil Survey of Great Britain (England and Wales), Technical Monograph No. 16, Harpenden.

Journaux, A. (ed.) (1978) *Carte de l'Environnement et de sa Dynamique*, Scale 1:50 000, Sheet Caen, Centre de Géomorphologie, CNRS, Caen, France.

Journaux, A. (ed.) (1987) *Integrated Environmental Cartography: A Tool for Research and Land Use Planning*, Unesco Man and Biosphere Programme Technical Notes 16, prepared in cooperation with the International Geographical Union, Paris.

Jungert, E., Fransson, J., Olssen, L., Toller, E., Borgefors, G., Lindgren, T. and **Roldan–Prado, R.** (1985) *Vega – a Geographical Information System* FOA Report D 30367–E1, National Defence Research Institute (FOA), Linköping, Sweden.

Jurdant, M., Belair, J.L., Gerardin, V. and **Duruc, J. P.** (1977) *L'Inventaire du Capital–nature*, Environnement Canada, Ottawa.

Justice, C. O. and **Townshend, J. R. G.** (1981) Integrating ground data with remote sensing, in Townshend, J.R.G. (ed.) *Terrain Analysis and Remote Sensing*, George Allen & Unwin, London, Ch. 3, pp. 38–58.

Justice, C. O., Townshend, J. R. G., Holben, B. N. and **Tucker, C. J.** (1985) Analysis of the phenology of global vegetation, using meteorological satellite data, *International Journal of Remote Sensing*, **6**(8), 1271–318.

References

Kahle, A. B., Schieldge, J. P., Abrams, M. A., Alley, R. E. and **Levine, C. J.** (1981) *Geological Applications of Thermal Inertia Imaging Using HCMM Data*, Jet Propulsion Laboratory Publication 81–55, NASA, Washington, DC.

Kandel, A. (1986) *Fuzzy Mathematical Techniques with Applications*, Addison–Wesley, California, Wokingham.

Kant, I. (1922) Lectures in Physical Geography, from article Kant in *Encyclopaedia Britannica*.

Kantey and Templer, Engineers (1959) *Geotechnical Map of the Proposed Trunk Route Mariental–Asab, Southwest Africa*, Cape Town.

Kaplan, R. (1975) Some methods and strategies in the prediction of preference, in Zube, E.H., Brush, R.O., and Fabos, J.G. (eds) *Landscape Assessment: Values, Perceptions, and Resources*, Dowden, Hutchinson & Ross, Stroudsburg, Pennsylvania, pp. 4–9.

Kiefer, R. W. (1967) Terrain analysis for metropolitan fringe area planning, *Journal of the Urban Planning and Development Division, Proceedings of the American Society of Civil Engineers*, UP4, **93**, 119–39.

King, L. (1962) *The Morphology of the Earth: A Study and Synthesis of World Scenery*, Oliver & Boyd, Edinburgh.

King, L. J. (1969) *Statistical Analysis in Geography*, Prentice-Hall, New York.

Kirkby, M. J. (ed.) (1978a) *Hillslope Hydrology*, John Wiley & Sons, Chichester.

Kirkby, M. J. (1978b) Implications for sediment transport, in Kirkby, M.J. (ed.) *Hillslope Hydrology*, John Wiley & Sons, Chichester, Ch. 9, pp. 325–64.

Kirkby, M. J. (1987) Modelling some influences of soil erosion, landslides and valley gradient on drainage density and hollow development, *Catena*, Supplement 10, Braunschweig, 1–14.

Kirkham, D. (1947) Studies of hillslope seepage in the Iowan drift area, *Proceedings of the Soil Science Society of America*, **12**, 73–80.

Klingebiel, A. A. and **Montgomery, P. H.** (1961) *Land Capability Classification*, USDA Soil Conservation Service, Agriculture Handbook No. 210, Washington, DC.

Klink, H. J. (1966) Die naturräumliche Gliederung als ein Forschungsgegenstand der Landeskunde, *Berichte zum Deutschen Landeskunde*, **36**, 223–46.

Kloosterman, B. and **Dumansky, J.** (1978) Data management capabilities of the Canada soil information system, in Sadovsky, A.N. and Bie, S,W. (eds.) *Development of Soil Information Systems*, Proceedings of the Second Meeting of the ISSS Working Group on Soil Information Systems, Varna/Sofia, Bulgaria; Wageningen, Netherlands.

Knapp, B. J. (1979) *Elements of Geographical Hydrology*, George Allen & Unwin, London.

Kondracki, J. (1964) Problems of physical geography and physicogeographical regionalization of Poland, *Geographica Polonica*, **1**.

Koppen, W. (1931) *Grundriss der Klimakunde*, Walter de Gruyter, Berlin and Leipzig.

Krajina, V. J. (1965) Bioclimatic zones and classification of British Columbia, in *Ecology of North America*, University of British Columbia Press, Vancouver, pp. 1–17.

Krinsley, D. (1970) *A Geomorphological and Paleoclimatological Study of the Playas of Iran*, US Geological Survey, 2 vols, Washington, DC.

Krumbein, W. C. and **Graybill, P. A.** (1965) *An Introduction to Statistical Methods in Geology*, McGraw-Hill, New York.

Kuchler, A. W. (1949) A physiognomic classification of vegetation, *Annals of the Association of American Geographers*, **39**, 201–10.

406

Kuchler, A. W. (1967) *Vegetation Mapping*, Ronald Press, New York.

LaBastille, A. (1981) Acid rain: how great a menace? *National Geographic Magazine*, **160**(5), 652–80.

Lacate, D. S. (1961) A review of land type classification and mapping, *Land Economics*, **37**, 271–8.

Lacate, D. S. (1969) *Guidelines for Biophysical Land Classification*, Canadian Forestry Service Publication No. 1264, Ottawa.

Landscape Research Group Symposium (1967) *Methods of Landscape Analysis*, London. See Childs; Clouston, Mott, and Tandy, (all 1967).

Langbein, W. B. and **Hoyt, W. G.** (1959) *Water Facts for the Nation's Future*, Ronald Press, New York.

Langdale-Brown, I. and **Spooner, R.** (1963) *Land Use Prospects in Northern Bechuanaland: A Reconnaissance Investigation*, Land Resources Development Centre (then Land Resources Division), Tolworth.

Langston, R. P. (1970) Maritime pollution, in AGARD *Information Analysis Centres*, North Atlantic Treaty Organization, Conference Proceedings No. 78, London.

Lapoukhine, N., Pront, N. A. and **Hirvonen, H.** (1978) *Ecological Land Classification of Labrador*, Ecological Land Classification Series No. 4, Environment Canada, Ottawa.

LARS (Laboratory for the Applications of Remote Sensing) (1968) *Remote Sensing in Agriculture*, 3rd Annual Report of LARS, Purdue University, West Lafayette, Indiana.

Laurie, I. C. (1975) Aesthetic factors in visual evaluation, in Zube, E. H., Brush, R. O., and Fabos, J. G. *Landscape Assessment: Values, Perceptions, and Resources*, Dowden, Hutchinson & Ross, Stroudsburg, Pennsylvania, pp. 102–17.

Laut, P., Heyligers, P. C., Heig, G., Loeffler, E., Margules, C. and **Scott, R. M.** (1977) *Environments of South Australia*, Handbook: Division of Land Resources, CSIRO, Canberra.

Laut, P. and **Nanninga, P. M.** (1985) *Landscape Data for Cattle Disease Eradication in Northern Australia*, CSIRO Division of Water and Land Resources Technical Paper No. 47, Melbourne.

Lawrance, C. J. (1966) *Block Diagrams of Landscapes for Terrain Classification*, MEXE Technical Note No. 6/66, Christchurch, Hampshire, England.

Lee, K. (1975) Ground investigations in support of remote sensing, in Reeves, R. G., (ed.) *Manual of Remote Sensing*, American Society of Photogrammetry, Falls Church, Virginia, pp. 805–56.

Leopold, L. B. (1969) Landscape aesthetics, *Natural History*, Oct., pp. 36–45. Also in Coates, D. R. (ed.) *Environmental Geomorphology and Landscape Conservation*, vol. III: *Non-Urban*, Dowden, Hutchinson & Ross, Stroudsburg, Pennsylvania, Ch. 26, 454–67.

Leopold, L. E., Wolman, M. G. and **Miller, J. P.** (1964) *Fluvial Processes in Geomorphology*, Freeman, San Francisco.

Leser, H. (1978) *Landschaftsökologie*, Uni-Taschenbücher 527, Verlag Eugen Ulmer, Stuttgart.

Lewis, D. (1974) Wind, wave, star, and bird, *National Geographic Magazine*, **146**(6), 747–55, and Map, Discoverers of the Pacific, Washington, DC.

Lewis, P. H. (1964) Quality corridors for Wisconsin, *Landscape Architecture*, **54**(2), 100–7.

Lillesand, T. M. and **Kiefer, R. W.** (1979) *Remote Sensing and Image Interpretation*, John Wiley & Sons, New York.

Lin Chao (1984) Development of regional geography in China, in Wu Chuanjun, Wang Nailiang, Lin Chao, and Zhao Songqaio (eds) *Geography in China*, Science Press, Beijing, pp. 147–63.

Linton, D. L. (1951) The delimitation of morphological regions, in Stamp, L. D. and Wooldridge, S. W. (eds) *London Essays in Geography*, Longman, London, pp. 199–218.

Linton, D. L. (1968) The assessment of scenery as a natural resource, *Scottish Geographical Magazine*, **84**, 219–38.

Lloyd, R. (1989) Cognitive maps: encoding and decoding information, *Annals of the Association of American Geographers*, **79**(1), 101–24.

Lobeck, A. K. (1923) *Physiographic Diagram of Europe*, Geography Press, Columbia University, New York.

Lobeck, A. K. (1939) *Geomorphology: An Introduction to the Study of Landscapes*, McGraw-Hill, New York.

Lobeck, A. K. (1958) *Block Diagrams*, Emerson-Trussel Book Company, New York.

Long, G. (1974) *Diagnostic Phyto-Ecologique et Aménagement du Territoire*, Collection d'Ecologie 4, (vol. 1), *Principes Généraux et Méthodes, Masson et Cie (Editeurs), Paris*.

Lotspeich, F. B. and **Platts, W. S.** (1982) An integrated land–aquatic classification system, *North American Journal of Fisheries Management*, **2**, 138–49.

Loucks, O. L. (1962) A forest classification system for the Maritime Provinces, Canadian Department of Forests reprint from *Proceedings of Nova Scotian Institute of Science*, **25**, Part 2.

Loughran, R. J. (1989) The measurement of soil erosion, *Progress in Physical Geography*, **13**(2), 216–33.

Low, F. K. (1967) Estimating potential erosion in developing countries, *Journal of Soil and Water Conservation*, **22**, 147–8.

Lowenthal, D. (1982) Revisiting valued landscapes, in Gold, J.R. and Burgess, Jacquelin (eds) *Valued Environments*, George Allen & Unwin, London, Boston, and Sydney, Ch. 5, pp. 74–99.

Lowry, W. P. and **Lowry, P. P.** (1989) *Fundamentals of Biometeorology*, volume 1 – The physical environment, Peavine Publications, Oregon, pp. 219.

Lynch, K. and **Hack, G.** (1984) *Site Planning*, 3rd edition MIT Press, Cambridge, Massachusetts.

Mabbutt, J. A. (1968) Review of concepts of land classification, in Stewart, G. A. (ed.) *Land Evaluation*, Macmillan of Australia, Melbourne, pp. 11–28.

Mabbutt, J. A. and **Stewart, G. A.** (1965) The application of geomorphology in resources surveys in Australia and New Guinea, *Revue de Géomorphologie Dynamique*, July–Sept., Nos 7–9, 97–109.

Macdonald Dettwiler (1987) *Frontiers in Digital Imaging*, and private communication, Richmond, BC, Canada.

McHarg, I. L. (1969) *Design with Nature*, Doubleday for the American Museum of Natural History, Garden City, New York.

Mackin, J. H. (1948) Concept of the graded river, *Bulletin of the Geological Society of America*, **59**, 463–512.

McNeil, G. (1967) *Terrain Evaluation: Data Storage*, MEXE Technical Note 5/67, Christchurch.

Maguire, J. (1989) *Computers in Geography*, Longman, London.

Mahler, P. J. (ed.) (1970) *Manual of Land Classification for Irrigation*, second approximation 1970, Ministry of Agriculture, Soil Institute of Iran, Tehran.

Mainguet, Monique M. (1972) *Le Modelé des Grès*, Institut Géographique National, Paris.

Makin, M. J., Kingham, T. J., Waddams, A. E., Birchall, C. J. and **Eavis, B. W.** (1976) *Prospects for Irrigation Development around Lake Zwai, Ethiopia*, Land Resource Study No. 26, LRDC, Tolworth.

Maletic, J. T. and **Hutchings, T. B.** (1967) Selection and classification of irrigable land, in Hagan, R.M. (ed.) *Irrigation of Agricultural Lands*, Agronomy 11, American Society of Agronomy, Madison, Wisconsin, pp. 125–73.

Mandelbrot, B. B. (1982) *The Fractal Geometry of Nature*, W. H. Freeman & Co., San Francisco.

Manley, G. (1962) *Climate and the British Scene*, Collins Fontana Library, London.

Marchesini, E. and **Pistolesi, A.** (1964) Landform maps of intermediate scale, *XX International Geographical Congress*, Section IX, London.

Mason, D. and **Cross, A.** (1985) *A Proposal for a UK Research Strategy for Integrated GIS*, Preliminary Report of the National Remote Sensing Centre Information Handling Working Group, Sub-Group on Geographic Information Systems, NERC Unit for Thematic Information Systems (NUTIS), Department of Geography, University of Reading.

Mather, P. M. (1976) *Computational Methods of Multivariate Analysis in Physical Geography*, John Wiley & Sons, London.

Mather, P. M. and **Doornkamp, J. C.** (1970) Multivariate analysis in geography with particular reference to drainage basin morphometry, *Transactions of the Institute of British Geographers*, **51**, 163–87.

Meier, R. L. (1965) *Development Planning*, McGraw-Hill, New York.

Meigs, P. (1957) *World Distribution of Arid and Semi-Arid Homoclimates*, with maps UN392 and UN393, Unesco, Paris.

Melton, M. A. (1958) *List of Sample Parameters of Quantitative Properties of Landforms: Their Use in Determining the Size of Geomorphic Experiments* Technical Report No. 16, Office of Naval Research, Dept of Geology, Columbia University.

Merriam, C. H. (1898) Life zones and crop zones of the United States, *Bulletin of the Division of Biological Survey*, USDA, **10**, 1–79.

MEXE (Military Engineering Experimental Establishment) (1966) *Quantitative Assessment of Terrain Conditions*, Technical Note No. 9/66, Christchurch.

MEXE (1968) *User Handbook for the Soil Assessment Cone Penetrometer*, Army Code No. 60285 (provisional Handbook (1966) was Army Code No. 14554), Christchurch.

Meyerhoff, H. A. (1940) Migration of erosional surfaces, *Annals of the Association of American Geographers*, **30**, 247–54.

Michigan University (1979) *Comprehensive Resource Inventory and Evaluation System*, CRIES Special Reports 1–5, Ann Arbor.

Mill, J. S. (1891) *A System of Logic*, 8th edn, Harper, New York, as quoted in Soil Survey Staff, USDA (1960), *Soil Classification. A Comprehensive System. Seventh Approximation*, Washington, DC.

Millard, R. S. (1967) Presentation in *Terrain Evaluation Symposium*, MEXE Report No. 1053, Christchurch, Hants, p. 9.

Miller, A. A. (1946) *Climatology*, Methuen, London.

Miller, J. P. (1961) *Solutes in Small Streams Draining Single Rock Types, Sangre de Cristo Range, New Mexico*, United States Geological Survey Water Supply Paper 1535–F, Washington, DC.

Miller, L. D., Pearson, R. L. and **Tucker, C. J.** (1976) A mobile field spectrometer laboratory, *Photogrammetric Engineering and Remote Sensing*, **42**, 569–72.

Miller, R. L. and **Kahn, R. S.** (1962) *Statistical Analysis in the Geological Sciences*, John Wiley & Sons, New York.

Miller, T. G. (1967) Recent studies in military geography (review article), *Geographical Journal*, **133**, 354–6.

Milne, G. (1935) Some suggested units of classification and mapping, particularly for East African soils, *Soil Research*, **4**(3), 183–98.

Milton, E. (1979) *Reading Band-Pass Radiometer User's Manual*, Remote Sensing Report, Reading University, UK.

Mishustin, Ye. N. (1983) V. V. Dokuchaev's natural zones and their reflection in the cenoses of microorganisms, *Soviet Soil Science*, **15**(3), 49–67.

Mitchell, C. W. (1971) An appraisal of a hierarchy of desert land units, *Geoforum*, **7**, 69–79.

Mitchell, C. W. (1973) *Terrain Evaluation*, 1st edn, Longman, London.

Mitchell, C. W. (1981) Soil degradation mapping from Landsat imagery in North Africa and the Middle East, in Allan, J.A. and Bradshaw, M. (eds.) *Geological and Terrain Analysis Studies by Remote Sensing*, Remote Sensing Society, Nottingham, pp. 49–69.

Mitchell, C. W. (1985) Landsat-based land system survey in northern Iraq, *Proceedings of the First National Symposium on Remote Sensing*, Baghdad, 26–28.10.85, vol.2, Space and Astronomy Research Centre, Baghdad, pp. 1–11 & 8 pages of illustrations.

Mitchell, C. W. (1987) Terrain evaluation, in Bell, F. G. (ed.) *Ground Engineers Reference Book*, Butterworths, London, pp. 23/1–23/8.

Mitchell, C. W. (1988) An international approach to GIS based on remote sensing and terrain classification, in *1988 International Geoscience and Remote Sensing Symposium* (IGARSS'88), European Space Agency, Paris, vol. 1, pp. 117–18.

Mitchell, C. W. and **Gavish, D.** (1980) Land on which battles are lost and won, *Geographical Magazine*, **LII**(12), 838–40.

Mitchell, C. W. and **Howard, J. A.** (1978) *Land System Classification. A Case History: Jordan*, FAO AGLT Bulletin 2/78, Rome.

Mitchell, C. W. and **King, R. B.** (1985) *Preliminary Assessment of Land Use Potential: Satellite Imagery Studies Related to the Preparation of a Comprehensive Economic Map of Namibia*, Report to accompany a 1:4 000 000 scale map of land use potential in Namibia, UNO/NAM/001/UNN (NAM/83/003), FAO, Rome.

Mitchell, C. W. and **Perrin, R. M. S.** (1966) The subdivision of hot deserts of the world into physiographic units, *Actes du IIe Symposium International de Photo–Interpretation*, Sorbonne, Paris, vol. IV, 1–89 to 106.

Mitchell, C. W., Webster, R., Beckett, P. H. T. and **Clifford, Barbara** (1979) An analysis of terrain classification for long-range prediction of conditions in deserts, *Geographical Journal*, **145**(1), 72–85.

Moore, E. (1989) Water management in early Cambodia: evidence from aerial photography, *Geographical Journal*, **155**, 204–14.

Morgan, R. P. C. (1979) *Soil Erosion*, Topics in Applied Geography, Longman, London.

Morgan, R. P. C. (1980) Soil erosion and conservation in Britain, *Progress in Physical Geography*, Edward Arnold, **4**(1), 24–47.

Morgan, R. P. C. (ed.) (1981) *Soil Conservation: Problems and Prospects*, John Wiley & Sons, Chichester, New York, Brisbane, Toronto.

Morgan, R. P. C. (1986) *Soil Erosion and Conservation*, Longman, London.

Moroney, M. J. (1965) *Facts from Figures*, 3rd edn, Penguin Books (Pelican), Harmondsworth.

Morozov, G. F. (1931) *Forest Science*, 6th edn, Sel'khozgiz, Moscow, quoted by Sukachev and Dylis, 1966.

Mott, P. G. (1967) Air photography as an aid to experimental planning, in *Landscape Research Group*, London, pp. 31–36.

Motts, W. S. (ed.) (1970) *Geology and Hydrology of Related Playas in Western United States*, University of Massachusetts, Amherst, Final Scientific Report for the Air Force Cambridge Research Laboratories.

Mountain, M. J. (1964) *Soils Engineering Map of an Area in the Immediate Vicinity of Etosha Pan*, Ovamboland, South West Africa, Kantey and Templer, Cape Town.

Mueller-Dumbois, D. and **Ellenberg, H.** (1974) *Aims and Methods of Vegetation Ecology*, John Wiley & Sons, New York.

Müller-Miny, H. (1958) Das Mittelrheingebiet und seine naturräumliche Gliederung, *Berichte z. Deutsch. Landeskunde*, **21**, 193–233.

Murdoch, G. (1972) *Views on Land Capability with Bibliography 1923–1970*, Miscellaneous Paper No. 132, Land Resources Division, Overseas Development Administration, Foreign and Commonwealth Office, London.

Murray, A. C. (1967) Power station siting: visual analysis, in Landscape Research Group, *Methods of Landscape Analysis*, London, pp. 20–3.

Murray, B. H. and **Niemann, B. J.** (1975) A landscape assessment optimization procedure for electric energy corridor selection, in Zube, E. H., Brush, R. O., and Fabos, J. G. (eds) *Landscape Assessment: Values, Perceptions, and Resources*, Dowden, Hutchinson & Ross, Stroudsburg, Pennsylvania, pp. 220–53.

Murphy, R. E. (1968) Landforms of the world, *Annals of the Association of American Geographers*, **58**(1), Map Supplement 9.

Nakano, T. (1962) Landform type analysis on aerial photographs, its principle and techniques, *Archives Internationales de Photogrammetrie*, **14**, Transactions of the Symposium on Photo-Interpretation, Delft, pp. 149–52.

Narasimham, T. N. (ed.) (1979) *Recent Trends in Hydrogeology'*, Geological Society of America Special Publication 189, Boulder, Colorado.

NASA (1982) *Heat Capacity Mapping Mission Data Users Bulletin*, **9**. NASA Goddard Space Flight Center, Greenbelt, Maryland.

Naturräumliche Gliederung Deutschlands mit Hoehenschichten (abbreviation NRGL) (1954) Westliches Blatt 1:1 Mio. *Bundesanstalt für Landeskunde und Zentralausschuss für Deutsche Landeskünde*, Bad Godesberg.

Neal, J. T. (ed.) (1965) *Geology, Mineralogy and Hydrology of US Playas*, US Army Air Force Cambridge Research Laboratories (AFCRL), Environmental Research Paper 96, Cambridge, Massachusetts.

Neal, J. T., Langer, A. M. and **Kerr, P. F.** (1968) Giant desiccation polygons of Great Basin playas, *Bulletin of the Geological Society of America*, **79**, 69–90.

Neef, E. (1963) Topologische und chorologische Arbeitsweisen in der Landschaftsforschung, *Petermanns Geographische Mitteilungen*, **107**, 249–59.

Newman, W. M. and **Sproul, R. F.** (1979) *Principles of Interactive Computer Graphics*, McGraw-Hill, Japan.

Ni Shao Xiang (1985) A preliminary review of research on the physiographic regionalization of China, *Area*, **17**(1), 19–24.

Nicod, J. (1982) *Phénomènes Karstiques*, vol. 3, Editions du Centre National de Recherche Scientifique, Paris.

Nossin, J. J. (ed.) (1977) *Surveys for Development*, Proceedings of ITC Symposium, Elsevier, Amsterdam.

Oke, T. R. (1978) *Boundary Layer Climates*, Methuen, London.

Ollier, C. D. (1975) *Weathering*, Oliver & Boyd, Edinburgh.

Olson, C. E. (1970) Multispectral remote sensing, *Symposium on Photo-Interpretation*, International Society of Photogrammetry, Commission 7, Dresden, vol. 2, pp. 678–88.

Olson, C. E. (1971) Collection and processing of multispectral imagery, International Union of Forest Research Organizations. Sect. 25. Joint Report by Working Group. Application of remote sensing in forestry.

Olson, G. W. (1974) Interpretive land classification in English-speaking countries, in FAO *Approaches to Land Classification*, FAO Soils Bulletin 22, Rome, pp. 1–25.

Oswald, E. T. and **Senyk, J. P.** (1977) *Ecoregions of Yukon Territory*, Canadian Forestry Service Publication No. BC–X–164, Environment Canada, Victoria, British Columbia.

Otremba, E. (1948) Die Grundsätze der naturräumlichen Gliederung Deutschlands, *Erdkunde*, 2, 156–67.

Oxley, N. C. (1974) Suspended sediment delivery rates and the solute concentration of stream discharge in two Welsh catchments, in Gregory, K. J. and Walling, D. E. (eds) *Fluvial Processes in Instrumented Watersheds*. Institute of British Geographers Special Publication No. 6, London, pp. 141–54.

Paffen, K. H. (1948) Ökologische Landschaftsgliederung, *Erdkunde*, 2, 167–73.

Paffen, K. H. (1953) Die natürliche Landschaft und ihre räumliche Gliederung. Eine methodische Untersuchung am Beispiel der Mittel-und Nieder–rheinlande, *Forsch. z. Deutsch. Landeskunde*, 68.

Pallister, J. W. (1956) Slope development in Buganda, *Geographical Journal*, 122(1), 80–7.

Palmer, E. and **Newton, C. W.** (eds) (1969) *Atmospheric Circulation Systems* (International Geophysics Series, Volume 13), Academic Press Inc., New York.

Palmer, R. C. and **Jarvis, M. G.** (1979) Land for winter playing fields, golf courses, fairways, and parks, in Jarvis, M.G. and Mackney, D. (eds) *Soil Survey Applications*, Soil Survey Technical Monograph No. 13, Harpenden, pp. 152–65.

Parry, J. T. and **Beswick, J. A.** (1973) The application of two morphometric terrain-classification systems using air-photo interpretation methods, *Photogrammetria*, 29, 153–86.

Parry, J. T., Heginbottom, J. A. and **Cowan, W. R.** (1968) Terrain analysis in mobility studies for military vehicles, in Stewart, G.A. (ed.) *Land Evaluation*, Macmillan of Australia, Melbourne, pp. 160–70.

Parsons, A. J. (1978) A technique for the classification of hill-slope forms, *Transactions of the Institute of British Geographers*, 3(4), 432–43.

Passarge, S. (1919) *Die Grundlagen der Lanschaftskunde*, L. Friedrichsen, Hamburg.

Passarge, S. (1926) Geomorphologie der Klimazonen oder Geomorphologie der Landschaftsgürtel, *Petermanns Mitteilungen*, 72, 173–5.

Pearson, A. R. (1979) An integrated terrain analysis system, in *Proceedings of the 13th International Symposium on Remote Sensing of Environment*, vol. 1, ERIM, Ann Arbor, Michigan, pp. 433–8.

Pecsi, M. and **Somogyi, S.** (1967) Physisch-geographische Regionen Ungarns, *Foldrajzi Kozlemenyak*, 15, 285–304.

Peitgen, H-O. and **Saupe, D.** (eds) (1988) *The Science of Fractal Images*, Springer-Verlag, New York.

Peltier, L. C. (1950) The geographic cycle in periglacial regions as it is related to climatic geomorphology, *Annals of the Association of American Geographers*, 40, 214–36.

Penck, W. (1927) *Die Morphologische Analyse*, Stuttgart.

Perrin, R. M. S. (1964) The use of drainage water analyses in soil studies, *Experimental Pedology*, Proceedings of 11th Easter School in Agricultural Science, University of Nottingham, 73–96.

Perrin, R. M. S. and **Mitchell, C. W.** (1970) *An Appraisal of Physiographic Units for Predicting Site Conditions in Arid Areas*, MEXE Report 1111, 2 vols, Christchurch.

Pettry, D. E. and **Coleman, C. S.** (1973) Two decades of urban soil interpretations in Fairfax County, Virginia, *Geoderma* **10**, 27–34.

Peuquet, D. J. (1984) Data structures for a knowledge-based geographic information system. *Proceedings of the International Symposium on Spatial Data Handling*, Zurich, vol.2, pp. 372–91.

Phillips, E. (1965) *Field Ecology, a Laboratory Block*, American Institute of Biological Sciences, Heath, Boston.

Photographic Survey Corporation Ltd (1956) *Landforms and Soils, West Pakistan*, scale 1:253 440, produced in cooperation with Central Soil Conservation Organization, Ministry of Food and Agriculture, Pakistan, Toronto, Canada.

Photo-Interprétation (1961–) Editions Technip, 7 rue Nelaton, Paris 15e.

Pielou, E. C. (1977) *Mathematical Ecology*, Wiley–Interscience, New York and London.

Pike, R. J. and **Rozema, W. J.** (1975) Spectral analysis of landforms, *Annals of the Association of American Geographers*, **65**(4), 499–516.

Plattner, R. M. (1975) The regional landscape concept for the Basel region, in Zube, E.H., Brush, R.O., and Fabos, J.G. (eds) *Landscape Assessment: Values, Perceptions, and Resources*, Dowden, Hutchinson & Ross, Stroudsburg, Pennsylvania, pp. 188–202.

Polakowski, K. J. (1975) Landscape assessment in the Upper Great Lakes Basin Resources: a macro-geomorphic and micro-composition analysis, in Zube, E.H., Brush, R.O., and Fabos, J.G. (eds) *Landscape Assessment: Values, Perceptions, and Resources*, Dowden, Hutchinson & Ross, Stroudsburg, Pennsylvania, pp. 203–19.

Poore, M. E. D. (1956) The use of phytosociological methods in ecological investigations, IV. General discussion of phytosociological problems, *Journal of Ecology*, **44**, 28–50.

Prince, D. R. (1980) Countryside interpretation in the North York Moors: a socio-psychological study, Ph.D. Thesis, University of Hull.

Prokaiev, V. I. (1962) The facies as a basic and smallest unit in landscape studies, *Soviet Geography: Review and Translation*, **3**(6), 21–9.

Putnam, W. C., Axelrod, D. I., Bailey, H. P. and **McGill, J. T.** (1960) *Natural Coastal Environments of the World*, University of California, Los Angeles.

Quinones-Garza, E. (1983) System of classification and physiographic survey, in Campos-Lopez, E. and Anderson, R.J. (eds) *Natural Resources and Development in Arid Regions*, Westview Press, Boulder, Colorado, vol. 7, pp. 85–98.

Ragg, M. (1960) *The Soils of the Country Round Kelso and Lauder* (Sheets 25 and 26), Memoirs of the Soil Survey of Great Britain (Scotland), HMSO, Edinburgh.

Raisz, E. (1938) Developments in the physiographic method of representing the landscape on maps, *Proceedings of the 15th International Geographical Congress*, Amsterdam, 2, Section 1, pp. 140–9.

Raisz, E. (1946) Landform, landscape, land use and land type maps, *Annals of the Association of American Geographers*, **26**(1), 102–3.

Raisz, E. (1962) *Principles of Cartography*, McGraw-Hill, New York, pp. 80–1.

Ranzinger, H. and **Ranzinger, M.** (1984) A geo-information expert system for synergetic use of map and image data, *Proceedings of the EARSeL/ESA Symposium on Integrative Approaches in Remote Sensing*, Guildford (ESA SP–214), pp. 263– 8.

Raunkiaer, C. (1934) *The Life Forms of Plants and Statistical Plant Geography*, Oxford University Press, New York.

Renwick, C. C. (1968) Land assessment for regional planning: the Hunter region of NSW as a case study in land evaluation, in Stewart, G.A. (ed.) *Land Evaluation*, Macmillan of Australia, Melbourne, pp. 171–9.

Reybold, W. U. and **Petersen, G. W.** (1987) *Soil Survey Techniques*, Soil Science Society of America Special Publication No. 20, Madison, Wisconsin.

Rhind, D. (1977) Computer-aided cartography, *Transactions of the Institute of British Geographers*, **2**(1), 71–97.

Rhind, D. (1985) Maps on the small screen, *Geographical Magazine*, **XVII**(1), 6.

Richards, L. A. (ed.) (1954) *Diagnosis and Improvement of Saline and Alkali Soils*, USDA Handbook No. 60, Washington, DC.

Richards, T. S. (1989) Evidence of ancient rainwater concentrating structures in northern Egypt as seen on Landsat imagery, *International Journal of Remote Sensing* **10**(6), 1135–40.

Richley, L. R. (1979) *Land Systems of Tasmania*, Department of Agriculture, Tasmania.

Richter, H. (1967) Naturräumliche Ordnung, in Neef, E. (ed.) *Probleme der Landschaftsokol. Erkundung und naturräumlichen Gliederung*, Wissenschaftlichen Abhandlungen der Geographische Gesellschaft der DDR, vol. 5, pp. 129–60.

Richter, H. (1968) Stand und Tendenzen der naturräumlichen Gliederung der DDR, *Geogr. Arb.* **69**, Warzawa, 63–79.

Riotte, R. J., Fabos, J. G. and **Zube, E. H.** (1975) Model for the evaluation of the visual–cultural resources of the southeastern New England region, in Zube, E. H., Brush, R. O., and Fabos, J. G. (eds) *Landscape Assessment: Values, Perceptions, and Resources*, Dowden, Hutchinson & Ross, Stroudsburg, Pennsylvania, pp. 254–73.

Riquier, J. (1978) A methodology for assessing soil degradation, in FAO *Report of the FAO/UNEP Expert Consultation on the Methodology for Assessing Soil Degradation*, Rome, pp. 25–59.

Robbins, R. G. (Compiler) (1976) *Lands of the Ramu–Madang Area, Papua New Guinea*, Land Research Series No. 37, CSIRO, Melbourne.

Roberts, C. R. and **Mitchell, C. W.** (1987) Spring mounds in southern Tunisia, in Frostick, L. E. and Reid, I. (eds) *Desert Sediments: Ancient and Modern*, Geological Society, Blackwell Scientific Publications, Oxford, pp. 321–34.

Robertson, V. C., Jewitt, T. N., Forbes, A. P. S. and **Law, R.** (1968) The assessment of land quality for primary production, in Stewart, G.A. (ed.) *Land Evaluation*, Macmillan of Australia, Melbourne, pp. 88–103.

Robinson, A. H., Sale, R. D., Morrison, J. L. and **Muehrcke, P. C.** (1984) *Elements of Cartography*, 5th edn, John Wiley & Sons, New York.

Robinson, A. R. (1983) Sediment yield as a function of upstream erosion, in Soil Science Society of America *Universal Soil Loss Equation: Past, Present, and Future*, Special Publication No. 8, Madison, Wisconsin, pp. 7–16.

Robinson, D. G., Wager, J. F., Laurie, I. C. and **Traill, A. L.** (eds) (1976) *Landscape Evaluation*, the Landscape Evaluation Research Project 1970–1975, Department of Town and Country Planning, Manchester University.

Robinson, G. W. (1949) *Soils, their Origin, Constitution, and Classification*, Thomas Murby, London.

Rodda, I. C. (1969) The flood hydrograph, in Chorley, R.J. (ed.) *Climate, Earth, and Man*, Methuen, London, Ch. 9.1, pp. 405–18.

Roose, E. J. (1975) Erosion et Ruisellement en Afrique de l'Ouest: Vingt Années de Mesures en Petites Parcelles Experimentales, ORSTOM, Abidjan, Ivory Coast (cyclostyled).

Rosayro, R. A. de (1959) The application of aerial photography to stock mapping and inventories in rainforest in Ceylon, *Empire Forestry Review*, **38**, pp. 141–74.

Rowe, J. S. (1972) *Forest Regions of Canada*, Canadian Forestry Service Publication 1300, Ottawa.

Rozov, N. N. and **Ivanova, E. N.** (1967) Classification of the soils of the USSR, *Soviet Soil Science*, **2**, 147–56.

Rubec, C. D. A. (1979) *Application of Ecological (Biophysical) Land Classification*, Ecological Land Classification Series No. 7, Environment Canada, Ottawa.

Rubenstein, H. M. (1969) *A Guide to Site and Environmental Planning*, John Wiley & Sons, New York.

Ruhe, R. V. (1960) Elements of the soil landscape, *Proceedings of the 7th International Congress of Soil Science*, Madison, Wisconsin, vol. 4, pp. 165–70.

Ruhe, R. V. (1975) *Geomorphology*, Houghton Mifflin, Boston.

Russell, E. W. (1978) *Soil Conditions and Plant Growth*, 10th edn, Longman, London. see Wild (1988).

Sabins, F. F. (1978) *Remote Sensing: Principles and Interpretation*, W.H. Freeman, San Francisco.

Sadovsky, A. N. and **Bie, S. W.** (eds)(1978) *Developments in Soil Information Systems*, Centre for Agricultural Publishing and Documentation, Wageningen.

Saint Onge, D. A. (1968) Geomorphic maps, in Fairbridge, R. W. (ed.) *Encyclopedia of Geomorphology*, Reinhold, New York, pp. 388–402.

Sampford, M. R. (1962) *An Introduction to Sampling Theory*, Oliver & Boyd, Edinburgh.

Savigear, R. A. G. (1952) Some observations on slope development in South Wales, *Transactions and Papers of the Institute of British Geographers*, **18**, 31–51.

Savigear, R. A. G. (1956) Techniques and terminology in the investigation of slope forms, *Premier Rapport de la Commission pour L'Etude des Versants'*, Union Géographique Internationale, pp. 66–75.

Savigear, R. A. G. (1960) Slopes and hills in West Africa, *Zeitschrift für Geomorphologie*, NF Supplement **1**, 156–71.

Savigear, R. A. G. (1962) Some observations on slope development in North Devon and North Cornwall, *Transactions and Papers of the Institute of British Géographers*, **31**, 23–42.

Savigear, R. A. G. (1965) A technique of morphological mapping, *Annals of the Association of American Geographers*, **55**(3), 514–38.

Schimper, A. F. W. (1903) *Plant Geography upon a Physiological Basis*, Oxford.

Schlesinger, J. Ripple, B., and **Loveland, T. R.** (1979) Land capability studies of the South Dakota automated geographic information system (AGIS), in *Computer Mapping in Natural Resources and the Environment*, Mapping Collection, Laboratory for Computer Graphics and Spatial Analysis, Harvard University.

Schmithusen, J. (1948) Fliesengefüge der Landschaft und Ökotopvorschlage zur begrifflichen Ordnung und zur Nomenklatur in der Landschaftsforschung, *Berichte zur Deutsch. Landeskunde*, **5**, 74–83.

Schmithusen, J. (1963) Der wissenschaftliche Landschaftsbegriff. *Mitteilungen d. Floristisch.- soziologischen Arbeitsgemeinsch.* NF **10**, 9–19.

Schmithusen, J. (1976) Allgemeine Synergetik, *Lehrbuch der Allgemeine Geographie*, **12**, Berlin.

Schneider, S. J. (1966) The contribution of geographical air-photo interpretation to problems of land division according to natural units, *Actes du IIe Symposium International de Photo-Interprétation*, Paris, VI 23–8.

Schneider, S. J. (1970) Die Verwendung der Luftbilder bei Problemen der Raumgliederung. *Bildmessung und Luftbildwesen*, Zeitschrift für Photogrammetrie, Photointerprétation und Luftbildwesen, **5**, 38 Jahrgang, Herbert Wichmann Verlag, Karlsruhe, pp. 295–301.

Schou, A. (1962) *The Construction and Drawing of Block Diagrams*, Nelson, London.

Schultze, J. H. (1955) Die naturbedingten Landschaften der Deutschen Demoktratischen Republik, *Petermanns Geographische Mitteilungen Erganzungsheft*. 257, Gotha.

Schultze, J. H. (1966) Landschaft, in Akademie f. Raumforschung u. Landesplanung *Handwörterbuch der Raumforschung und Raumordnung*, Hannover, pp. 1819–39.

Schumm, S. A. (1977) *The Fluvial System*, John Wiley & Sons, New York.

Schumm, S. A. (1979) Geomorphic thresholds: the concept and its applications, *Transactions of the Institute of British Geographers*, New Series **4**(4), 485–515.

Schwab, G. O., Frevert, R. K., Edminster, T. W., and Barnes, K. B. (1966) *Soil and Water Conservation Engineering*, John Wiley & Sons, New York.

Sharpe, G. W. (1976) *Interpreting the Environment*, John Wiley & Sons, New York.

Short, N. M. and Blair, R. W. (eds) (1986) *Geomorphology from Space: A Global Overview of Landforms*, NASA Special Publication SP486, Washington, DC.

Siegal, B. S. and Gillespie, A. R. (1980) *Remote Sensing in Geology*, John Wiley & Sons, New York.

Siegal, S. (1956) *Non-parametric Statistics for the Behavioural Sciences*, McGraw-Hill, New York.

Simonds, J. O., (1961) *Landscape Architecture: The Shaping of Man's Natural Environment*, McGraw-Hill, New York.

Simpson, J. E. (1964) Sea-breeze fronts in Hampshire, *Weather*, **19**, 208–20.

Smardon, R. C. (1975) Assessing visual–cultural values of inland wetlands in Massachusetts, in Zube, E. H., Brush, R. O., and Fabos, J. G. (eds) *Landscape Assessment: Values, Perceptions, and Resources*, Dowden, Hutchinson & Ross, Stroudsburg, Pennsylvania, pp. 289–318.

Smith, D. I. and Newson, M. D. (1974) The dynamics of solutional and mechanical erosion in limestone catchments on the Mendip Hills, Somerset, in Gregory, K. J. and Walling, D. E. (eds) *Fluvial Processes in Instrumented Watersheds*, Institute of British Geographers, Special Publication No. 6, London, pp. 155–68.

Smith, D. M. (1975) *Patterns in Human Geography*, David & Charles, Newton Abbot.

Smith, J. (1949) *Distribution of Tree Species in the Sudan in Relation to Rainfall and Soil Texture*, Sudan Government Ministry of Agriculture Bulletin No. 4, Khartoum.

Smith, K. G. (1950) Standards for grading texture of erosional topography, *American Journal of Science*, **248**, 655–68.

Smith, L. P. (1984) *The Agricultural Climate of England and Wales*, Great Britain: MAFF Technical Bulletin 35, Reference Book 435, HMSO, London.

Smith, T. R. (1985) Notes for *RSS/CERMA GIS Workshop*, at School of Oriental and African Studies, London. Remote Sensing Society, Nottingham.

Smyth, A. J. (1966) *The Selection of Soils for Cocoa*, Soils Bulletin 5, FAO, Rome.

Snedecor, G. and Cochran, W. G. (1978) *Statistical Methods*, 6th edn, Iowa State University Press, Ames, Iowa.

Socava, V. B. (1974) Das Systemparadigma in der Geographie, *Petermanns Geographische Mitteilungen*, **118**, 161–6.

Soil Science Society of America (1983) *Universal Soil Loss Equation: Past, Present and Future*, Special Publication No. 8, Madison, Wisconsin.

Soil Survey of England and Wales (1977) *Winter Rain Acceptance Potential*, Ordnance Survey, Southampton.

Soil Survey of England and Wales (1978) *Bioclimatic Classification of England and Wales at 1:625,000*, Ordnance Survey, Southampton.

Soil Survey of England and Wales (1979) *Land Use Capability Map* at 1:1 000 000, Harpenden.

Soil Survey Institute (Netherlands) (1960) *Soil Map of the Netherlands* at 1:50 000, Wageningen.

Solntsev, N. A. (1962) Basic problems in Soviet landscape science, *Soviet Geography: Review and Translation*, 3(6), 3–15.

Stamp, L. D. and **Beaver, S. H.** (1933) *The British Isles: A Geographic and Economic Survey*, Longmans Green and Co., London.

Stamp, L. D. and **Willatts, E. C.** (1934) *The Land Utilization Survey of Great Britain: An Outline of the First Twelve One Inch Maps*, Land Utilization Survey of Britain, London School of Economics, London.

Steele, J. G. (1967) *Soil Survey Interpretation and its Use*, FAO Soils Bulletin No. 5, Rome.

Steinitz, C. F. (1970) Landscape resource analysis: the state of the art, *Landscape Architecture*, 60(2), 101–4.

Steur, G. G. L., Paas, W. and **Bakker, H. de** (1984) The soil maps of the German–Netherlands border-area project: A comparison of maps, mapping methods, and classifications, in Haans, J. C. F. M., Steur, G. G. L., and Heide, G. (eds) *Progress in Land Evaluation*, Proceedings of a Seminar on Soil Survey and Land Evaluation, Wageningen, Netherlands, A.A. Balkema, Rotterdam and Boston, pp. 143–76.

Stewart, G. A. (ed.) (1968) *Land Evaluation*, Macmillan of Australia, Melbourne.

Stocking, M. A. and **Elwell, H. A.** (1973) Soil erosion hazard in Rhodesia, *Rhodesian Agricultural Journal*, 70, 93–110.

Stocking, M. and **Elwell, H. A.** (1976) Rainfall erosivity over Rhodesia, *Transactions of the Institute of British Geographers*, New Series 1, 231–45.

Stoddart, D. R. (1971) Climatic geomorphology: review and assessment, in Board, C., Chorley, R. J., Haggett, P., and Stoddart, D.R. (eds) *Progress in Geography: International Reviews on Current Research*, vol. 1, Edward Arnold, London, pp. 159–222.

Stone, R. O. and **Dugundji, J.** (1965) A study of micro-relief, its mapping, classification and quantification by means of Fourier analysis, *Engineering Geology*, 1(2), 89–187.

Storie, R. E. (1954) Land classification as used in California for the appraisal of land for taxation purposes, *Fifth International Congress of Soil Science*, Commission VI, Leopoldville, pp. 407–12.

Storie, R. E. (1964) Soil and land classification for irrigation development, *Proceedings of Eighth International Congress of Soil Science*, vol. V, pp. 873–82.

Story, R., Galloway, R. W., van der Graaff, R. H. M. and **Tweedie, A. D.** (1963) *General Report on the Lands of the Hunter Valley*, Land Research Series No. 8, CSIRO, Melbourne.

Strahler, A. N. (1957) Quantitative analysis of watershed geomorphology, *Transactions of the American Geophysical Union*, 38, 913–20.

Strahler, A. N. (1969) *Physical Geography*, 3rd edn, John Wiley & Sons, New York.

Sudan, Republic of the (1963) Ministry of Agriculture *Roseires Soil Survey*, Report No. 1. *Gezira Extension Area. Soil Survey and Land Classification*, vol. 1, Sir Murdoch Macdonald and Partners by Hunting Technical Services Limited, Boreham Wood.

Sukachev, V. and **Dylis, N.** (1966) *Fundamentals of Forest Biogeocoenology*, translated by J. M. McLennan, Oliver & Boyd, Edinburgh.

Sweeting, Marjorie M. (1972) *Karst Landforms*, Macmillan, London.

Sys, C. and **Frankart, R.** (1971) Land capability classification in the humid tropics, *Sols Africains*, **16**, 153–75.

Ta Liang (1964) *Tropical Soils: Characteristics and Airphoto Interpretation*, School of Civil Engineering, Cornell University, prepared for Air Force Cambridge Research Laboratories.

Tandy, C. R. V. (1967) The isovist method of landscape survey, in Landscape Research Group, *Methods of Landscape Analysis*, London, pp. 9–10.

Tanner, W. F. (1961) An alternative approach to morphogenetic climates, *Southeastern Geologist*, **2**(4), 251–7.

Tansley, A. G. (1935) The use and misuse of vegetational terms and concepts, *Ecology*, **16**, 284–307.

Tansley, A. G. (1953) *The British Islands and their Vegetation*, 2 vols, Cambridge University Press.

Tarran, Ann E. (1984) Suitability mapping: first early potatoes, in Haans, J.C.F.M., Steur, G.G.L., and Heide, G. (eds) *Progress in Land Evaluation*. Proceedings of a Seminar on Soil Survey and Land Evaluation, Wageningen, Netherlands, A.A. Balkema, Rotterdam and Boston, pp. 27–42.

Taylor, B. W. (1959) Ecological land use surveys in Nicaragua, *Estudios Ecologicos*, **1**.

Teaci, D. and **Burt, M.** (1974) Land evaluation and classification in east European countries, in *Approaches to Land Classification*, FAO Soils Bulletin 22, Rome, pp. 35–46.

Terrell, T. T. (1979) *Physical Regionalization of Coastal Ecosystems of the United States and its Territories*, US Fish and Wildlife Service FWS/OBS–78/80, Washington, DC.

Thie, J., Chaetrand, N. and **Mills, G.** (1979) Interpretation of a data base using the Canada Land Data System, in Rubec, C.D.A. (ed.) *Applications of Ecological (Biophysical) Land Classification in Canada*, Ecological Land Classification Series No. 7, Lands Directorate Environment Canada, pp. 351–60.

Thirlaway, H. I. S. (1959) *Annual Report of the Unesco Arid Zone Geophysical Research Project in Pakistan*, Paris.

Thomas, I. L. and **Stewart, N. J.** (1985) *GIS Survey Conclusion Report*, General Technology Systems, London. Available through Space Branch, Department of Trade and Industry, London.

Thomas, M. F. and **Thorp, M. B.** (1985) Environmental change and environmental etchplanation in the humid tropics of Sierra Leone: the Koidu etchplain, in Douglas, I. and Spencer, T. (eds) *Environmental Change and Tropical Geomorphology*, George Allen & Unwin, London, pp. 239–67.

Thompson, R. D. (1973) The influence of relief on local temperature: data from New South Wales, Australia, *Weather*, **28**(9), 377–82.

Thompson, R. D., Mannion, A. M., Mitchell, C. W., Parry, M. and **Townshend, J. R. G.** (1986) *Processes in Physical Geography*, Longman, London.

Thompson, T. R. E. (1979) Soil surveys and wildlife conservation in agricultural landscapes, in Jarvis, M. and Mackney, D. (eds) *Soil Survey Applications*, Soil Survey Monograph No. 13, Harpenden, pp. 184–92.

Thornbury, W. D. (1954) *Principles of Geomorphology*, John Wiley & Sons, New York.

Thornthwaite, C. W. (1948) An approach towards a rational classification of climate, *Geographical Review*, **38**, 85–94.

Thorp, J. (1931) The effects of vegetation and climate upon soil profiles in northern and northwestern Wyoming, *Soil Science*, **32**(4), 283–301.

Tomlinson, R. F. (1968) A geographic information system for regional planning, in Stewart, G. A. (ed) *Land Evaluation*, Macmillan of Australia, Melbourne pp. 200–10.

Tomlinson, R. F. (1984) Geographic information systems: a new frontier. Keynote address to the *International Symposium on Spatial Data Handling*, Zurich.

Tomlinson, R. F., Calkins, H. W. and **Marble, D. F.** (1976). *Computer Handling of Geographical Data: An Examination of Selected Geographical Information Systems*, Unesco, Paris.

Tothill, J. D. (ed) (1952) *Agriculture in the Sudan*, Oxford University Press.

Townshend, J. R. G. (ed) (1981) *Terrain Analysis and Remote Sensing*, George Allen & Unwin, London.

Travaglia, C. and **Mitchell, C. W.** (1982) *Applications of Satellite Remote Sensing for Land and Water Resources Appraisal: People's Democratic Republic of Yemen*, FAO Report RSC Series 9 TCP/PDY/0104 (mi), Rome.

Tricart, J. (1965a) *Le Modelé des Régions Chaudes: Forêts et Savanes*, Traité de Géomorphologie, vol. 5, CEDES, Paris.

Tricart, J. (1965b) *Principes et Méthodes de la Géomorphologie*, Masson, Paris.

Tricart, J. and **Cailleux, A.** (1972) *Introduction to Climatic Geomorphology*, translated from the French by Conrad J. Kiewiet de Jonge, Longman, London.

Troll, C. (1939) Luftbildplan und ökologische Bodenforschung, *Zeitschrift d. Gesellschaft f. Erdkunde*, Berlin, 241–98.

Troll, C. (1971) Landscape ecology (geoecology) and biogeocenology, a terminological study, *Geoforum*, **8/71**, 43–6.

Trudgill, S. T. (1976) Rock weathering and climate: quantitative and experimental aspects, in Derbyshire, E. (ed.) *Geomorphology and Climate*, John Wiley & Sons, New York and London, pp. 55–99.

Trudgill, S. and **Briggs, D. J.** (1981) Soil and land potential, *Progress in Physical Geography* **5**(2), 274–85.

Tuan, Yi-Fu (1974) *Topophilia: A Study of Environmental Perceptions, Attitudes, and Values*, Prentice-Hall, Engelwood Cliffs, New Jersey.

Tucker, C. J. (1978) *An Evaluation of the First Four Landsat-D Thematic Mapper Reflective Sensors for Monitoring Vegetation: A Comparison with Other Satellite Sensor Systems*, NASA Technical Memorandum 79617, Goddard Space Flight Center, Greenbelt, Maryland.

Turner, F. T. and **Weiss, L. E.** (1963) *Structural Analysis of Metamorphic Tectonites*, McGraw-Hill, New York.

Twidale, C. R. (1982) *Granite Landforms*, Elsevier, Amsterdam, New York, and Oxford.

Unesco (1967) *Quaternary Maps* of Europe, Africa, etc. at 1:2 500 000, Paris.

Unesco (1968) *International Tectonic Map of Africa, 1:5 000 000* in association with Association of African Geological Surveys (ASGA), Paris.

Unesco (1970) *International Hydrogeological Map of Europe*, 1:500,000, Paris.

Unesco (1971) International Co-ordinating Council of the Programme on Man and the Biosphere (MAB), *Final Report*, Paris.

Unesco (1973a) *International Classification and Mapping of Vegetation*, Series 6, Ecology and Conservation, Paris.

References

Unesco (1973b) Programme on Man and the Biosphere (MAB): Expert Panel on Project 13: Perception of Environmental Quality (1973) *Final Report*, Paris.
Unesco (1974) *Metamorphic Map of Europe, 1:2 500 000*, Paris.
Unesco (1975) *Legends for Geohydrochemical Maps*, Paris.
Unesco (1976) *Engineering Geological Maps: A Guide to their Preparation*, Paris.
Unesco (1979) *Map of the World Distribution of Arid Regions*, Paris.
United Kingdom National Remote Sensing Centre (1987) *Data Users Guide*, Farnborough.
United Kingdom Natural Environmental Research Council (NERC) (Department of Education and Science) (1983) *Geographic Information System Developments at the Thematic Information Services, NERC, Swindon.*
Unstead, J. F. (1916) A synthetic method for determining geographical regions, *Geographical Journal*, **48**, 230–49.
Unstead, J. F. (1933) A system of regional geography, *Geography*, **18**, 175–87.
Unstead, J. F. (1937) Classification of the regions of the world, *Geography*, **22**, 253–82.
USAEWES (US Army Engineer Waterways Experiment Station) (1959) *Handbook: A Technique for Preparing Desert Terrain Analogs*, Technical Report No. 3–506, Vicksburg.
USAEWES (1962) *Desert Terrain Evaluations. 1/4 Ton Utility Truck at Yuma*, Vicksburg.
USAEWES (1963a) *Forecasting Trafficability of Soils: Airphoto Approach*, Technical Memorandum No. 3–331, vol. 2, Vicksburg.
USAEWES (1963b) *Environmental Factors Affecting Ground Mobility in Thailand*, Technical Report No. 5–625, Appendix C, Trafficability, Vicksburg.
USDA (US Department of Agriculture) (1954) *Diagnosis and Improvement of Saline and Alkali Soils*, Agriculture Handbook No. 60, Washington, DC.
USDA Soil Conservation Service (1971) *Guide for Interpreting Engineering Uses of Soils*, US Government Printing Office, Washington, D.C.
USDA Soil Survey Staff (1962) *Soil Survey Manual*, Handbook No. 18, Washington, DC.
USDA Soil Survey Staff (1976) *Soil Taxonomy*. Soil Conservation Service, USDA Handbook 436, US Government Printing Office, Washington, DC.
US Department of the Army (1979) *Remote Sensing Applications Guide*, Engineer Pamphlet 70–1–1, Washington, DC.
US Department of the Interior (1951) Bureau of Reclamation Manual, vol. 5, *Irrigated Land Use* Washington, DC.
US Geological Survey (1970) *Apollo 6 Photomaps of the NW Corridor from the Pacific Ocean to Northern Louisiana, 1:500 000*, Washington, DC.
van Beers, W. F. J. (1958) *The Auger Hole Method for the Field Measurement of Hydraulic Conductivity*, International Institute for Land Reclamation and Improvement, Bulletin No. 1, Wageningen, Netherlands.
van Valkenburg, S. A. and **Huntington, E.** (1935) *Europe*, John Wiley & Sons, New York.
van Zuidam, R. (1985) *Aerial Photo Interpretation in Terrain Analysis and Geomorphologic Mapping*, Smits, The Hague.
van Zuidam, R. A. and **van Zuidam-Cancellado, F. I.** (1978) *Terrain Analysis and Classification Using Aerial Photographs: A Geomorphological Approach*, ITC Textbook of Photo-Interpretation, vol. 1, Ch. 6, ITC, Enschede.
Veatch, J. O. (1933) *Agricultural Classification and Land Types of Michigan*, Michigan Agricultural Experiment Station, Special Bulletin No. 231.

Verstappen, H. Th. (1983) *Applied Geomorphology*, Geomorphological Surveys for Environmental Development, Elsevier, Amsterdam.

Verstappen, H. Th. and **van Zuidam, R. A.** (1968) ITC system of geomorphological survey, in *ITC Textbook of Photo-Interpretation*, Delft, Ch. 7.

Verstappen, H. Th. and **van Zuidam, R. A.** (1970) Orbital photography and the geosciences – a geomorphological example from the central Sahara, *Geoforum*, **2**, 33–47.

Vink, A. P. A. (1975) *Land Use in Advancing Agriculture*, Advances in Agronomy, vol. 1, Springer-Verlag, Berlin/Heidelberg.

Vink, A. P. A. (1983) *Landscape Ecology and Land Use*, Longman, London.

Vinogradov, B. V., Gerenchuk, K. I., Isachenko, A. G., Raman, K. G. and **Teselchuk, Yu N.** (1962) Basic principles of landscape mapping, *Soviet Geography: Review and Translation*, **3**(6), 15–20.

Vita-Finzi, C. (1978) *Archaeological Sites in their Setting*, Thames & Hudson, London.

von Engeln, D. D. (1942) *Geomorphology*, Macmillan, New York.

von Humboldt, A. (1856) *Cosmos: A Sketch of a Physical Description of the Universe* 3 vols, Longman, Brown, Green, and Longmans, London.

Wainwright, A. and **Brabbs, D.** (1985) *Fell Walking with Wainwright*, Michael Joseph, London.

Walker, P. H. and **Butler, B. E.** (1983) Fluvial process, in Australia, CSIRO *Soils: An Australian Viewpoint* Ch. 6. pp. 83–90.

Wallace, A. R. (1876) *The Geographical Distribution of Animals*, Macmillan, London.

Walter, H. and **Box, E.** (1976) Global Classification of natural terrestrial ecosystems, *Vegetatio*, **32**, 75–81.

Walters, A. A. (1975) *Noise and Prices*, Clarendon Press, Oxford.

Waltz, J. P (1969) Groundwater, in Chorley, R. J. (ed.) *Water, Earth, and Man*, Methuen, London, Part 6, Ch. I pp. 259–67.

Ward, R. G. (1989) Earth's empty quarter: Pacific islands in a Pacific century, *Geographical Journal*, **155**(2), 235–46.

Waters, R. S. (1958) Morphological mapping, *Geography*, **43** 10–17.

Watts, S. H. (1975) Mound springs, *Australian Geographer*, **13**, 52–3.

Way, D. S. (1968) *Airphoto Interpretation for Land Planning*, Harvard University Department of Landscape Architecture, Cambridge, Massachusetts.

Way, D. S. (1973) *Terrain Analysis: A Guide to Site Selection using Aerial Photographic Interpretation*, Dowden, Hutchinson & Ross, Stroudsburg, Pennsylvania.

Webster, R. (1977) *Quantitative and Numerical Methods in Soil Classification and Survey*, Clarendon Press, Oxford.

Webster, R. and **Beckett, P. H. T.** (1964) A study of the agronomic value of soil maps interpreted from aerial photographs, *Transactions of the 8th International Congress of Soil Science*, **5**, 795–803.

Weiers, C. J. (1975) Soil Classification and land evaluation, *Town and Country Planning*, **43**, 390–3.

Weiers, C. J. and **Reid, I. G.** (1974) *Soil Classification, Land Valuation, and Taxation, the German Experience*, Centre for European Agricultural Studies, Wye, Ashford, Kent.

Welch, D. M. (1978) *Land/Water Classification*, Ecological Land Classification Series No. 5, Environment Canada, Ottawa.

Westermann (1970–) *Landformen im Kartenbild*, 1:25000 scale terrain maps of the main landform regions of Germany: Gruppe I: Norddeutsches Flachland; II and III: Mittelgebirge; IV: An Vulkanismus Gebundene Formen; V: Alpenvorland; VI: Alpen Nordliche Flysch – und Kalk–alpen; VII: Alpen – Zentralalpen.

References

Westerveld, G. J. W. and **Van den Hurk, J. A.** (1973) Application of soil and interpretive maps to non–agricultural land use in the Netherlands, *Geoderma*, **10**, 47–66.

Whipkey, R. Z. (1965) Subsurface stormflow on forested slopes, *Bulletin of the International Association of Scientific Hydrology*, **10**(2), 74–85.

Whittlesey, D. (1936) Major agricultural regions of the earth, *Annals of the Association of American Geographers*, **26**, 199–240.

Whittow, J. B. (1979) Landscape perceptions, in Goodall, B. and Kirby, A. *Resources and Planning*, Pergamon Press, Oxford, Ch. 12, pp. 243–62.

Whittow, J. B. (1984) *The Penguin Dictionary of Physical Geography*, Harmondsworth, England.

Wiken, E. B. (1979) Rationale and methods of ecological land surveys: an overview of Canadian approaches, in Taylor, G.D. (ed.) *Land/Wildlife Integration*, Ecological Land Classification Series No. 11, Lands Directorate Environment Canada, Ottawa, pp. 11–19.

Wiken, E. B., Welch, D. M., Ironside, G. R. and **Taylor, D. G.** (1981) *The northern Yukon: An Ecological Land Survey*, Ecological Land Classification Series No. 6, Lands Directorate Environment Canada, Ottawa.

Wild, A. (ed.) (1988) *Russell's Soil Conditions and Plant Growth*, 11th edn, Longman Scientific and Technical, Harlow, UK.

Williams, D. F. (1981) Integrated land survey methods for the prediction of gully erosion, in Townshend, J.R.G. (ed.) *Terrain Analysis and Remote Sensing*, George Allen & Unwin, London, pp. 154–68.

Williams, G. E. (1964) Some aspects of the aeolian saltation load, *Sedimentology*, **3**, 257–87.

Wilson, I. G. (1971) Desert sandflow basins and a model for the development of ergs, *Geographical Journal*, **137**, 180–99.

Wischmeier, W. H. (1976) The use and misuse of the universal soil loss equation, *Journal of Soil and Water Conservation*, **31**(1), 5–9.

Wischmeier, W. H., Johnson, C. B. and **Cross, B. V.** (1969) A soil erodibility nomograph for farmland and construction sites, *Journal of Soil and Water Conservation*, **26**, 189–93.

Wischmeier, W. H. and **Smith, D. D.** (1965) *Predicting Rainfall–erosion Losses from Cropland East of the Rocky Mountains*, Agricultural Handbook 282, USDA, Washington, DC.

Wischmeier, W. H. and **Smith, D. D.** (1978) *Predicting Rainfall Erosion Losses: a Guide to Conservation Planning*, USDA Handbook No. 537, Washington, DC.

Wischmeier, W. H., Smith, D. D. and **Uhland, R. E.** (1958) Evaluation of factors in the soil loss equation, *Agricultural Engineering*, **39**(8), 458–62, 474.

Wood, W. F. and **Snell, J. B.** (1957) *The Dispersion of Geomorphic Data Around Measures of a Central Tendency and its Application*, HQ Quartermaster Research and Engineering Command, Technical Report EA–8, Natick, Massachusetts.

Wood, W. F. and **Snell, J. B.** (1959) *Predictive Methods in Topographic Analysis. I. Relief, Slope, and Dissection on Inch–to–the–Mile Maps in the USA*, HQ Quartermaster Research and Engineering Command, Technical Report EP–112, Natick, Massachusetts.

Wood, W. F. and **Snell, J. B.** (1960) *A Quantitative System for Classifying Landforms*, HQ Quartermaster Research and Engineering Command, Technical Report EP–124, Natick, Massachusetts.

Wooldridge, S. W. (1932) The cycle of erosion and the representation of relief, *Scottish Geographical Magazine*, **48**, 30–6.

Working Party (1982) Land surface evaluation for engineering practice, by Working Party under the Auspices of the Geological Society, *Quarterly Journal of Engineering Geology*, **15**, 265–316.

Worthington, E. B. (ed.) (1977) *Arid Land Irrigation in Developing Countries: Environmental Problems and Effects*, Pergamon Press, Oxford.

Wright, R. L. (1964) *Unesco Geomorphology Mission, Pakistan*, Report to Unesco, Paris.

Wright, R. L. (1967) A geomorphological approach to land classification', Ph. D. Thesis, University of Sheffield.

Yates, F. (1960) *Sampling Methods for Censuses and Surveys*, 3rd edn, Griffin, London.

Young, A. (1963) Some field observations on slope form and regolith and their relation to slope development, *Transactions and Papers of the Institute of British Geographers*, **32**, 1–29.

Young, A. (1971) Slope profile analysis: the system of best units, in Institute of British Geographers: *Slopes: Form and Process*, Special Publication No. 3, pp. 1–13.

Young, A. (1976) *Tropical Soils and Soil Survey*, Cambridge University Press.

Young, J. A. T. (1986) A UK geographic information system for environmental monitoring, *Resource Planning and Management Capable of Integrating and Using Satellite Remotely Sensed Data*, Remote Sensing Society Monograph No. 1, Department of Geography, Nottingham University.

Youngs, E. G. (1987) Estimating hydraulic conductivity values from ring infiltrometer measurements, *Journal of Soil Science*, **38**, 623–32.

Yule, G. G. and **Kendall, M. G.** (1950) *An Introduction to the Theory of Statistics*, 14th edn, Griffin, London.

Zhao Songqiao (1984) Comprehensive physical geography in China, in Wu Chuanjun, Wang Nailiang, Lin Chao, and Zhao Songqiao (eds) *Geography in China*, Science Press, Beijing, pp. 1–16.

Zhao Songqiao (1986) *Physical Geography of China*, John Wiley & Sons, New York.

Zimmermann, R. C. (1969) *Plant Ecology of an Arid Basin Tres Alamos–Redington Area Southeastern Arizona*, US Geological Survey Professional Paper 485–D, Washington, DC.

Zimmermann, R. C. and **Thom, B. G.** (1982) Physiographic plant geography, *Progress in Physical Geography*, **6**(1), 45–59.

Zoltai, S. C., Pollett, F. C., Jeglum, J. K. and **Adams, G. D.** (1975) Developing a wetland classification for Canada, In *Forest Soils and Forest Land Management*, Les Presses de l'Université Laval, Quebec.

Zube, E. H. (1976) Landscape aesthetics: policy and planning in the US, in Appleton, J. (ed.) *The Aesthetics of Landscape* Rural Planning Services Publication No. 7, Department of Geography, University of Hull, pp. 69–79.

Zube, E. H., Brush, R. O. and **Fabos, J. G.** (1975) *Landscape Assessment: Values, Perceptions, and Resources*, Dowden, Hutchinson & Ross, Stroudsburg, Pennsylvania.

Zube, E. H. and **Carlozzi, C. A.** (1967) *An Inventory and Interpretation – Selected Resources of the Island of Nantucket*, Cooperative Extension Service Publication 4, University of Massachusetts, Amherst, Massachusetts.

INDEX